Living in Digital Worlds

Living in Digital Worlds investigates the relationship between human society and technology, as our private and particularly our public lives are increasingly undertaken in spaces that are inherently digital: digital public spaces.

The book unpicks why digital technology is such an inextricable part of modern society, first by examining the historical relationship between technological development and the early progression of human sociality. This is then followed by an examination of the ways in which modern life is currently being impacted by the expansion of digital information and devices into multiple aspects of our lives, including focuses on privacy, bias and ownership in digital spaces. Finally, it explores potential future developments and their implications, and proposes that it is crucial to consider the design of technology and systems in order to support a positive and beneficial direction of change.

Each chapter includes case studies, primarily drawn from The Creative Exchange, a five-year programme which ran from 2012 to 2016 to explore the notion of the digital public space through collaborative cross-sector research.

Naomi Jacobs is a Research Fellow whose various work focuses on interaction in digital and physical spaces. Previous topics of research by Dr Jacobs have included interdisciplinarity and knowledge exchange, fan practices of sharing online and the effect on communities of internet of things technologies.

Rachel Cooper OBE is Distinguished Professor of Design Management and Policy at Lancaster University. Professor Cooper's research covers: design thinking; design management; design policy, across all sectors of industry. She has a specific interest in design for wellbeing and socially responsible design, and has published widely on these topics.

Living in Digital Worlds

Designing the Digital Public Space

Naomi Jacobs and Rachel Cooper

Routledge
Taylor & Francis Group
LONDON AND NEW YORK

First published 2018
by Routledge
2 Park Square, Milton Park, Abingdon, Oxon OX14 4RN

and by Routledge
711 Third Avenue, New York, NY 10017

Routledge is an imprint of the Taylor & Francis Group, an informa business

British Library Cataloguing-in-Publication Data
A catalogue record for this book is available from the British Library

Library of Congress Cataloging-in-Publication Data
A catalog record for this book has been requested

ISBN: 978-1-4724-5283-2 (hbk)
ISBN: 978-1-315-59278-7 (ebk)

Typeset in Times New Roman
by Cenveo Publisher Services

Contents

Figures and tables

Figures

Tables

Case studies

Acknowledgements

We would like to thank all of our colleagues and expert advisors who worked with us on the AHRC Creative Exchange Hub Project between 2012 and 2016. Their insight, energy, and contribution to issues around the digital public space inspired and informed the text. We would of course like to thank our funders the Arts and Humanities Research Council without whose funding we would not have been able to lead and engage with the research undertaken by our 21 doctoral students and their supervisors.

We would also like to thank those who have kept us company in various digital public spaces inhabited over the years, particularly the Agincourt Agitators for punctuation advice, and being interested. Finally, to Ruth Templeton for helping us deliver!

Introduction

Digital is a term which is over-used, abused and in some senses becoming meaningless. Digital is everywhere and nowhere, indistinguishable from other aspects of the material world and the way we live in it – at least for most people in modern developed societies.

This book results from a five-year research programme called The Creative Exchange (CX), which focused its attention on the 'digital public space'. Initially this was defined as:

> Where anyone, anywhere, anytime can access, explore and create with digital content.

The concept of digital public space was originally developed by the BBC and taken up as a topic for investigation by the Creative Exchange, particularly with regard to the relationship of the creative industries to academic research. However this book will explore more closely the origins and aspects of how we can talk about digital public space, and also focus on broader interpretations and implications.

Our work takes a particular lens to this question, drawing on the authors' background and focus. It does not claim to be taking an exhaustive or privileged position, but complements other work on digital public space produced during the Creative Exchange research process, which includes the theses of 21 students whose doctoral work took place during the project.

Because digital technology is becoming ubiquitous, and is now an integral part of all of our lives, we need to understand it. However the systems are so complex that it is becoming more and more difficult to fully grasp all aspects and the impacts they can have on our lives and our possible futures. For this reason, we want to return initially to the basics of what we mean by digital public space, and how humans have developed in terms of their interactions with each other and the world. This is followed by an analysis of the physical attributes and effects, and the intangible non-physical attributes and effects of this concept of digital public space. It is clear from considering these dimensions of digital public space that there are numerous consequences which we must understand. An understanding of this context will enable us to think more carefully about our future in the digital public space and how we design and create within it. We have therefore structured the book to lead us through this journey.

Section 1: How did we get here?

- Chapter 1: We examine the history of the digital public space as a concept, unpick the definition, and explore what and who is encompassed (and is not) by the term.
- Chapter 2: We give a brief overview of the evolutionary history of human culture and cognition, and relate this to the concept of 'technological revolutions'; points beyond

which a particular innovation is irrevocably bound to the way we live our lives so that it cannot be removed without major societal collapse.

Section 2: What are the attributes and effects of digital public space?

- Chapter 3: We explore the ways in which physicality has affected the development of digital public spaces. This includes: virtual spaces which are built on templates in the physical world, telepresence allowing physical embodiment at a distance, hybrid digital/physical interfaces and spaces, and how pervasive computing might enhance our physical environment.
- Chapter 4: We propose that a key part of the digital public space is the *information space* that pervades our modern connected lives. This affects our cognitive processes, society and interactions, and conception of ourselves as individuals.

Section 3: What are the consequences of digital public space?

This section looks at specific contexts in which digital public spaces affect behaviour and interaction, and at the challenges posed by the integration of these technologies into our societies. Trust is a critical aspect in each of these challenge areas.

- Chapter 5: We discuss aspects of ownership and transaction, and new forms which are being developed for digital public space.
- Chapter 6: We consider privacy and security in the digital public space, what the differences between these are, and how it might be impacting on our wellbeing and interactions.
- Chapter 7: We discuss different forms of bias that can affect the digitally connected world.

Section 4: How do we design digital futures?

- Chapter 8: We look forward to futures of digital public spaces, emphasising that we do not aim to predict, but rather speculate and ask what would be the consequences of possible endpoints of current technology trends, and how this might impact on design decisions.
- Chapter 9: Our final chapter foregrounds design principles and processes, focusing on why these are important when developing new technologies and systems in digital public spaces. We highlight digital public spaces as emergent systems, and the importance of the user in considerations of design.

It is clear that this topic is fast-moving and dynamic. We realise that we are writing at a particular moment in time and referencing contemporary technologies and thinking that might rapidly change. However, we think it is important to understand the opportunities and challenges for design and designers that are arising from and being signalled by the consequences of these technologies.

Section 1

How did we get here?

1 Defining the digital public space

Today our world is made up of colliding parallel dimensions; the digital world is as ubiquitous as our physical world, and there is space in both worlds that we navigate and inhabit. This chapter discusses the origin of the term 'digital public space', how it is evolving in general practice and application, and how it might relate to these parallel worlds.

The Digital Public Space (DPS) project: BBC origins

The phrase 'digital public space' is becoming more widely known. There are however a variety of different ways in which it can be used and interpreted. As with any new term, there are divergent ideas of what it means. Therefore, before it can be explored in any great detail we must look at definitions of digital public space, and where the term originated.

Although a few isolated instances of the phrase can be found earlier, the first real use of the term comes from work by the BBC, announced in 2011. Earlier uses are mostly in the context either of use of the digital in cities (Frenchman & Rojas, 2006; Graham & Aurigi, 1997) or curated online spaces for the public to use (Hinssen, 1995). From May 2012 until March 2013, the BBC ran an initiative called 'The Space'. Extended from an initial six-month pop-up service, this experimented with the idea of making the corporation's extensive archives shareable, both within the BBC and with external organisations which might be able to make use of them. The Space was described as a service which 'knits together BBC technology expertise and content from the corporation, BFI and UK arts bodies with £3.5m in Arts Council commissioning funding' (Kiss, 2013). Its strapline was 'the arts – live, free and on demand'.

Interaction was a key part of this early venture, as explained by Alan Davey[1], Chief Executive of Arts Council England: 'What's really exciting about The Space is that it will provide a communal playground for arts and cultural organisations, for technology wizards, and for audiences – anyone who's open to new ways to connect with culture – to come in, to be creative, and to feed back about their experience.'

Key players in the development of this project were Tony Ageh (Controller of Archive Development) and Bill Thompson (the archive's Head of Partnership Development). Following on from The Space, they led the DPS project which centred on a shared technical platform for indexing, searching and publishing material in partnership with other UK cultural organisations.

An article in JISC Inform[2] in 2013 sets out the following principles:

- Long-term sustainability of the material – it needs to be future-proofed against formats becoming obsolete.

- Detailed metadata needs to be available for every object.
- Freedom of access and use in education and research.
- Open technical standards.
- Trustworthy management of user data.

Arguably the most critical of these is the one which refers to metadata being available for every object. It is disingenuous to think of the DPS project as simply making the BBC's previously broadcast programmes available online. As Brody and Fass describe it:

> The Digital Public Space (the DPS) originally began life as a way of thinking about how the BBC archive could be made available and accessible to all. This was swiftly followed by the realisation that surrounding the BBC content was a huge seam of additional data and a new context for digitally mediated cultural experiences.
>
> (Brody & Fass, 2013)

The archive is not just the programmes which have been commissioned, but every bit of information that was recorded and stored which surrounds those programmes. This might contain content including but not limited to scripts, soundtracks, contracts, unused footage and concurrent news. It is the surrounding content and links to other BBC materials and holdings that makes the possibilities of this connectivity so enthralling. This is made possible due to metadata: information about an item that allows it to be placed in a context. For example, traditional cataloguing information, such as the Dewey decimal system or ISBN numbers, allows books to be identified and placed within categories. If you go to the shelf to look for a particular book and the books have been shelved with others which fit in that category, you may find another relevant book which you did not previously realise existed. Metadata for digital content can be even more important, because you cannot easily 'browse the shelves' of the huge amounts of available information; information which may also be in many various forms and stored in different media. The connections between items are often going to be automatically extracted. This therefore requires correct categorisation and labelling, allowing links to be drawn between connected items, whether they are connected because they are created by the same artist, or have the same historical building or event as their subject.

Members of the DPS project hope that by creating a consistent system by which their collection is stored and catalogued, this can be extended out and linked with other similar systems. To this end, the working group led by the BBC includes partners such as the BFI, British Museum, Tate, the British Library, and many others. By pooling collective resources and strengthening the power of the way they are indexed, a truly comprehensive public archival service can be imagined. As Tony Ageh puts it: 'The Digital Public Space is not a product or a service, but an arrangement of shared technologies, standards and processes that will be collaboratively developed and commonly applied, to deliver a set of principles, objectives and purposes against which collective enterprise can be evaluated' (Ageh, 2013).

More recently, in a speech at Royal Holloway University in 2015, he described his vision of the DPS as 'a secure and universally accessible public sphere through which every person, regardless of age, income, ability or disability, can gain access to an ever growing library of permanently available media and data held on behalf of the public by our enduring institutions'. These principles can be seen to be extremely aligned with the aims of the wider BBC. In 2006, Bill Moggridge wrote:

> A policy goal of the BBC is to help people become engaged in the digital world, so they want to build engaging services for people who are not used to the Internet, or to any other digital media. For the design of the BBCi homepage, that means trying to make it really intuitive to explore the site, making sure that people have the opportunity to find out what they want easily. It also means helping them engage in dialogue by offering a simple way to comment on something they've seen, or send an email. Live chats and message boards are offered and connected with the BBC programs, so there is more two-way traffic.
>
> (Moggridge, 2006)

This focus on audience engagement and interaction indicates that the BBC is interested not just in allowing people to access fixed content, but to create additional content of their own and contribute to the body of information.

The digital public space in this context therefore, appears to particularly pertain to open archives whereby existing data, metadata and digital objects can be shared in a meaningful way with and amongst the public. This worthy initiative is one that cultural institutions around the world are more and more seeing as a priority. Examples of this include the Tate's 'Insight' project[3] to digitise their collection, which has been running since 1998, and digitisation projects at many national collections including the Wellcome Library[4] and Trinity College Cambridge.[5]

Larger international initiatives such as Google's Cultural Institute, and Europeana, act not directly as stores for the digital artefacts themselves, but as curated spaces where archives from many different sources can be connected and displayed. In this way, they can be searched and linked more easily, by encouraging the use of metadata as described above. Clearly, we are moving more and more towards our cultural heritage not just being transferred to a digital form, but stored in such a way that it can be easily navigated.

The broader definition of digital public space

The BBC's digital public space is clearly an exciting concept, but the term is too useful to leave in this specific context. More and more content and information is being made available online, and put in an associated framework where related strands can be brought together at the request of the user. But this is not limited to traditional 'objects' that may have been stored in a library, archive or museum. The ease of information transfer means that new information is constantly being created and shared. This may take the form of commentary, mashups, blogs or other new content created in a purely digital space by users as a form of expression. Increasingly it may also be information that previously would have been temporarily exchanged by individuals before vanishing, but is now retained in the online space forever – such as conversation and memories shared via social media platforms.

The objective of the BBC's DPS project is about giving people access to digital content. Taking this a step further, the facility to access and manipulate is allowing people to *interact* with digital content. We often use metaphors of place and space when we 'visit a website' or 'retrieve a file'. This may not purely be a habit of expression, but might be more fundamentally representative of the way that we manipulate information mentally, and indicative of something critical to the way people naturally interact online.

Ageh (2015) talks about the digital public space as a location, but then discusses mainly objects which exist within the space (metadata, archive material, created objects) without considering in detail the nature of the space in which these objects are located. There are various ways of considering digital public space – as a collection or archive, a single object made up of a collection, or alternatively as the vessel in which this collection is located. It is important to clearly define how digital public space will be used as a phrase going forward in this book, because there does not so far appear to be a clear consensus for this as a concept and so avoiding confusion with any other definition is paramount.

To start off with, let us look at each of the words individually. Each of the three words 'digital' 'public' and 'space' are ones we use regularly, but put together there are a variety of meanings, all of which have a great potential for exploration.

Digital

Although digital is a word used widely, from policy makers to equipment manufacturers, it is actually quite non-specific. Going back to the very precise definition, 'digital' describes information which is measured in discrete units, such as binary (on/off) states. Only values corresponding to one of these can be measured. The word comes from the Latin 'digitalis' which means finger – when you are counting on your fingers you can only use whole numbers. You cannot count to 4.6 using your fingers, only four or five. By contrast, an analogue scale or measure is smooth and gradated, and measurements can be taken at any point. An example of this would be a child who stands against a wall and makes a pencil mark once a month to measure their height.

However, these days when people use the word digital they are usually referring to the use of computers or other higher technology. Computers are by their nature digital, because they transpose information into a form that can be understood by a machine – using 1s and 0s of binary code. An image displayed on a computer screen has been translated into series of points, or pixels, which are either on or off. But because the use of computers is so pervasive in our daily lives, using the word 'digital' is becoming less and less specific. Phrases like 'digital economy', 'digital strategy' or 'digital experience' are trying to convey a much broader category of new technologies which cannot be identified simply as 'using computers', and so it becomes very difficult to pin down exactly what is meant by them.

When talking about digital public space, the term is going to be used to distinguish from non-digital public space, meaning that which exists free of high technological influence. Though as we will see, that does not exclude physical spaces. 'Digital' is a place characterised by the underlying principle of translating everything to 1s and 0s for storage, analysis, reuse and requisition. This brings with it implications of transferability and persistence: digital information is easy to copy, transmit and manipulate and this brings many implications for its use.

Public

'The general public' is a phrase that is used to mean, basically, everyone. It might seem that 'public space' then is something which has an obvious common sense meaning, as being available for usage by all. We might think about public toilets, or public gardens. This equates to the notion of the 'commons' or commonly held resources.

But dig a little deeper and it is a complex social construct. Public bodies are those which are funded, usually via tax, by the population of a nation to provide services for all. Part of this social contract is the understanding that those who have more resources available can contribute on behalf of those who might not be able to afford it. In that way, a society can make critical services available for everyone. Jill Cousins of Europeana, talking about the digital commons, describes the foundation of the concept of the commons as follows: 'Underpinning the foundation of the commons is a set of resources in the public domain that are owned collectively or 'held in common' and shared openly among a community. The key feature is that, unlike private property, the ownership of resources held in common is inherently inclusive'(Cousins, 2012). For this to exist in terms of the digital realm, there needs to be a public social agreement in place to establish these 'common resources' as described above.

When talking about a digital version of this public space then, we are assuming that there is some overarching body which governs digital content. This is not the case – it is made up of a network of private computers which, although publically accessible, are owned. James Bridle sums it up as follows: 'Digital space is always owned in some way: there is no true commons online' (Bridle, 2012). Projects such as Europeana and DPS are attempting to construct this shared public openness by creating a common system of shared metadata. These are funded by the public. But the Google Cultural Institute has similar aims, and is a private company. The lines become blurred.

Perhaps rather than looking at questions of ownership, we should look at accessibility. When considering how one uploads information to the internet, a consideration is often whether something should be 'public' or 'private'. Public in this sense does not relate to the notion of publically owned, but rather publically available. If it is available for anybody to view, then it is in public in the same way that public behaviour refers to that which can be viewed by unrelated strangers who happen to be passing. The smoking ban in England is generally considered to be a ban on smoking in 'public places' (one is allowed to smoke in a private home) but includes owned spaces such as restaurants, hotels and workplaces. The difference is that individuals can freely enter and exit these spaces, although the proprietor usually retains the right to exclude particular individuals should they wish, and could in theory withdraw the entire service without notice.

The key aspect here is control. Control of access to a space or object determines whether it is private or public. As an example, a continuum can exist in the object of a book; from a notebook which an individual does not let anybody else access (private) to a copy of a novel which they own but lend out to friends (private, but more widely shared) to a copy of the same novel on a shelf in a cafe which can be borrowed by anyone who wishes to read it (public). The defining feature is who has access to the book. Similarly with spaces – in a private house, the only people who have access are the inhabitants with keys, and invited guests. A private party may have a large attendance list, but only those who are on the guest list may come in. This is still private even though there may be a large number of people there. A public telephone box can only be used by one person at any time, but can be used by anyone and is therefore public. Corey Doctorow, science fiction author, activist, journalist and blogger, states it thusly in an article about how young people deal with privacy in the digital age: '"Privacy" doesn't mean that no one in the world knows about your business. It means that you get to choose who knows about your business' (Doctorow, 2014).

Another factor which influences how 'public' or 'private' something is generally considered, is cultural context. For example, although a public picnic area in a park can be used by anybody, there may be delineated 'private' areas defined by how it is being used. If I come and spread my picnic blanket in one corner of a large park, it would seem unusual for another member of the public to set up their picnic two feet away from me when a much larger empty space is available. By laying open claim to a piece of 'public' space, I have indicated that using the area in direct proximity is invading my privacy. Of course, this area adjacent to 'mine' becomes public again if the park is very busy (say, because of a music festival) and the amount of available space is reduced. In that instance, a far smaller space becomes defined as 'mine' and unavailable for others to make use of. This is a shared created context – the construction of spaces that have particular rules. The behaviour appropriate to these spaces may vary as well. danah boyd[6] (2007) gives the example of a beach versus a lecture theatre. Both of these are considered to be public, but you would only wear a bathing suit, wear sun-cream and stretch out on a towel in one of them. Social contract determines usage. We create similar context and spaces virtually. boyd talks about mediated publics which must necessarily be accessed using technology. In these situations the context and the level of privacy can be complex.

If a private space is one which has been created or claimed by an individual or group for a specific purpose and usage, either through actual restrictions or social understanding, a digital example of this might be a private Facebook group. This is an interesting layering of public and private. Facebook is a private company, and 'owns' the space in which all interactions on Facebook are made. However, the infrastructure is perceived as a 'public' space, which can be visited and contributed to by anyone. By restricting access to specific parts of this area, the users create a private space that curates their own constructed audience and context. Within this space, they may feel safe from those they do not wish to be visible to – although 'ownership' of this space could be revoked by the true owners (Facebook) at any time. Issues can arise if the perceived audience is not the same as the actual audience, and content can spread in an unintended manner. This, says boyd, is why participants in these public social online spaces may not be as safe as they think they are. Users may make the assumption that their privacy is protected by 'security through obscurity'; that the sheer weight of information means nobody will find the 'uninteresting' thing they are saying. She points out that 'this puts all oppressed and controlled populations (including teenagers) at risk because it just takes one motivated explorer to track down even the most obscure networked public presence' (boyd, 2007).

boyd further notes four key features which she says define networked publics that are mediated by technology: persistence, searchability, replicability and invisible audiences. The first three of these criteria can be thought of as applying to 'items of information'. Persistent items are or can be stored indefinitely, as opposed to a fleeting conversation offline. Searchable items are indexed and can be retrieved using applicable search terms, or discovered by following threads of data. Replicability is the phenomenon that (arguably) perfect copies can be created of digital objects. The fourth point, invisible audiences, is a consequence of the first three. Because these records persist, and can be copied and shared, then one must be aware at their creation that the potential audience is infinite.

While the features above are given in the context of online social spaces, this private/public dichotomy can also apply to digital objects. Locking items to make them private is usually implemented by making them encrypted (therefore requiring some kind of password or authentication to access). At the other end of the scale are objects which are fully indexable, that have been categorised and can be easily found via standard search tools. These may have been augmented with characteristics that make them well suited to being shared and can be easily linked, duplicated and transmitted. This is often encouraged in online social media tools which allow mechanisms for sharing such as retweets and 'likes'. Objects may be created with the use of these tools in mind and could be described as being 'social', but if something is not intended to be used in this way, the issue of replicability may have implications for copyright and ownership.

One final aspect of public space we must consider is the power of democratic freedom that the populace has to speak and be heard. This encompasses both the network of communicating points of view, and the process of information filtering, structure and synthesis 'in such a way they coalesce into bundles of topically specified public opinions'; described in this way by Habermas (1996) who calls it the public *sphere*. A key feature of modern digital communication is that broadcast media is no longer pre-eminent: everyone has the capability to write a blog post or build a website and put their voice in the public sphere so it can be heard by all. One possible effect of this, according to Yochai Benkler, is an increase in individual autonomy. 'This gives individuals a significantly greater role in authoring their own lives, by enabling them to perceive a broader range of possibilities, and by providing them a richer baseline against which to measure the choices they in fact make' (Benkler, 2006). This might be because information can be spread and multiplied with less possibility of control by authoritarian regimes, or simply an increased opportunity to contribute to debate and be a part of the public conversation. Our ability to communicate any aspects of our lives to a wide potential audience, in a way that was not previously possible, means that some things we might have previously considered private now become the domain of the public. This will be explored further in Chapter 6. Concerns of information flow, such as fragmentation and polarisation of discourse, digital divides, authoritarian censorship and the obscuring effects of information overload, may result in an ineffective public sphere despite the use of such technology. But the nature of public discourse is something that is important to consider in discussions of digital public space.

It is difficult to draw clear lines around what is public and what is private in digital space. In this book we will take the broadest meaning of public, that an object or space is public if anyone, anywhere *could* access and use it. The fact that any individual has an equal potential use and access does not however mean that every person does use it, or will.

Space

Above we suggested that digital public space could mean a collection or archive, an object made up of a collection, or the vessel in which this collection is located. To us, the use of the word 'space' implies thinking in terms of this latter. However if we are thinking about a space in this sense, it begs questions of what kind of objects sit within it, and also whether there are other ways that the space can be used. Can things other than objects exist there? Is there a map? How do we navigate within it? This may also not be a purely metaphorical construct. It is a conceptual space, but one which can be accessed and visited in a real sense. We may speak of it in terms of spatial correlates, even though these may not relate to the four dimensions (three of space and one of time) which we are used to considering. The spatial implications of non-physical environments are discussed further in Chapters 4 and 9.

Literal digital analogues to space have been tried, including virtual worlds modelled on physical spaces such as *Second Life* and those of 'MMORPG' multi-user games such as *World of Warcraft*. The use of virtual reality to go to an 'alternate' space that is traversed in the same way as physical space is one that has been explored by many speculative fiction writers, particularly in the genre of cyberpunk. The word 'cyberspace' was coined by William Gibson in his seminal novel *Neuromancer* (1984) which was one of the original works in this genre. In that novel, as well as *Snow Crash* by Neal Stephenson (1992), the characters plug into headsets which allows them to access a shared digital world which is experienced as if it were physical, and where transactions can be carried out and people can meet and socialise. These books predated the general use of actual shared virtual worlds by several years. But such virtual worlds do not seem to have taken off in the wider sense, and are used by a relatively small subset mostly for entertainment purposes.

Rather than asking us to enter a digital realm modelled on the real one, digital interactions have started invading the physical space. A key example of this is 'backchannels' which are enabled by mobile wireless and smartphones as well as other mechanisms. By giving people portable access to online connectivity, it allows the sharing of information between people who are collocated in physical space, or participating in the same activity. As predicted by McCarthy and boyd in 2005, the majority of academic and industry conferences these days will provide participants with a hashtag which can be used on social media services such as Twitter and Facebook. This collects all communication regarding the event as a single resource which can be viewed and searched by other participants. In this way, commentary and feedback about the event can be made in real time, allowing a secondary conversation to emerge alongside that in the 'real' world and as part of the linear discourse of the presenters. This also occurs for large shared events such as sports, for example the Olympics, broadcasts of particular television programmes or during national disasters such as hurricanes. Co-ordinating online messages in a searchable way means that anyone can join the conversation by 'eavesdropping' on the public statements that are made, and physical space collapses as people across vast distances share experiences.

We also have digital infiltration of the space that we inhabit in the form of information screens that pervade our daily lives. We no longer pay attention to screens unless we want something from them; they have become a 'pull' technology. Cities are full of digital content: from advertisement hoardings, to bus and train timetables, to live traffic updates. Some of this is not directly obvious to those walking down a street but may be integral to the way the city functions (such as algorithmically controlled traffic lights), or sustain digital behaviour in those who know it is there (such as widespread wireless or mobile data coverage accessed by those using mobile devices).

Augmented reality proposes that by using devices almost as 'magic windows', you can view the digital overlay on the real world. This technology was proposed as being game changing, has been slower to take off than predicted, only recently beginning to break out beyond rather specific implementations. Perhaps it will take technologies which directly include these in users' field of view (such as Google Glass or 'smart contact lenses') to bring such things into the mainstream. But this should not obscure the fact that we are already living in a hybrid space that has digital information woven through it, and that it is a 'space' with more than the three usual spatial dimensions. The digital public space gives us a term which can be used to describe this new realm of information that is held in media other than the traditional minds, books or physical objects. This then, is the digital public space that we will talk about in this book and explore in more depth.

Case study 1.1: The Creative Exchange and the many definitions of digital public space(s)

Throughout this book, we will highlight a series of case studies describing novel research exploring some of the issues particularly relevant to digital public space, its impacts, implications, and potential. The majority of these have been the result of work conducted as part of the Creative Exchange project, a collaboration between Lancaster

University, Newcastle University and the Royal College of Art; one of the UK Arts and Humanities Research Council's four 'Knowledge Exchange Hubs'.

The four hubs, spread across the UK, were set up in 2012 with the aim of furthering connections between creative industries, and arts and humanities academics. The Creative Exchange focused specifically on the topic of digital public space as a rich area in which to develop new collaborative work, inspired in part by the work of the BBC, Tony Ageh and Bill Thompson in examining the future of digital archives of public content.

At the heart of the knowledge exchange methods used by the project was a cohort of 21 PhD students, who were recruited from interdisciplinary backgrounds to undertake a new form of practice-based PhD around the theme of digital public space. An initial scoping exercise involved guided interviews with a range of creative industry contacts and academics with an interest in the area, and identified six key cluster topics within digital public space which were either areas of current research and work, or opportunities for development. These were:

- Public Service Innovation and Democracy
- Making the Digital Physical
- Performance, Liveness and Participation
- Stories, Archives and Living Heritage
- Rethinking Working Life
- Building Social Communities: Dynamic Structures for Growth

A 'Creative Exchange Lab' workshop event was held for each of these topics, with a range of academic, industry and third sector partners invited to join the Creative Exchange team and the PhD researchers to add their expertise and develop collaborative projects in these topic areas. Collaborative teams then worked on these projects to investigate issues relating to digital public space and develop products or services within the sector. Over fifty projects were carried out, which were extremely wide ranging as can be seen both from some of the examples used throughout this book, and others which are not included here: from best practice guidelines for 'good procurement' in the digital sector, through digital capturing and sonification of craft practices such as crochet, to a hybrid digital-physical co-working space.

The PhD researchers built their research theses around the research carried out during these projects, and some in the second phase of the Creative Exchange initiated their own digital public space related projects in conjunction with academics and industry partners. Through the research they produced, many different aspects of digital public space and digital public spaces were investigated.

The work of the Creative Exchange has not resulted in a single definition of digital public space: in fact, it has expanded it beyond existing definitions and exposed the plurality of concepts that are associated with these new technologies in our public spaces. Rather than answering a question, the projects which have taken place across the work of the Hub have begun asking new ones that in turn might expose the changes that are taking place in our cultural lives and how we might use design to address and shape these in a way that is socially responsible.

What exists in the digital public space?

> The Internet, then, forms an unstable digital field, a potential space between the archive and the encyclopedia, which we have termed the encyclomedia.
>
> (Featherstone, 2009)

As we have discussed earlier in this chapter, a large part of what people immediately start to think of when discussing the digital public space is archives. Archives are generally thought of as being collections from which items (documents or objects) can be retrieved: an accumulation of primary sources; a 'repository of the national memory' (Featherstone, 2000). The critical difference between digital archives and physical archives is that because of computational processing power, it is much easier to find and retrieve objects quickly from a much larger stock. Even with a complicated indexing system, it would take a while to find the correct book you wanted from a library of ten thousand books, walk to the correct place in the stacks and retrieve it. Yet in a library of digital books you can almost instantly find out how many books use a particular word and then have them immediately available for you to read. Another aspect is that the physical space required to store these objects is incomparably reduced. Those ten thousand books can easily fit not just in your pocket, but on your fingernail, on a microchip. These two facts combined mean that digital archives can be enormously large, and may therefore potentially have less attention paid to selection and curation of the 'most worthy' items. On top of this, as we have described earlier, many large cultural organisations are also now making large efforts to digitise their collections and make them available online, to allow a greater reach and significantly easier access.

When you can keep everything, it makes sense to do so. Where pictures were previously precious and only those most promising were developed from negatives and then chosen to be put into an album, now you can store hundreds of digital photos from every day of your life. And people are increasingly doing so.[7] This may seem frivolous, but it may be valuable data for social scientists and historians of the future. As Featherstone puts it 'should we not seek to extend the walls of the archive to place it around the everyday, the world? If everything can potentially be of significance shouldn't part of the archive fever be to record and document everything, as it could one day be useful?' (Featherstone, 2000). The internet has allowed anyone to 'publish' their writing or make it available simply through starting a blog. The practice of 'life logging' is becoming more widespread, and technology is making it easier to collect data from all aspects of your life to add to your 'personal record'. The mechanisms for this may range from a camera that hangs around your neck or is fixed on your glasses (such as film maker and artist Alan Kwan has been wearing since November 2011) to more abstract 'quantified self' applications such as the Fitbit which record information about activity, sleep patterns or other aspects of one's daily life.

Of course, while this information is all in the digital space, it is not necessarily in the digital *public* space. But increasingly, such personal archives are not stored locally on private machines, either because distributed (cloud) storage is more convenient, or because sharing has become entangled with storage on services such as Flickr and Facebook. People like to share, even data that might have once been considered extremely private; there are many discussion forums where those in the quantified-self movement share and compare data they have gathered about daily habits. Additionally, the person digital information relates to is not always the person who creates and holds it. Businesses now collect large amounts of information on us. Some of this is gathered through our interaction with websites and online services, particularly social or linked services or those who make their money through knowing which adverts we might find relevant. Some may be gathered via our physical actions in the real world, such as CCTV cameras or use of loyalty cards. The Open Data movement advocates making as much data as possible from governmental or publically funded service available for open use. The reasoning for this is that doing so gives greater power to data analysis that can be used to benefit wider society. All of this adds up to a cloud of information trailing off us into the digital space all the time.

With this much information available, we must consider two critical aspects of its storage and sharing; the overall collection and its management, and how objects are accessed and retrieved in way that is meaningful. This process of retrieval from such a large store means that selection processes must be used, and these can lead to conscious or unconscious biases in what is selected. This will be explored in more detail later in Chapter 7.

Beer and Burrows (2013) identify four interrelated categories of archive that are not mutually exclusive, but which categorise and describe the way that this 'sea of data' and online information can be thought of. These are: transactional archives, archives of the everyday, viewpoint or opinion archives, and crowdsourcing archives.

Transactional archives are perhaps the closest to traditional libraries, in that they contain materials which have been produced or put online specifically so that they can be purchased or selected, and downloaded or viewed. This would include examples such as iTunes and Amazon or the BBC iPlayer. While money does not necessarily have to be exchanged to obtain something from these archives, there is a transaction taking place in that, in return for being given access to the content, information is provided as to who is accessing it. This may simply be a record of the number of times that it is downloaded, or may be more complex user data. For example, it could be connected to profiles (which could include details of demographics or location) and other purchasing behaviour. This archive of transactional information may be critical for commercial organisations that use this to create revenue.

The second category, *archives of the everyday*, covers documentation of mundane daily life, and include much of the self-generated information that we have described above; either the things that people share and post on social networking sites such as Facebook and Twitter about their activities, or via lifelogging and other self-quantification methods. These are not necessarily produced with the direct intentions of being a persistent archive, but may arise out of transient 'status updates' to be shared in a particular moment to connect with others. However, the overall collection of this data when gathered together can be used for powerful ends such as assisting anthropologists and sociologists in analysing 'everyday' behaviour.

The third type of archive identified by Beer and Burrows is the *viewpoint or opinion archive*. In this they include blogging and microblogging which is providing an opinion on something or giving insight into the viewpoint of the creator. They also include in this category the practice of commenting, rating and reviewing – this might be structured reviews on sites that collect feedback for a product or service, but also comment systems on blogs or forums, or perhaps even the practice of clicking a 'like' or 'favourite' button. Although these very simple interactions might not seem to contain much information, they may reveal surprisingly more in a networked space. For example, a study from Cambridge University showed that individual traits such as sexuality could be predicted to quite a high accuracy level by aggregating 'likes' on Facebook (Kosinski et al, 2013).

The final category of archive in Beer and Burrows' classification is that of *crowdsourcing archives*. These are communally constructed archives such as wikis that build up not only contributions from many authors, but also associated metadata about when (and sometimes why) changes and updates were made. This arguably might also include certain contributory comments sections on blog posts which can (especially on systems which allow nested comments) become comprehensive discussions significantly longer than the original material. Twitter hashtags aimed specifically at collecting examples rather than being tied to a specific event also create crowdsourced archives; these allow contributions under a particular topic to be grouped together and found and read by others who are interested in the same topic. They can be used for diverse purposes from humour (#movieprequels) to political activism (#blacklivesmatter) to creating a crowdsourced real-time information system about the weather (#uksnow).

It is unclear whether Beer and Burrows would include collaborative creation in the category of crowdsourcing archives; not only commentary as mentioned above but adaptation of existing digital artifacts. The culture of creation of 'mashup' songs and videos combining various popular media, and the sharing and building upon of memes is widely known. This 'remixing' can be facilitated by removing the necessity for technical skill. For example, during Barack Obama's 2008 presidential campaign, it became popular to share personalised versions of his iconic 'Hope' campaign poster, including altered text or images. Various sites were created[8] which allowed contributors to easily re-adapt it to their version. The poster continues to be widely used as inspiration for such transformative work, for example many variations were seen during the 2016 US presidential elections.

Figure 1.1 Original 'Hope' poster for the Barack Obama 2008 presidential campaign

Figure 1.2 A 2016 example of a remixed variation as part of an ironic protest against Donald Trump, adapted to use the image of actor Misha Collins (who was at the time campaigning for Hilary Clinton)

Sources: *Figure 1.1* Obama Hope Poster by Shepard Fairey. Photograph originally posted to Flickr by Lord Jim at http://flickr.com/photos/25028863@N00/2245362705 and reproduced under a Creative Commons Attribution 2.0 Generic license. Figure 1.2 Misha Collins Nope Poster by Tracy McCusker and Lua James. Reproduced with permission.

We could also consider fanfiction an example of something which might fit in this category; creative production and adaptation in shared world writing, such as writing new stories about favourite characters in a television series or film. While this is not a new phenomenon and has arguably been around for centuries or longer (think of Virgil's *Aeneid* as his own version of what happened during parts of Homer's *Iliad*, or *Wide Sargasso Sea* by Jean Rhys which is a prequel to *Jane Eyre*), the internet has increased the speed and durability of sharing and thus the visible volume of such works. 'Archive of Our Own' is a site designed and maintained by volunteers to provide a searchable archive of such works and currently hosts over two million contributions.[9]

This is not an exclusive list of categories of archive, and there may be many others not covered under this classification. For example, community archives may be used to store communal public knowledge. Professional archives may exist within businesses and organisations which are not public, but can be accessed in specific ways by the public. Specific sectors may have unique categories of archive, for example medical data generated in vast quantities. Such archives can be subdivided and categorised based on topic and precise content, or other categorisations might emerge. One of the key properties of large data archives is that the data can be stored, sorted and partitioned in multiple ways; users can then procure it and reuse it for their own purposes and re-categorise it.

In the discussion about the 'space' aspect of digital public space above, we suggested that the digital layering of information can provide an alternate venue for interaction. This is distinctive from traditional mechanisms because the interactions which take place there will often be, as with other digital content, replicable, searchable and persistent rather than ephemeral. They may therefore make up a component of the archival nature of digital public space, and we must consider in our categorisation of 'stuff' within the digital public space that a large part of it might be these seemingly inconsequential interactions.

Recently, public awareness has been raised about the fact that information online does persist, and that this is not always desirable. With accusations of 'spying' levelled at major governments, and the realisation of a teenage generation that seemingly private discussions online may be used against them in the future, there has recently emerged a market for artificially ephemeral online spaces. An example of this is Snapchat, which posts images fleetingly for sharing, which are subsequently removed. A high-profile court case led to Google providing the 'right to be forgotten', that is, enabling individuals to ask for misleading or outdated information on them to be removed (Lee, 2014).

There is also an argument that not every experience can be or should be digitised, and that some of the value of the connectivity of tangible experiences is in that which cannot be stored. The National Theatre Live project is innovative in that it allows a much wider audience to experience their theatre performances by live streaming them directly to cinemas around the world. But the National Theatre is very aware of the importance of the 'experience' of a live performance, so has limited the number of screenings which take place a significant time after the original performance, and does not sell the recordings on DVDs or other ownable media. To quote the 'lessons learned':[10] 'There may be limits to the "anytime, anywhere" attitude to the consumption of content. The response to the live aspect of NT Live would suggest that there is a right time and a right place for some cultural experiences'. You can only go and see the show 'live', even if this liveness is remote or delayed by a week or two. This keeps the 'specialness' of the occasion and event, and attempts to replicate the experience of being one of a limited number of audience members present in a theatre at the time.

These examples highlight the temporal aspects of the digital public space; the timing of content is critical to the experience, and may also affect the relevance of the digital content.

More recent items may have newer content. Whether they are more accurate depends on the nature of the content; consider for example that revision can be made for political ends not just for noble ones. It may be possible to find the original versions if you dig deep enough, but if the information management systems that we mentioned above preferentially bring up newer versions, then it is the information from those that is going to become the prevailing story.

The discussion above has been focusing on things that exist purely in the online part of the digital public space. But as mentioned earlier in this chapter, the digital public space can often be a hybrid realm that creeps out into the physical. As well as these purely online spaces, we should consider the fact that the digital public space includes access points which cross over with the physical, be it via augmented reality or an interactive screen. The 'digital mile' project in Zaragoza, Spain,[11] was envisaged as an urban space fully integrated with digital technology such as interactive fountains, displays of the 'mood' of the city, and free wifi access covering the entirety of the public space.

Recently a new category of physical objects has come to exist within the digital public space. Objects in the real world can be 'tagged' and linked to other items, be they physical or digital. A cereal packet can link to the toys that come free with the contents, and can link to the website promoting the brand. A public bench can record how many people sit on it, and send information to the council which will tell them whether it is worth maintaining a bench in that particular spot. This is the burgeoning realm of the 'internet of things' and is the natural extension of the digital public space into a more tangible reality. Later in this book we will discuss why this tangibility is so important. Giving things a name, a unique identifier that can be tabulated with other 'things', allows us to harness the power of computers to almost instantly catalogue and find anything. By doing so we have the power to concrete its place in a web, like that of a spider: vibrating an alert when a connected thread is touched. Another aspect of this is that objects can become intelligent – like the bench mentioned above, or the oft-used example of a fridge which knows that you have almost used up all of the milk, so puts in an order with the supermarket for more to be delivered the next morning. Trov, a start-up based in San Ramon, California, gives people the ability to record and catalogue all their possessions, creating a 'digital locker'. The physical objects become digital objects in our connected environment.

When thinking about what exists within the digital public space we should be expansive rather than restrictive. If everything we do and everywhere we go has digital information hiding behind it in the digital realm, then anything can potentially exist within the digital public space if we want it to. This can and is transforming the way that we live our lives, and the implications of this will be explored in further chapters. The flip side of this is to remember that if anything can exist, then that can include things we might not want to be retained forever and connected digitally.

Who exists in the digital public space?

There is an effect called the 'Malkovich Bias' (Glusman, 2010), named after the film *Being John Malkovich* which includes, as the name suggests, sequences where the whole world appears to become the actor in question. The Malkovich Bias suggests that there is a tendency to assume that everyone uses technology in the same way that we do. This could be extended to the fact that we sometimes forget that the use of technology is not evenly distributed. We can talk about the digital public space as if it were a consistent entity which is available equally to everyone, and used in the same way everywhere that it is available. But this is not necessarily the case. We must be careful making assumptions about how ubiquitous this situation really is.

Firstly, we must take into account that not everybody has equal access to digital technology and computers. The pervasive digital culture that is discussed in this book is not shared worldwide, and access to such technology may be limited either by economic or political pressures. Technology can be taken up at different rates in different parts of the world, and political force and economic status can play a huge role. It may seem like every human on the planet is having their lives transformed by technology. But this may be a form of the Malkovich Bias in action. There are different levels of interaction with the digital public space for different cultural contexts.

A very broad overview of differences can be seen simply by looking at the numbers of internet users in different regions of the world, while recognising that internet does not mean existence in the digital public space. There is a strong variance in the percentage of the population in each region which uses the internet, and significant differences can also be seen in the rate of growth between 2000 and 2015. The lowest growth is in those regions which already have high levels of internet penetration, which could imply that other regions are starting to 'catch up'. It is very difficult though to draw any conclusions about specific use, because the regions listed are so large and varied (*see* Table 1.1).

The 2012 Eurostat report on internet use in households and by individuals (Seybert, 2012), makes it clear that there are differences by location within Europe in how and how pervasively people use the internet. The percentage of individuals who access the internet through portable or handheld devices while away from home was found to be above 50% in six Member States (Denmark, Ireland, Luxembourg, the Netherlands, Finland and Sweden) but also below 20% in five Member States (Bulgaria, Italy, Hungary, Lithuania and Romania).

Much has been written about the 'digital divide' between the developing and developed world.

Table 1.1 World internet usage and population statistics, November 15, 2015

World regions	*Population (2015 est.)*	*Population (% of world)*	*Internet users (latest data)*	*Penetration (% of population)*	*Growth 2000–2015*	*Users (% of table)*
Africa	1,158,355,663	16.0%	**327,145,889**	28.2%	7,146.7%	9.8%
Asia	4,032,466,882	55.5%	**1,611,048,215**	40.0%	1,309.4%	48.1%
Europe	821,555,904	11.3%	**604,147,280**	73.5%	474.9%	18.1%
Middle East	236,137,235	3.3%	**123,172,132**	52.2%	3,649.8%	3.7%
North America	357,178,284	4.9%	**313,867,363**	87.9%	190.4%	9.4%
Latin America/ Caribbean	617,049,712	8.5%	**339,251,363**	55.0%	1,777.5%	10.1%
Oceania/ Australia	37,158,563	0.5%	**27,200,530**	73.2%	256.9%	0.8%
World total	**7,259,902,243**	**100.0 %**	**3,345,832,772**	**46.1%**	**826.9%**	**100.0%**

Source: www.internetworldstats.com/stats.htm

Developing countries tend to have far lower 'internet populations', and this can be for many reasons including infrastructure (such as the density of telephone lines) education levels, income, and national legal development and regulations (Chinn & Fairlie, 2010). Some initiatives such as the One Laptop Per Child programme which developed and distributed extremely cheap computers, have tried to address the challenges of connectivity in the developing world, but there has been criticism about their efficacy and methods. It has been said that access to information technology will be a cornerstone for the progression of developing countries by providing better quality of life through greater access to education, health care and government; but the paradox is that without these things it can be difficult to enable. Political will can often also affect the way in which these technologies are distributed: for example, large areas of the globe have access only to a portion of the wider digital space because of censorship and restrictions on access such as the 'Great Firewall of China'. This may create biases in the digital public space; this concept will be explored further in Chapter 7.

But this is not to say that there is a complete absence of digital public space in the developing world, or that education is necessarily a restriction on usage of such technologies. A recent experiment by the One Laptop Per Child organisation dropped tablet computers (powered by solar rechargers and pre-loaded with software) in remote villages where the children had, according to Nicholas Negroponte (founder of OLPC), never previously seen printed materials, road signs, or even packaging that had words on them (Talbot, 2012). Within five days the children were using 47 apps per child, and five months later had hacked the android system to restore functionality to some aspects of the devices that had been disabled, such as the camera. Digital technology is often extremely intuitive to use and can be picked up extremely easily by young people, therefore suggesting it is availability rather than education which acts as a barrier.

Those in the developing world may use technology within the infrastructure limitations in novel, ingenious ways due to different challenges and goals. Alternative uses may be made of technology that we in the developed world consider 'low tech' or 'old fashioned', for extremely creative purposes which rival the 'high tech' solutions used elsewhere. In rural Africa, SMS apps (which function on all mobile phones, smart or not) are transforming lives by offering services such as MedAfrica. 'Part phone book and part basic medical reference text, the app has a symptom checker and lists nearby doctors and hospitals based on the user's location' (Fellett, 2011). By sharing information using the SMS system and apps which collate this data, agriculture can be improved by monitoring for crop health and outbreaks of pest species. Having access to even the simplest digital technology can change behaviour, for example allowing more accurate time keeping and scheduling which increases efficiency by allowing people to not have to wait lengthy amounts of time for appointments. Digital banking and money transfer services allow phones to be used as a digital wallet, opening up new income streams for those who truly need it, for example by allowing instant payment for goods which are sold by a remote agent elsewhere (Vince, 2010).

The introduction of digital services seems to be a priority worldwide and is the focus of many national and international initiatives. Clearly it is believed that by reducing the divide and increasing the spread of technology, the lives of many worldwide can be improved. Despite this, the digital space that they join may not be the same as that currently inhabited by those in areas such as North America, Asia and Europe (which may also differ from each other).

As well as the gaps in digital connectivity due to geographic economic differences, there is not a consistently high level of internet usage across the board even within single regions. Many studies have looked at differences in usage based on age. Children born after the invention of the World Wide Web in 1991 have never lived in a world without online accessibility, and these children are now old enough to have children of their own. Those belonging to the generation born after 1980 have often been described as 'digital natives' (Prensky, 2001) and this perhaps overused phrase gives lie to what is the real issue – we live our lives by interaction with the world and other inhabitants of it, and digital technology gives a novel avenue for this interaction which becomes more familiar the longer you have been immersed in it. The United Nations Economic Commission for Europe (UNECE) Statistical Database[12] has data from 2013 for prevalence of weekly internet use in 16-24 year olds for 30 countries, and out of these only 5 show levels of usage at less than 90% (Bulgaria, Greece, Italy, Romania and Turkey). Three countries, Iceland, the Netherlands and Luxembourg, show figures of 100% indicating that every 16-24 year old is online regularly.

Prensky postulates that the brains of this generation are fundamentally different because of the environment in which they have grown up. While older people may be 'digital immigrants', the young are 'digital natives' who are fluent in the use of technology because they have grown up with it. There is much disagreement surrounding this. Opponents of his viewpoint argue that it is not the brains of the younger generation that are different, but the tools that are available to them and the uses that brains are being put to. This is even more the case as mobile devices become more and more prevalent, and young people in developed countries are exposed to a culture of 'always available' connected digital spaces. A survey in March 2011[13] indicated that ownership of mobile devices is growing year on year in the UK, and found that 94% of children in the UK aged 12-15 owned a mobile phone. A different more detailed survey (Masters, 2010) found that 52% of all children owned a mobile phone, with 11% owning smartphones, and 35% owning laptops (with the figure at 50% for 11-16 year olds).

A 2016 report from the Office of National Statistics (ONS) in the UK (ONS, 2016) also notes that when analysed by age, distinct differences in use of the internet in the UK become apparent. Adults in the two youngest age groups (16-24 and 25-34) were found to be proportionately the largest users of many of the 'internet activities' surveyed. These included (among several other categories) social networking, uploading content and internet banking. This rise in digital technology integration is not just found in the UK. The 2012 Eurostat report (Seybert, 2012) found that about 60% of the young EU27 population (aged 16-24 years) used the internet on the move, and between 2006 and 2012 the proportion of individuals in the EU27 who never used the internet declined from 42% to just 23%. That means almost three quarters of individuals across Europe have used the internet in some form.

With this attention on the internet changing the culture of the young, you might be forgiven for thinking that those individuals not using the internet are mostly in the older age groups, and that this 'new online world' will soon be available to everyone under the age of 30, seeping backwards up to older people as they learn how to make use of it at a slower pace. The ONS report (2016) notes that in 2016, 41.8 million adults in Great Britain (82%) accessed the internet every day, compared with 78% (39.3 million) in 2015 and 35% (16.2 million) in 2006 when directly comparable records began. They found that 23.7 million households (89%) had internet access, and that access to the internet using a mobile device almost doubled between 2011 and 2016, from 36% to 70%.

But even in the developed world, differing ways of using technology is not purely a generational divide. Specific demographic sectors such as those with lower income or education levels, or the disabled, may not have as much access to digital technology as others. Helsper and Eynon (2010) critique Prensky's assertion that it is the age of the person that contributes primarily to whether or not they are a 'digital native'. Instead, they propose a number of digital activities that indicate digital nativeness, and discuss which types of people might display these. They point to breadth of use, experience, self-efficacy and education as the critical factors in the digital divide. Various studies have looked at how social demographic factors affect the distribution of use of technology, though authors van Duersen and van Dijk (2014) argue that it is usage which is different across socio-economic variables, and not access, with those who are better educated using the internet in a more productive way.

This economic/educational digital divide may lead to social disadvantages for sections of the population. If certain services become primarily provided via online methods, then those who are not familiar with receiving services and information in this way may be cut off from them. Lack of access to preferentially digital systems such as internet banking may mean having to spend longer, for example, while queuing at the bank in person. In April 2013, the UK Government announced its intention that the 'Universal Credit' benefits system should be accessed primarily online, potentially disadvantaging those who do not have regular access (Peachey, 2015).

Crang et al (2006) note that individuals from certain social demographics may be disenfranchised if they are excluded from digitally mediated links to services and infrastructure that have become normalised for those with access to them. These people 'will be distanced from the complex and multiscale constellations of increasingly normalised electronic flows, transactions, and exchanges which operate within, through and beyond their neighbourhoods'. When digital infrastructure becomes fundamental to complex lifestyles, the disadvantages for those without become more significant. Some people may even find that they are cut off from services which they previously relied upon, because due to the normalisation of digital access in the daily routines of the more affluent and socially privileged, the non-digital access is removed.

Despite this, we should not ignore the fact that even those who never use a computer or access the internet are not, unless they are located in an extremely remote rural area, living in a completely non-digital world. More and more, services are provided in a broadcast sense which does not require individual connectivity, and as discussed above, the hybrid digital-physical space is becoming wider and thus including those who might not own their own dedicated internet capable devices. This may be something as broad as the digital information boards which are used at services such as train stations and bus stops, or a specifically targeted single-function digital service such as digital television. As of October 2012, analogue television services in the UK were turned off, so that all those who wanted to watch television were required to switch over to a digital signal. This allows concurrent digital services such as the BBC's 'red button' which allows a higher level of interaction.

Some individuals who are not counted in traditional statistics of internet usage may also have sporadic access if they use the internet on devices that belong to other people rather than having access themselves. This might not be counted in traditional surveys of internet use that focus on daily access to owned devices. If someone who does not claim to use the internet themselves asks a relative to look up the best holiday options for them, or buy a washing machine online to get a better deal, they are indirectly using the technology and therefore still have a connection, albeit second hand, to the digital public space. Equally, as digital penetration gets more ubiquitous and the population ages, there will be fewer people who do not have the skills to cross the digital divide.

In our consideration of the cultural and social impact of the digital public space and the design implications therein, we will mainly in this book be focusing on those who are on the 'top' side of the digital divide; those who are fully immersed in the digital world, and live their lives in the hybrid landscape of the digital public space. In the language used by Crang et al (2006), we are interested in constant users rather than episodic ones; those who are in constant connection with the digital space rather than those who see the internet as something that they occasionally visit for a specific purpose. This constancy is considerably facilitated by mobile internet access. Mobile telephones were not equivalent to fixed line telephones even before the advent of smartphones. SMS text messages were originally envisaged only as a means for engineers to contact each other, but quickly became a critical means of contact when talking was not suitable or only short non-time-sensitive messages were required. Mobiles are fashion items, treated as jewellery by some, and as status symbols by others who must have the 'latest model' as soon as possible. This ubiquity of connectivity has already changed our perceptions of the world, for example by allowing meetings to be flexible, so that you can change arrangements on the fly and be in touch if you are running late or need to change the venue. The rise of smartphones and the relatively massive computing power of a handheld device mean that mobile technology is now a critical portal to the digital public space, making it something that is available everywhere if you have means to use it, rather than something fixed to a particular time and place. Those who have access to this new world may live in it in very different ways.

Summary of principles

The original notion of the digital public space arose from concepts of making national and international cultural archives digitally accessible, searchable and shareable for the benefit of the wider public. Use of digital storage and metadata enables unprecedented connectivity and cross media referencing. Engagement and interaction are critical components of this vision: not just a static archive but a living 'cultural memory'.

However, because digital storage and sharing are not restricted to traditional cultural artifacts, use of the term 'digital public space' can be extended out more widely to describe the conceptual space created by publicly shared digital information. This space may be purely information based or be interwoven with physical correlates, and may in fact be made up of a variety of different digital public spaces which are connected and co-dependent. Being an inhabitant of these spaces may significantly impact on our culture and social interactions. How and why this occurs is a question which will be explored throughout this book.

To define digital public spaces in this context, key criteria we must consider are that digital public space:

– includes digitally stored and shared information

Digital information transfer conveys the properties of transferability, persistence and searchability.

– is accessible to the public

True digital commons is unfeasible because digital networks and archives are made up of privately owned areas. Instead, the key consideration of the digital public space is accessibility: accessible in principle to anyone, anywhere.

Digital public spaces may have varying behaviour norms dictated by the social construction of spaces and cultural context. They may be within privately owned contexts but accessible to the public, or may have areas restricted to a limited number or type of individuals, but who represent 'the public'.

– is space within which information flows

We are treating the digital public space as a container for digital information flow and content which gives rise to notions of space, with navigable qualities. Digital public space may collapse physical distances or may be interwoven with physical space.

– includes both archives of content and venues for interaction

Large volumes of content are possible in the digital public space because storage is trivial. Many different types of archive can be included, including content intended to be both long lasting and ephemeral, and that which is derived from other materials.

– has temporal aspects: not everything may be persistent, some is time specific

There may be some content which should be removed from digital public space, or reduced in prominence over time.

– is potentially accessible to all, but may not be evenly distributed

A digital divide exists which may be based on geographic or economic factors (large or small scale), or on age or other social considerations and pressures. Limited access may lead to novel uses of connected technology. Being a 'digital native' brought up with connected technology may affect lifestyles and culture significantly. If digital public space usage becomes ubiquitous, lack of connectivity may negatively impact those excluded for whatever reason.

Having now defined digital public space as this openly accessible hybrid digital/physical realm in which information is transmitted and interactions take place, what will follow is a discussion of the form in which it currently exists and how it is used, and how it interacts with the daily lives of those who live with and within it. Digital public space can then be put in the wider context of other ways in which experience of, and interaction with, the environment and our social communities has co-developed with technology.

With these principles in place, we can begin to explore the effects that digital public space is having on our lives: from how the historical development of communication methods has influenced how we use digital technology, through the nature of spatiality in regard to digital dimensions, to challenges and opportunities for digital living, and looking forward into the future of digital public space. We will look at the role which design has in building and developing digital public space, or mitigating some of the issues and challenges which might arise out of it as emergent processes.

Key points

- The digital public space as a concept first came to prominence as an idea and project spearheaded by the BBC, as a publicly accessible archive of public digital content including metadata.
- An increasing volume and importance of online content and interaction means that broader definitions of digital public space are necessary.
- Digital will be used here to denote computer-based information transfer with associated implications of transferability and persistence.
- There are various axes of public and private: control of access to information is critical, as well as cultural context for usage.
- While it is difficult to clearly delineate public and private, this book will use the public criteria of universal access in potentiality if not in fact.

- As well as the growth of virtual spaces for interaction, digital information is infiltrating the physical world to create hybrid spaces.
- The lack of physical space needed for digital storage means that many new types of archives can be kept, of content that might have previously been deemed worthless.
- This may include records of mundanity, created 'opinion' content such as blogs and or feedback, crowdsourced records, and collaborative creation, remixing and transformative works.
- Increasingly, the content of these archives is shared, and information relating to individuals may be collected or held by others.
- With so much content, management and retrieval strategies become critical.
- Persistent digital information may not always be desirable (if it can be used for unwanted purposes) and some experiences may not transfer easily to a digital format.
- Temporality is critical to the digital public space: timing of content is an intrinsic part of the experience.
- Access to digital public space may not be equally distributed, and may be limited by political, geographic or economic pressures, as well as demographic aspects such as age.
- Utility of different aspects of digital technology may depend on context and local factors.
- Lack of access to digital services may disempower groups to whom it is not available, particularly if non-digital versions are withdrawn.
- Mobile digital access is causing major societal changes.
- In defining or considering digital public space, there are certain key aspects which must be considered.

Notes

1 www.bbc.co.uk/mediacentre/latestnews/141111space.html
2 Accessible at www.webarchive.org.uk/wayback/archive/20130607091841/www.jisc.ac.uk/inform/inform36/DigitalPublicSpace.html#.V-P_xlgVDIU
3 www.tate.org.uk/about/projects/insight-digitisation-tate-collection
4 http://wellcomelibrary.org/about-us/projects/digitisation/
5 www.trin.cam.ac.uk/index.php?pageid=1328
6 boyd prefers her name to be spelled with no capitalisation.
7 See a variety of examples at www.newscientist.com/special/lifelogging
8 For example www.obama-me.com/; http://obamapostermaker.com/
9 http://archiveofourown.org/
10 www.nesta.org.uk/sites/default/files/nt_live.pdf
11 www.interactivearchitecture.org/zaragoza-digital-mile.html
12 http://w3.unece.org/
13 www.gsma.com/connectedliving/wp-content/uploads/2012/03/uk110811interactive1.pdf

References

Ageh, T., 2013. Why the digital public space matters. *In:* Hemment, D., Thompson, B., de Vicente, J.L. and Cooper, R., eds. *Digital Public Space.* FutureEverything Publications. Available from: http://futureeverything. org/publications/digital-public-spaces/ [Accessed 23 September 2016].

Ageh, T., 2015. *The BBC, the licence fee and the digital public space.* Open lecture presented at Royal Holloway University of London. Available from: www.royalholloway.ac.uk/harc/documents/pdf/tonyageh.pdf [Accessed 23 September 2016].

Beer, D. and Burrows, R., 2013. Popular culture, digital archives and the new social life of data. *Theory, Culture & Society*, *30*(4), pp.47–71.
Benkler, Y., 2006. *The wealth of networks: how social production transforms markets and freedom*. Yale University Press, p.9.
boyd, d., 2007. Social network sites: public, private, or what. *Knowledge Tree*, *13*(1), pp.1–7.
Brody, N. and Fass, J., Digital Public Space and the Creative Exchange. Available from: http://researchonline.rca.ac.uk/1389/19/BrodyFass_DigitalPublicSpace_2013.pdf [Accessed 23 September 2016].
Bridle, J., 2012. The library and the forum. *In:* Hemment, D., Thompson, B., de Vicente, J.L. and Cooper, R., eds. *Digital Public Space.* FutureEverything Publications. Available from: http://futureeverything. org/publications/digital-public-spaces/ [Accessed 23 September 2016].
Chinn, M.D. and Fairlie, R.W., 2010. ICT use in the developing world: an analysis of differences in computer and internet penetration. *Review of International Economics*, *18*(1), pp.153–167.
Cousins, J., 2012. Creating the backbone. *In:* Hemment, D., Thompson, B., de Vicente, J.L. and Cooper, R., eds. *Digital Public Space.* FutureEverything Publications. Available from: http://futureeverything.org/publications/digital-public-spaces/ [Accessed 23 September 2016].
Crang, M., Crosbie, T. and Graham, S., 2006. Variable geometries of connection: Urban digital divides and the uses of information technology. *Urban Studies*, *43*(13), pp.2551–2570.
Doctorow, C., 2014. *You are not a digital native: privacy in the age of the internet.* [online] Available at: www.tor.com/2014/05/27/you-are-not-a-digital-native-privacy-in-the-age-of-the-internet/ [Accessed 26 September 2016].
Featherstone, M., 2000. Archiving cultures. *The British Journal of Sociology*, *51*(1), pp.161–184.
Featherstone, M., 2009. Ubiquitous media: an introduction. *Theory, Culture & Society*, *26*(2-3), pp.1–22.
Fellett, M., 2011. Phone tech transforms African business and healthcare. *New Scientist*, 2833
Frenchman, D. and Rojas, F., 2006. Zaragoza's Digital Mile: place-making in a new public realm [Media and the city]. *Places*, *18*(2).
Gibson, W., 2000. *Neuromancer*. Penguin.
Glusman, A., 2010. *The Malkovich Bias.* [online] Available at: http://glusman.blogspot.co.uk/2010/02/malcovich-bias-over-years-ive-noticed.html [Accessed 26 September 2016].
Graham, S. and Aurigi, A., 1997. Virtual cities, social polarization, and the crisis in urban public space. *The Journal of Urban Technology*, *4*(1), pp.19–52.
Habermas, J., 1996. *Between facts and norms: contributions to a discourse theory of law and democracy.* Cambridge: MIT Press, p.360.
Helsper, E.J. and Eynon, R., 2010. Digital natives: where is the evidence? *British Educational Research Journal*, 36(3), pp.503–520.
Hinssen, P., 1995. Life in the digital city. [online] *Wired Magazine*, *3*(6) Available at: www.wired.com/1995/06/digcity/ [Accessed 26 September 2016]
Kiss, J., 2013. BBC makes space for cultural history. *The Guardian,* 6 January 2013. Available from www.theguardian.com/media/2013/jan/06/bbc-digital-public-space-archive [Accessed 23 September 2016].
Kosinski, M., Stillwell, D. and Graepel, T., 2013. Private traits and attributes are predictable from digital records of human behavior. *Proceedings of the National Academy of Sciences*, *110*(15), pp.5802–5805.
Lee, D., 2014. *What is the 'right to be forgotten'?* [online] Available at: www.bbc.co.uk/news/technology-27394751 [Accessed 26 September 2016].
McCarthy, J.F. and boyd, d. 2005. Digital backchannels in shared physical spaces: experiences at an academic conference. In: *CHI'05 extended abstracts on human factors in computing systems* (pp. 1641-1644). ACM.
Masters, S., 2010. *High tech kids cost parents £537 a year.* [online] Available at: www.parentallychallenged.co.uk/news/according-to-a-survey/high-tech-kids-cost-parents-537-a-year/ [Accessed 26 September 2016].
Moggridge, B., 2006. Adopting technology. *In: Designing interactions*, pp.237–317.

ONS, 2016. *Office for National Statistics. Statistical bulletin internet access – households and individuals, 2016*. Available at: www.ons.gov.uk/peoplepopulationandcommunity/householdcharacteristics/homeinternetandsocialmediausage/bulletins/internetaccesshouseholdsandindividuals/2016 [Accessed 26 September 2016].

Peachey, K., 2015. *Q&A: universal credit and the benefits overhaul*. [online] Available at: www.bbc.co.uk/news/business-11735673 [Accessed 26 September 2016].

Prensky, M., 2001. Digital natives, digital immigrants part 1. *On the Horizon*, 9(5), pp.1–6.

Seybert, H., 2012. Internet use in households and by individuals in 2012. *Eurostat*. Available at: http://ec.europa.eu/eurostat/documents/3433488/5585460/KS-SF-12-050-EN.PDF [Accessed 23 September 2016].

Stephenson, N., 1992. *Snow crash*. Bantam Books.

Talbot, D., 2012. Given tablets but no teachers, Ethiopian children teach themselves. *Technology Review*. Available at: www.technologyreview.com/news/506466/given-tablets-but-no-teachers-ethiopian-children-teach-themselves/ [Accessed 23 September 2016].

van Deursen, A.J. and Van Dijk, J.A., 2014. The digital divide shifts to differences in usage. *New Media & Society*, *16*(3), pp.507–526.

Vince, G., 2010. Who needs banks if you have a mobile phone? *New Scientist*, 2748.

2 Digital public space for the evolved mind

In order to understand how being an inhabitant of the digital public space influences us, and how we influence and shape it as an environment, it is useful to trace the evolution of the human mind, how it has emerged as it is, and how that is an interesting variable in the lens of the way in which we interact with the digital public space.

We will see from a historical perspective, that the ability to live in social groups with support networks of others related to us has been critical to our development, and that other technologies have built upon the human propensity to hold shared intellectual resources, communicate and share information. Digital public space is an extension of the informational public space that we already share.

In order to understand why people interact with digital information networks in particular ways, it is useful to look back through evolutionary history at the origins of humans as a species, what differences in our brains separate us from other species, and to examine how the long-term relationship that we have with tools, technology and social information has shaped our behaviour.

The evolutionary history of human sociality

As established in Chapter 1, our definition of digital public space is based upon information being shared digitally between groups or individuals across a public, communal space where humans transfer ideas and objects. This concept of existing in a highly interconnected information network is however not restricted to the digital environment, but rather something that appears to be the case for all humans. This is true in whatever culture they live – be it hunter-gatherers with little or no access to digital technology, or individuals surrounded by advanced information technology in the developed world. Indeed, it is part of the very definition of 'culture'. But, arguably, this is not something which is true of any other species on this planet. So how did we come to be so intensely reliant on the sharing of information? We can examine this by looking at the factors which contributed to the development of humans as a social species, going back to pre-human ancestors who began to develop the traits which lead to our uniquely human characteristics.

Relatively speaking, we do not know a huge amount about how our very early ancestors lived. The field of palaeoanthropology, the study of fossil remains of humans and human ancestors, is hampered by the fact that (because of the nature of how remains are preserved) the fossil record only yields a small number of examples. This means that each new discovery can rapidly change our picture of the historical record and theories of the evolutionary history of our species. For this reason, this chapter will not dwell on the specifics of the human evolutionary tree, because to do so would run the risk of becoming outdated quickly.

New, significant, previously undiscovered fossil species that feature in the history of humanity, either as direct ancestors or related lineages, are still discovered every few years; for example *Austrolepithecus sediba* in 2008, or the well-publicised *Homo floriensis* or 'hobbit' in 2003. In 2015, while this book was in progress, a large discovery of hominid fossils was discovered in a cave in South Africa and assigned as a new species: *Homo naledi* (Berger et al, 2015). At the time of writing, the remains had yet to be precisely dated, and the relationship of this species to *Homo sapiens* is still ambiguous. Each new find prompts a rewriting of the already murky 'family tree'. Often, these species are 'identified' through one or two individuals, and not even complete skeletons but a fragment of fossil – a tooth or a bone. But this does not mean we cannot infer lots of details about both anatomy and behaviour, as well as, more recently, their relationship to us in terms of genetics. Modern techniques of genetic sequencing have allowed us to peer more closely at the relationships that the owner of these sparse remains might have had with modern humans. A good example of this is the recently discovered 'Denisovans'. In 2010, a group of scientists found a tooth and a finger bone in the Denisova cave in southern Siberia (Reich et al, 2010). Through genetic sequencing, they have been able to determine that not only were these from the same species of hominin, but that this new species, dubbed Denisovans after the cave, were both distinct from Neanderthals and modern humans, and also had contributed parts of their genome to certain groups of living humans – in other words, our ancestors likely interbred with theirs.

In order to find the origin of differences between other ape species and humans, it is useful to look at what happened between our common ape-like ancestors, and modern humans. Although, as described above, our understanding of early human ancestors is not complete, what seems clear is that at some point about 6 million years ago, a particular species of ape that lived in Africa began walking upright and started down the evolutionary path that eventually led to us. These were called Australopithecines, and did not yet possess language[1] or use complex tools. However, it is believed that they lived in larger groups than other ape species, a fact which gave them an advantage against predators but which meant that there were additional constraints on how they lived. These constraints meant adaptations for social living were necessary.

Of the two main groups of Australopithecines (known as robust and gracile), it was the gracile species that, in East Africa, gave rise to species in which we see the first clear signs of higher mental capacities. The first clear evidence of simple 'Olduvan' stone tools appears around 2.6 million years ago, usually associated with the species *Homo habilis,* whose name means 'handy man'. This use of tools should not be overestimated as a major shift towards particularly human capabilities, since other species such as crows and octopi use complex tools, and chimpanzees have been taught to create stone tools similar to those of the Olduvans. However in later parts of what is known as the Early Stone Age (which lasted from the period of the first Olduvan tools up until around 250,000 years ago), more complex tools can be seen to have developed which included skilfully shaped 'Acheulean' cutting tools. The species these tools relate to is called *Homo erectus,* which had a much larger brain than previous species, and was around for a very long time – around one and a half million years. During this period there was another steady but modest increase in hominin brain size.

Authors such as Stout et al (2008) note that these were parallel trends of brain expansion and technological elaboration but mention that the link between them is controversial and poorly understood. Several theories have been put forward to explain how the changes we can see in brain size, social complexity and evidence for increasingly complex tools, might be connected. That they occur in such close proximity suggests something more than coincidence, and that these aspects must be linked – although which, if any of them, are the initial cause or whether they are reactions to some other event, is not something that is yet clear. One suggestion is that increases in parts of the brain relating to dexterity and manual manipulation were related to the ability to hold and work with tools, but also led to brain lateralisation (different tasks allocated to different sides of the brain). By taking measurements of the brains of people using Stone Age techniques to produce stone tools, Stout et al saw evidence that the brain regions used when experts created Acheulian style stone tools had overlaps with those lateralised areas known to be involved in language.

Some researchers (Dunbar, 2014) suggest that the critical difference in the *Homo erectus* species might have been control of fire – allowing larger brain size not just because it gave access to cooked meat (which is easier to digest and therefore more nutritious) but because the light from evening fires would extend the usable time during each day and allow groups to carry out social bonding during this period. This in turn would allow larger groups to form, needing larger (or at least more complex) brains to keep track of the ensuing social relationships.

These species were still not really humans as we think of them. It seems unlikely, for example, that they used language to communicate with each other. Some new theories do however suggest that they might have used a new invention to release endorphins communally and cement their social groups – laughter. And it may have been this in combination with the brain lateralisation mentioned above that preceded the major transition into anatomically modern humans – the ability to use language.

Around 500,000 years ago, a new set of species emerged out of the *Homo erectus* lineage, and these were the first archaic humans. These species, such as *Homo heidelbergensis*, show dramatic and rapid increase in brain size, and signs of far more complex material culture and use of tools. Although early migration patterns are still contested, current leading theories suggest they began to spread out into wider regions of the world around this time, and were the ancestors of species such as the Denisovans and Neanderthals. The Neanderthals shared several features with us, including larger brains, but appear to have used their brains differently with a greater proportion devoted to vision rather than the cortex (used for complex cognition).

In Africa, the *Homo erectus* lineage evolved separately into anatomically modern humans, *Homo sapiens,* around 200,000 years ago. These humans would have had the cognitive capacities of those who live today; however the archaeological evidence of tool use stays similar to that of archaic humans for a long period. It appears that these early humans left Africa around 60,000 years ago, replacing the other pre-existing species in Europe and Asia (perhaps with some interbreeding). It is at some point during the period of 60,000–80,000 years ago that significant changes begin to be seen in the types of artifacts left behind – significantly, the use of symbols can be seen at sites such as the Blombos cave in South Africa (Henshilwood, 2007) and objects can be found that do not appear to be purely utilitarian.

Around 40,000 years ago, anatomically modern humans began to spread out and colonise Europe and it is at this point that we see a major shift in the archaeological record; a major explosion of more complex and symbolic artifacts. These range from the paintings that are found on cave walls, to beautiful items of statuary and jewellery. It seems clear these people appreciated art, and had complex internal lives which might have involved religion or symbolic thinking – but certainly involved living in community groups with other individuals. This is known as the Upper Paleolithic Revolution, and is the start of what can truly be called human mentality. Although we have developed culturally since this point, this is where brain volume slowed its onward increase, and where we can say with a fair amount of confidence that these were people who thought 'like us'.

There are certain trends which can be seen across this evolutionary history, which give evidence that modern human culture is a result of evolution and influences of the selection pressures on our ancestors. A key change that is often mentioned is the increased brain size; as a result of this, humans have brains two to three times larger than you would expect given our body size, based upon close relatives such as chimpanzees. But larger brains are costly, and must have evolved for a specific reason. A leading current theory is that this reason was to facilitate living in larger groups.

Robin Dunbar has built upon the work of Byrne and Whiten (1989) to develop the social brain hypothesis, and proposes that the main constraint on group size is the time which needs to be spent on keeping track of the social hierarchies in the group. This is needed to alleviate some of the stresses that accompany group living. He argues that while it is favourable for social primate species to live in groups, there are costs as well. This includes direct costs, indirect costs and 'freeloading' (Dunbar, 2014). Direct costs are those which come from conflicts with other members of the group, for example fighting over territory or food. Indirect costs include the fact that certain aspects of communal living demand more time or energy; for example group members must spend more time on foraging and travelling, because a larger group consumes more food, may have to travel further to find new sources, and can only travel as fast as the slowest member. Freeloading costs are those incurred by members of the group who are not providing as much support as they take. By living in a group, time needed for certain tasks increases. This includes things like the aforementioned foraging and travelling but also the time for social bonding, which is necessary to negate some of the costs associated with conflict, which would otherwise become untenable. Keeping track of the pecking order in your group, who owes you favours and what the relationship is between other members, takes more brain power and higher complexity of thought. Therefore the means to cope with this was an expansion in brain size to keep track of these social hierarchies, starting a trend which would enable other mental abilities.

The legacy of this can be seen in modern behaviours. We are an intrinsically social species, not happy on our own, and we put an inordinate amount of effort into knowing what other people are doing. Humans do not have highly evolved defensive or offensive capabilities – we do not have sharp claws or run fast, nor are we powerfully jawed predators. But what we do have is flexibility, and co-operation. Language is a critical factor in being able to track social interactions in a group.

Table 2.1 (*see* p. 32) summarises these different stages in the history of human evolution, and allows us to compare how different physiological and cognitive traits emerged in tandem. Some of these traits, such as tool use and interaction, will be explored in more detail and summarised in Tables 2.2 (*see* p. 34) and 2.3 (*see* p. 38) later in this chapter.

Theory of mind

The story above is useful in relation to the digital public space because it tells us how critical social interconnectedness is to understanding both our evolutionary history and the way we live now. In fact, since there is a strong argument that our larger brains specifically evolved in order to be able to cope with managing these relationships, it makes sense that they are adapted to highlight and prioritise them. This may be why aspects of social connectedness like Facebook dominate the 'web 2.0' and current trends in digital technology. But what exactly did these large brains offer which enabled us to track this connectedness? Some current theories of evolutionary developmental psychology suggest that the critical skill these larger brains afforded was that of mentalising, or theory of mind. This is the ability to conceptualise experiences of others as being different to our own, and at greater removes.

Theory of mind can be understood in terms of intentionality; an ability to comprehend intentions at different levels. First order intentionality is the awareness of oneself, knowing

Table 2.1 Stages in the history of human evolution

	Date	*Brain size*	*Group size***	*Tool use*	*Hypothesised group interaction*
Apes	Last common ancestor 6.5mya	~50cc	<50	Limited	One-on-one or small kin groups
Austrolopithecines	6mya		<50	Limited	Laughter
Homo habilis	2.4mya		50	Olduwan tools	Laughter
*Homo erectus**	1.65mya		90	Acheulean tools	Laughter, song, dance
Archaic humans	~500kya		130	Some evidence of complex tools	Proto-language
Anatomically modern humans	200kya		150	Complex tools and early technology	Language
Humans with symbolic culture	50kya	200cc	150	Evidence for symbolic thinking	Culture

*We are also including here *Homo ergaster*, which *Homo erectus* is known as earlier during the lifetime of the species.
**From Dunbar (2014)

your own mind. It is widely believed that most conscious organisms do this in some form. Second order intentionality is where theory of mind comes in, and is the ability to understand that someone else has different thoughts. In children this can be seen to arise at around age five (depending on the development of the child) and can be tested by experiments, such as that where a box of crayons is presented to the child in the presence of another person (Bill, say), emptied out of sight of Bill and the crayons hidden elsewhere, following which the child is asked where Bill thinks the crayons are. Children without a fully developed theory of mind will indicate that Bill knows where the crayons are hidden, because they themselves are in possession of that knowledge. Third order intentionality is the understanding of this experiment – I understand that the child thinks Bill knows where the crayons are. Adult humans can generally cope with, on average, up to five levels of intentionality though some people can handle one fewer or greater. This appears to be higher than that possible for most primates, which can only get as far as first order intentionality. Chimpanzees can achieve second order intentionality, but they struggle with it and are generally not as good as five year old children. Dunbar has plotted intentionality ability against frontal lobe capacity, and suggests that it has been enabled by brain expansion over human evolution and is linked to the ability to comprehend more complex social inter-linkages. According to these calculations, Neanderthals would have had the frontal lobe capacity to only deal with third order intentionality – not enough to use a fully developed language in the way that we use it. Table 2.2 opposite shows the possible development of these orders of intentionality across species.

A side effect of these social abilities and ability to think at several levels of remove may have been a level of flexibility not possessed by earlier species. Kathleen R. Gibson (1993) describes the property of hierarchical mental constructional capacity, being able to break concepts and perceptions into fine units and recombine them into new ones, and suggests that humans have a greater capacity for this. She links this capacity to such skills as tool making,

Table 2.2 Stages in the history of human evolution: orders of intentionality

	Date	*Group size***	***Orders of intentionality*****	*Tool use*	*Hypothesised group interaction*
Apes	Last common ancestor – 6.5mya	<50	**1/2**	Limited	One-on-one or small kin groups
Austrolopithecines	6mya	<50	**2**	Limited	Laughter
Homo habilis	2.4mya	50	**2**	Olduwan tools	Laughter
*Homo erectus**	1.65mya	90	**3**	Acheulean tools	Laughter, song, dance
Archaic humans	~500kya	130	**4**	Some evidence of complex tools	Proto-language
Anatomically modern humans	200kya	150	**5**	Complex tools and early technology	Language
Humans with symbolic culture	50kya	150	**5**	Evidence for symbolic thinking	Culture

*We are also including here *Homo ergaster,* which *Homo erectus* is known as earlier during the lifetime of the species.
**From Dunbar (2014)
***Some of these values are surmised based on evidence of brain size, group size and taxonomic relatives. From Dunbar (2014)

language and social intelligence. When you can think about what your friend thinks about your other friend, you can also compare different scenarios, and try out different options mentally without having to put them all into practice. We will come back later to how this ability might be important for the design of technology. Perhaps it is this flexibility which allowed humans to succeed and adapt to changing conditions when other species, such as the Neanderthals could not.

Tools and technology

> Technology is the active human interface with the material world. But the word is consistently misused to mean only the enormously complex and specialised technologies of the past few decades.
>
> (Ursula LeGuin, 2004)

In discussing the development of human ancestors, tools were repeatedly mentioned as a key development in increased mental capacity, and an important facet of archaeological evidence for sociality and lifestyle. But in discussing modern humans and digital public space, we usually talk about technology rather than tools. Working on the assumption that it is technology, not tool use, that is a uniquely human trait, what has not yet been discussed is when tools become technology, and what exactly this means.

We must make an aside here to emphasise that to restrict our thinking of ancient tools (and hence technology) to stone tools would be wrong. So much weight is placed on evidence from stone tools and objects, that it is easy to think of them as being a particularly significant innovation in human culture, and the only tools being used at the time. But they are simply extant and persistent, because stone lasts. They were perhaps one of many different tool types used by these hominins. The rest of their technologies may simply not be preserved, thus giving us no way of knowing what they were, although some indirect evidence exists for technologies such as clothing (Toups et al, 2011).

The use of tools is not in and of itself unique to humans. However humans have elevated this act of creation to another level. The Acheulean stone tools show evidence of detailed forethought and planning. A *Homo erectus* making a hand axe must have been able to know before they started what they wanted the outcome to be. It is not simply the case that a task needed to be done and therefore an object was located that could be manipulated to complete that task. Instead, the hominin must have first visualised a tool (which did not exist) to solve a problem, before making changes to the objects they had available to create the tool.

This higher order tool development is the basis of technology – forward thinking that allows creation of second order tools which then solve the problem or can sometimes be used to create other tools. To create these tools it might even be necessary to take steps that initially make it look like you are going away from the end goal. This ability, while not exclusive to humans, is rare in animals and seen only in specific groups such as chimpanzees and corvids (Hunt, 1996); those which also live in highly social groups and which are thought to achieve second order intentionality.

These second order tools, though complex, are still not technology. Bruce Sterling (2005) makes a distinction between *artifacts* and *machines*. In this schema, the tools we have been discussing so far would be artifacts, since machines are defined as: 'complex, precisely proportioned artifacts with many integral moving parts that have tapped some non-human, non-animal power source [and] require specialized support structures for engineering skills, distribution and finance'. The missing step to bridge this gap may be *cumulative* culture and innovation. Some (for example, Dean et al, 2014) have theorised that the transmission and accumulation of ideas is at the root of the increasingly complex levels of tool use seen as early humans developed. This differs from currently extant non-human species, which for the most part appear to use only the short term copying of behaviours and the re-invention of skills, without the ability to adapt or improve upon these with further experimentation.

In *The Prehistory of Mind* (1996), Steven Mithen discusses the work of Bill McGrew, 'author of the most comprehensive study of chimpanzee material culture', which measures complexity in terms of 'technounits'. These are the individual components of any individual tool. Mithen, as an example, compares a hoe, made of three components ('shaft, a blade and a binding') and therefore afforded a value of three technounits, to computerised robots which have millions of components and therefore a similar technounit value. He does however note that the skill required to use these different tools may not be in proportion to their technounit value, with the more 'basic' hoe requiring far more skill and expertise to use effectively than a robot operated with the push of a button. The key here is that it would be almost impossible for a single individual to create something of more than four or five technounits, because a significant amount of innovation is needed for each one, and the mechanism to use them together. It requires individuals to take existing components and learn how to combine and improve on them, to create higher technounit tools.

It is these cumulatively developed complex aids to living (which would include 'machines' but also high-technology objects and systems which require a depth of knowledge built upon culture) that we are referring to when we talk about technology. More specifically, while tools are items which serve a specific purpose and solve a particular problem (like cracking a nut or stretching a hide), technologies exist in a context of wider problem solving; they are built upon previous tools and technologies, require skill and knowledge to use (or construct) effectively, and are unlikely to be something that a single individual could have independently created since they rest upon the context of the wider society and culture as well as previous developments. It may be that this is related to the mental capabilities of species to think at higher orders of intentionality. Is the complexity of human tool use possible because higher order intentionality allows understanding of what someone else meant for an invention to do, and the ability to mentally manipulate the options further of how we might improve it?

Another advantage we have over other species is that our faculties of invention themselves are much greater. Mithen (1996) talks about the 'technical intelligence' of humans, who can look at a problem and devise a simple solution; this solution may be found independently many times, so simple behaviours will be seen in various places and 're-invented' by children, who explore their environment. Conversely, while inventive behaviours and tool use have been seen to be 'culturally transmitted' in certain animal species, these ideation events are rare and only passed to individuals directly connected to the original 'inventor' (Krützen et al, 2005). There is an argument that one of the uniquely human skills that contributes to our success as a species is an intrinsic drive to copy others' behaviour, mimic it ourselves – and learn rapidly from doing so.

We do not simply copy others, but add our own ideas. Over time technology may be improved upon by individuals, leading to a 'ratchet' effect of cumulative culture (Tomasello, 1999). We also are very much influenced by what others around us do, and feel a strong urge to conform to group expectations and values. Sociality is a key part of our human behaviour, and is bound into such evolved traits as our long childhoods (and our helplessness as babies). Behaviours that may not have an obvious immediate advantage but are socially acceptable within the group may be copied and maintained within our social communities, leading to culture.

These then are the two factors that lead to the critical distinction between tools and technology – technology is created to solve problems, and therefore requires forward thinking and complex cognitive thought at higher order levels of 'theory of mind'. But it is also created through the cumulative innovation of cultural transmission, perhaps allowable by conceptualising what the previous owner or inventor of a piece of technology meant it to be used for.

Of course, this is far easier if the creator of a technology can explain to you what it is for, using language or symbolic representation. It is language that has really allowed technology to blossom within our species.

Language, storytelling and games

It is sometimes tempting to talk about the 'invention' of language, and discuss language as if it were a tool or technology itself. However this is not the case; language is a behaviour, for which parts of our brain and bodies are uniquely adapted. Children do not spontaneously create particular tools (though they do show the experimentation and exploration needed to *invent* tools). They will however start to use language in all but the most extreme circumstances. There are many examples of groups of children with parents who could communicate with each other only in rudimentary forms, who create their own grammatically structured language between them. These include deaf children who spontaneously create their own sign languages, and the development of creoles in children brought up where people without a common language are brought together. The parents will speak pidgin languages that are crude and ungrammatical, but their children will transform this vocabulary into a new, true language.[2]

Language is a key feature that allows us to socially interact with other humans. A grammatical system of speech allows behaviours that are not possible without it – for example utterances can contain meaning that is transferred robustly between individuals. It can encompass displacement; talking about things that are not present, either in space or in time (that is, happened in the past or have not yet happened). It allows the production of novel utterances, things that have never been said before (Hulit & Howard, 1993). This ability may have co-evolved with the development of higher orders of intentionality as discussed above. Constructing speech to be understood by another requires internal representations of what someone else is thinking; their motivations and the reasoning behind their purpose. This requires the mental capacity to consider the mind of others, the ability to construct a plausible hypothetical sequence of events, and also working memory capacity to hold these chains of occurrences in mind. Internal representations can also be created of things that are not yet in existence – allowing for planning and innovation. Language is a complex system involving both physical adaptations and mental agility, and only emerged at the later stages of hominin development as outlined above.

Wynn and Coolidge (2007) highlight working memory as necessary for allowing recursion of language and thus complexity of sentences. They highlight it as a key component of intellectual development, linked to executive brain functions. Wynn (quoted by Wayman, 2012) uses the example of building a trap for animals, which requires the designer to 'think up a device that can snag and hold an animal and then return later to see whether it worked'. This ability to mentally consider a problem and how it could be solved if certain steps were taken requires 'visualisation' of the problem. This ability to plan is critical for design, and is seen even at the very early stages of the evolutionary history told above. Imagining the process behind a tool requires not just visualising the steps, but putting them in sequential order: as a narrative. This ability could also be related to language, which has the additional benefit of allowing sharing of these plans and narratives.

Language may not be the only evolved social behaviour whose purpose relates to group cohesion. As mentioned above, laughter and jokes may act as a bonding exercise, a 'chorusing' behaviour that releases endorphins, especially when it occurs as part of a group (Dunbar, 2004). Laughter is also well known to be 'contagious', moving quickly within a group, and more easily experienced in company. Sharing experiences may create bonds by aligning the mental 'expectations' of the world between members of a group. Dunbar suggests that singing and dancing may have followed laughter as shared behaviours, as well as jokes once language developed. Shared 'in-jokes' create close ties between a community. Theories of humour and laughter suggest that it may result from something unexpected occurring (the 'incongruity theory'), or the release of tension after a risk is taken (the 'relief theory'). Both of these theories require the ability to understand a narrative of connected, successive actions.

In order for something to be 'unexpected', we must have had an 'expected' version – a plan in our minds of how we predicted things to unfold. Some researchers such as Jeff Hawkins (see Hawkins & Blakesee, 2007) have suggested that part of the way our brains function is to have a constantly unfolding 'map' of predictions about the world. Attention is drawn only to those things that differ from the predicted version, so that the brain does not have to be constantly overloaded with every piece of information coming in – just the ones that are new. This also supports the idea that larger brains are necessary for living in larger groups when they are tasked with creating highly detailed predictions for social interactions, which are more complex and involve higher orders of intentionality. To make these predictions, a narrative must be constructed: 'If I do this, then he will do that and she will think this'.

Control of fire would have extended the 'usable' day by allowing groups to socialise in the evenings. Once language developed, it is not such a great leap to imagine that this fuelled our propensity for storytelling; a group huddled around a fire telling tales, perhaps scary stories that would also facilitate the release of endorphins in those listening. This might be a romanticised notion, but narrative and the telling of stories plays a crucial role in how we think and behave, and this can be seen in the way in which we use and conceive of digital public space – narrative is a critical component.

Narrative and storytelling allows not only relating of activities that have taken place previously, but also hypotheses about what might happen in future. Kim Shaw-Williams (2014) has suggested links between storytelling and tracking animals, describing personal experience of mentally juggling different options that are possible given the track-signs and evidence. He relates these skills to narrative cognition, which he elaborates as meaning 'imagining being "in the body-and-mind" of absentee agents, including past and future selves. This enables mental time travel (Suddendorf & Corballis, 2007) and mind-reading (Tomasello et al, 2005) of other agents in the present.'

This drive to create narratives and stories can be seen in many aspects of human behaviour. We use stories to justify and shape our beliefs about the world, as demonstrated via experiments with sufferers of obsessive compulsive disorder. These people provided 'justification' for their habitual behaviour, citing reasons that they had previously shown they knew not to be true (Gillan, 2014). Storytelling begins very young, with children making up stories and playing 'let's pretend'. Many different species play, in that youngsters can be seen to act out adult behaviours (such as fighting) in a way that has fewer consequences. Such widespread behaviours are generally considered to have a useful function, and it has long been suggested that play allows practice of adult behaviours in a 'safe' space. It may also be related to social bonding or safely exploring flexibility, giving an advantage in adaptive behaviour.

Children imagine themselves in scenarios which are based on copying what they have seen in adults, but also combining these things in new ways. Very young children engage in sensorimotor play, simply exploring their bodies and other objects. From the ages of two to six however, children undertake symbolic/representational play involving dressing up, imaginary games, make-believe with toys, and so on. Note that this is the age range in which theory of mind skills begin to develop, and this symbolic play has been linked to the development of these skills (Varga, 2011). Dunbar (2004) links both symbolic play of this sort and storytelling to higher orders of intentionality and theory of mind. As they grow slightly older and get to school-age, other forms of play emerge such as games with rules, which are introduced by adults as practice for the complex social rule-based interactions which govern adulthood.

Games are enjoyed by adult humans as well as children, and it may be that this is practice for how to function with rules and social interaction. We continue to enjoy 'playing' with stories (through films, novels and other forms of fiction), and with rules, throughout adulthood. 'Gamification', the practice of setting semi-arbitrary goals and rewards for desirable behaviour, is another expression of this, as may be our propensity to put things in groups and ordered taxons. In fact Steven Mithen (1996) has speculated that this tendency to organise and categorise is a basic biological feature of our language, and relates to giving things names. By creating a name for something and placing it within a taxon (relatable to other items in that category) we make the concept retrievable and understandable. Relationships

Table 2.3 Stages in the history of human evolution: hypothesised group interaction

	Date	*Group size***	*Orders of intenti-onality****	*Tool use*	*Hypothesised group interaction*
Apes	Last common ancestor – 6.5mya	<50	1/2	Limited **1st order**	One-on-one or small kin groups
Austrolopithecines	6mya	<50	2	Limited **1st order**	Laughter **Basic understanding of surprise and causality**
Homo habilis	2.4mya	50	2	Olduwan tools **1st/2nd order**	Laughter **Basic understanding of surprise and causality**
*Homo erectus**	1.65mya	90	3	Acheulean tools **2nd order**	Laughter, song, dance
Archaic humans	~500kya	130	4	Some evidence of complex tools	Proto-language **Expression of displacement or fiction**
Anatomically modern humans	200kya	150	5	Complex tools and early technology **(e.g. clothes, dwellings)**	Language **Narrative**
Humans with symbolic culture	50kya	150	5	Evidence for symbolic thinking **Cumulative culture/ technology**	Culture **Communication via objects Information technology**

*We are also including here *Homo ergaster,* which *Homo erectus* is known as earlier during the lifetime of the species.
**From Dunbar (2014)
***Some of these values are surmised based on evidence of brain size, group size and taxonomic relatives. From Dunbar (2014)

are built with other words, objects and concepts. Rules are also important for social living, and are closely related to our need to exist within culture. Adherence to rules is rewarded, and those who do not adhere to them are punished; this may sound harsh, but it is how society functions and prevents 'freeloaders' from taking advantage of the benefits without having to pay the costs. It is possible to consider ritual behaviour in this light. This might include shared stories of 'what happens to those who are bad' (either through moralistic fairy tales or religious teachings) or ritualised rule-following that cement members of a community, such as the challenges offered during rituals of initiation. Understanding this wider belief requires the development of an 'internal world' with consistent beliefs held across many individuals, something which Dunbar suggests is only possible with fifth order intentionality (Dunbar, 2014). This has great power for social cohesion.

The technology of communication is intricately linked with these two principles of categorisation and telling stories. For example, Kuhn and Stiner (2007) suggest that the beads which are commonly found in archaeological early human settings are a form of information technology, in that they communicate information about the wearer. But this information could also be said to be in the form of a narrative; where they come from, how much wealth they own, and who their family and ancestors are. This last is very important, if the purpose of such communication is to understand hierarchical structures and influence how much you should trust and support any individual. Cultural context is necessary to understand such informational markers and shorthand, and ritualistic behaviour and use of specific language can act to demonstrate to others that you are part of the same hierarchy by undertaking behaviours that on the surface do not seem to serve a purpose. In the modern world, these rituals might form part of our online behaviour.

In considering the digital public space, we should think about what narratives are intrinsic to the ways in which people use it, and how structures encourage certain narratives to develop. These can be either helpful or harmful, and will be explored further in later chapters. Equally, if the aim is to invoke or enable certain behaviours, we should consider how the design of technologies takes into account behaviours around narrative and play – and by using rule and reward based systems in a way that is positive rather than harmful, and paying attention to the cultural practices that form alongside them.

Culture, cognition and technology

Several times now we have mentioned culture as being an evolved behaviour, like language, that is a vital part of 'being a human'. Culture is made up of an accumulation of behaviours that are taught to members of the group (usually as children) and form the communal social behaviour expectations. Technology may form an integral part of this cultural behaviour, such as the use of body decoration to send cultural signals about status and heritage, as discussed above. Although the last major change in terms of brain enlargement and development happened with the advent of modern humans around forty thousand years ago (and in fact there is some evidence that human brains have become *smaller* in the last 20,000 years[3]), there has been significant change in cultures and their structure. When technological novelty leads to a major structural change in the nature of culture for all those who are exposed to it, this may be a called a revolution. Revolutions shift our cultural life in ways that become ingrained and embedded; the building blocks of 'cumulative culture' have reached a point at which all further technology is impacted by the level which went before, and it is almost impossible to return to the previous state. However, the transition to this ubiquitous use will always take some time.

Weizenbaum wrote in 1976 that technologies such as the clock or the map become part of 'the very stuff out of which man builds his world.' The ratchet affect means that technologies accrue, but it also means that it is difficult to separate them from the structures of culture; which by necessity require conformity and uniform acceptance of practices. This makes it almost impossible to remove such an integrated technology once it has become a critical part of society. A revolution could therefore be said to describe a technology which 'becomes an indispensable component of any structure once it is so thoroughly integrated with the structure, so enmeshed in various vital substructures, that it can no longer be factored out without fatally impairing the whole structure' (Weizenbaum, 1976). Abandoning such a technology, or even disruption to move to a 'better' one, would cause major chaos and confusion.

The earliest of these we can only speculate about, but they may have included such things as clothing, doctrinal religions (as opposed to personal shamanistic ones which do not impose external rule systems), agriculture and cities. There are several key factors a technology needs to have to be firmly embedded in a culture to the extent that its use is almost ubiquitous; primarily, there must be a benefit to using it, either for the individual or the larger society or both. For example, the introduction of agriculture would have meant a more predictable and reliable source of food for the group, allowing it to increase in size. Clothing must have played a key part in the expansion of humans outside of the equatorial regions, by allowing early hominins to keep warm at night or in more northern or southerly climes – especially since it is likely they had by then lost the covering of body hair as an insulator. In a sense, this technology allowed us to adapt the functionality of our body. In Chapter 3 we will examine trends in digital technology that have radically changed our relationship with physical space.

Some critical technology is useful for reducing cognitive load, rather than giving such direct benefits; adapting the functionality of our minds rather than our bodies. A useful example of this might be the development of writing. To read a piece of writing and learn from it, you do not ever need to meet the person who wrote it. This therefore allows transmission of information at significantly greater removes of both space and time (younger generations can even read it when the original author has died). Writing also solidifies information and reduces the amount of error in transmission, compared to spoken communication which must be remembered. Multiple copies can be made, to assist spread, and something written can be amended or added to by the original author or others. Being able to write things down also has cognitive benefits for the individual, for example by storing things in a written list rather than in our memory. Writing may even improve the logical reasoning process, allowing more complex and structured works which are refined and developed on paper over time.

Unlike language, writing is a technology rather than a behaviour. However once it is learned it becomes highly integrated into our mental processes. This is possible because our brains have a high degree of flexibility, known as neuroplasticity. This can achieve remarkable things including restructuring of brain usage in those with sensory impairments. Neuroplasticity is not necessarily restricted to early development. Sensory substitution can be learned, such as in the case of individuals who use echolocation to 'see' surroundings by producing clicks with their mouth. Most often this is studied in those who are visually impaired, particularly from birth, but it seems from recent work that this is a skill that can be learned by anybody, to varying degrees of success (Teng & Whitney, 2011).

Technological developments which integrate with mental processes do not have to be 'easy' to learn, but should give benefits once they are. This is evidence that the nature of cumulative culture allows cognitive gains from a technology to be postponed. Culture can support the learning of difficult skills as children (and in fact we as humans have an exceptionally long childhood) which can overall improve cognitive efficacy of our lifestyle. As children, we are very flexible and adapted not to have particular skills, but the ability to learn new ones and pick them up from our teachers and peers.

There is a certain rhetoric that modern technology is 'changing our brains', often imagined to be for the worse. For example, Nicholas Carr (2011) has expressed concern that capacity for 'concentration and contemplation' is reduced by the high-speed information stream of the internet. However, even if this were the case – and some (for example Smart, 2010) have argued it is not – it is individual brains that are changing due to neuroplasticity and the cognitive learning process, and not our species as a whole. Like learning the skill of reading and writing, which allows us to externalise memory and not have to possess the skills of rote memorisation of oral histories, some skills are being prioritised over others. And the fact that the skills to which we give priority differ, may not necessarily be a negative phenomenon, if newly prioritised skills are more suited for our modern environment. For example, experiments in computer science in interaction design have found that individuals will refrain from memorising information content if there is a sufficiently low cost of accessing the information from the environment (for example, Gray & Fu, 2004, Waldron et al, 2007). In a world where large amounts of information are easily accessible at the touch of a button, perhaps we are instead prioritising the ability to search efficiently through large amounts of information to locate the correct item that we want to retrieve?

It is also worth considering the fact that by utilising technology of this sort, we may be extending the capabilities of our brains in the same way that binoculars or a microscope extend the capabilities of our eyes. Writing allowed us to store information externally by writing it down, and digital public space allows the information to be replicated and shared almost infinitely, very rapidly. This is critical to the concept of distributed cognition; that our intelligence is not just related to what goes on inside our heads, but also what goes on in the space between them, where we interact. Fischer (2000) discusses two different types of distributed cognition, between ourselves and objects (as demonstrated by creating a written shopping list), and between a number of individual human minds. The latter highlights how our disinclination to memorise information that can be retrieved elsewhere does not just apply to technology. Clive Thompson, in his book *Smarter Than You Think* (2013) describes a similar phenomenon where spouses will 'share' the information that they recall, not bothering to memorise things that they know their wife or husband will be able to 'remember for them'. This same transactive memory functions when we know that the computer will 'remember it' for us. Fischer makes note however of the fact that when knowledge is shared between several individuals rather than one individual and inanimate tools, there is no 'head', and thus externalisations – a record of shared mental efforts – are needed to undertake joint social creativity.

Along with the binoculars and microscopes mentioned above, plenty of other technological inventions fit into the category of support or augmentation for our bodies, such as glasses which correct failing sight, pacemakers which improve heart rhythm, or artificial prostheses for missing limbs. But rather than think of these technologies as a separate category from brain augmentation, we should think about the intrinsic relationship between the functioning of our bodies and how our brains interpret and interface with the world.

Brains and bodies

Like the propensity to focus on transfer of social information, other evolutionary based elements of how our brains operate, beyond culture, can influence technological development. The way that our senses work is also fundamentally tied to our evolutionary history. This includes our visual and auditory mechanisms, but also aspects of perception tied to the tangible awareness of our body within physical space. This is something that increasingly is being seen to have an impact on digital interaction technologies, which are beginning to move away from the 'screen based' purely visual context, to one which involves more of our senses.

Quirks of our sensory systems can mean we interpret incoming information in particular ways. An example of this is the fact that we are highly attuned to movement. Our visual system evolved in a context where spotting something in the corner of your vision may have meant a predator lurking. This is true of most animals, but we should be aware of this bias when considering the use of digital technology. Many designers already make use of such biases, for example the proliferation of internet advertisements which use moving images to draw the attention. The way that we perceive shapes is influenced by expectations that the mind makes, which leads to visual illusions such as the one in Figure 2.1.

Figure 2.1 Visual illusion
Source: Akiyoshi Kitaoka (2009), www.ritsumei.ac.jp

The image consists of concentric circles, and not a continuous spiral as it appears. Our brains take the pattern of light and shade and extrapolate too far to tell us that there is a continuous spiral, even though that is not the case.

Disconnect between our perception and reality can be seen on deeper and more surprising levels. As described above, it appears that our brains create a predicted version of the world and attend only to things which do not fit this map. Sometimes details that are different to what we have mapped out are not perceived at all, especially if we are focusing on something else; a factor in many road traffic accidents involving cyclists and motorcyclists whom the drivers 'did not see'. This is a phenomenon known as change blindness: when we are not expecting something to change, we do not always perceive it; particularly if the exact moment of change is 'masked' by something, our attention is focused on a different aspect of what we are looking at, or the change happens during visual saccades. The most famous example of this is the 'gorilla experiment' of Simons and Chabris (1999), in which 50% of participants watching a video of a basketball match and asked to focus on the number of passes being made, failed to notice a person in a gorilla suit walking across the pitch. The participants who did not notice the gorilla were frequently very surprised when told what they had not 'seen' and refused to believe it until they were shown the video again.

A related version of this experiment was done more recently with radiologists, who were asked to examine x-rays of lungs for nodules (Drew et al, 2013). On the final scan, a small image of a gorilla was placed in the upper quadrant of the lungs, but 83% of the radiologists[4] failed to notice it, because they were focused on examining the lungs for the nodules. When we are focused on looking for something in particular, we do not perceive the rest of the things within our field of view that we expect to remain the same.

What these studies demonstrate is that just because something happens in our environment, in front of our eyes, it does not mean that we take in what is actually happening. Therefore interaction designers who are developing technology requiring attention must bear this in mind. It is all very well putting lots of information into the digital public space, for example on screens displayed in public areas, but this does not necessarily mean that they will be seen by people walking through that space, as can be seen from the case study of TILO (see Chapter 3). Physiological differences may also need to be taken into account to ensure accessibility and prevent discrimination: not only in terms of those with sensory impairments but more subtle differences between groups. As an example of this, some preliminary work by danah boyd (2014) suggests that there may be a difference between the primary way that men and women process 3D vision (motion parallax versus shape-from-shading). This difference may mean that women are more likely to experience nausea when using 3D technology such as the Oculus Rift that relies on using just one of these systems.

We must also be careful not to think of our brains as disconnected from our bodies, with sensory information being restricted to visual and audio streams as if from a camera feed entering a black box. The physicality of our bodies is a key part of how we experience the world, and the theory of embodied cognition argues that awareness and consciousness would not be possible without physical experience. Some evidence which reinforces this point of view relates to how we learn. Many people will say that they find it easier to learn something if they can actually practice it physically rather than just watch. Consider how you track back through your movements to remember where you were when you were using a lost item, to enable you to find it again. Changes in the body can cause changes in the mind, and vice versa. Examples include studies finding that moving the facial muscles into the position of a smile causes cartoons to be rated as funnier (Strack et al, 1988) and that we feel colder when we are lonely (Zhong & Leonardelli, 2008). Based on theories of embodied cognition, new AI research is being undertaken which rather than attempting to model intelligence in a purely computer based system, allows it to emerge taking into account physical hardware and sensors (Mullins, 2011).

The integration of physical experience and our mental processes also extends to the tools which we use. When you are entirely comfortable and familiar with the use of something, it can almost feel like an 'extension of yourself'. Part of this can be attributable to habit – many people will relate to looking at an empty wrist for a forgotten watch, as described by James Cooray-Smith on Twitter:

> I left my watch at home & it's really, really bothering me. I feel really, profoundly uncomfortable. Like I've lost a finger.
>
> (Cooray-Smith, 2014)

However, these feelings are not always just related to habit, or just metaphorical. Brain imaging studies have shown that the mental mapping of yourself, or 'body schema' can be 'extended' onto a tool; for example, extending the reach of the arm (Maravita & Iriki, 2004, Sposito et al, 2012). This integration can lead to remarkable phenomena such as the 'rubber hand illusion' (Botvinick & Cohen, 1998; Ehrsson et al, 2004; Tsakiris & Haggard, 2005) in which participants are shown a prosthetic rubber hand being stroked simultaneously with their own (hidden) hand, and experience the prosthetic as part of their body, feeling shock if it is struck. Such phenomena have been used to treat phantom limb pain, allowing amputees to 'feel' an artificial arm as their own, 'unclench' the painful phantom fist and relieve the pain. Although a level of identification as a 'real' limb is needed for this to work (for example it does not work with a wooden stick: Tsakiris et al 2008, de Preester & Tsakiris 2009), enough personal connection and expression can lead to, for example, musicians who feel their instruments are 'part of themselves'.

Linking this back to the way in which culturally ubiquitous technologies form the basis of cultural revolutions, we should consider whether digital tools extending mental capabilities (such as memory) can be incorporated as an extension of the body. All the evidence we have seen here shows that it is very difficult to separate how we experience the world physically, from how we think about it. This is the basis of material engagement theory, proposed by Malafouris (2004), which hypothesises that human cognition, sapient 'minds', arise synergistically out of brains, bodies and objects. By developing new technology which acts upon our brains and bodies, we are therefore affecting the shape of our minds.

The fact that multiple senses are tied into our perception, including physical tangibility, is one that is important to bear in mind when considering how we use and design technology. For example, it is often considered difficult for individuals to understand large amounts of data which may now be available. Using sensory modalities other than visual, for example being able to 'feel' the data, may be able to make this experience easier. Future systems should make use of our full range of senses, to give a more integrated experience and one which is closer to our everyday experience of the world based on our evolutionary history.

Case study 2.1: Northpaw

Can digital technology give us new physical experiences of the world, and extend our existing senses? The neuroplasticity of our brain means that it can adapt to a wide variety of tools which give augmentation – such as the stick used by the visually impaired – but this does not have to be restricted to replacing or even enhancing sensory experiences which already exist. Glasses help us see as well as someone with perfect 20:20 vision, and a periscope helps us see around a corner. But can we extend this even further to more novel experiences? Some people are experimenting with technology that goes in this direction. The Northpaw device, created by Sensebridge, aims to utilise the brain's neuroplasticity to give an additional sense – that of magnetoreception. This exists in some species such as birds, who can navigate based on their sense of magnetic north, but does not (as far as we are aware) exist in humans because we do not possess the necessary receptors. The Northpaw consists of an anklet containing vibrating motors of the sort found in mobile phones. The motor closest to magnetic north will vibrate, giving the wearer a constant 'sense' of where north is. However the aim of the device is not just to be able feel the vibration and use it to find north. Because the signal is constant input, the brain may theoretically learn to interpret this in the same way as senses such as sight or smell, and allow the wearer to 'know' where north is in the same way that you 'know' whether you are in the light or the dark.

Jacobs and Huck (2017) undertook an investigation of how such sensory directional information might be incorporated into everyday life, by adapting the Northpaw into a belt which was worn for a period of a month. Jacobs undertook a series of navigational tests both before wearing the belt, and having worn it for several weeks. The results were surprising: in part because the tests were completed successfully both with and without the belt. Navigational ability was seen both without the intervention, and during it, although it was not conscious – in each case the tasks were successfully completed despite a feeling of 'lostness'. It seems that directional 'sense' is something that we do have even when we are not aware of it, and it is relating this to map locations which may be a cognitive hurdle. There were some indications that being aware of the directionality, via the sensation of the belt as it became familiar, increased the feeling of 'knowing' where you were; at least in terms of improving the 'sense of direction' which appears to be highly variable between individuals. The main effect that was observed seemed to be in terms of mindfulness – a heightened awareness of both sensory input and specifically directionality as a result of being constantly reminded that it is a factor.

While there are no definitive answers here about incorporation of 'novel' senses, it does seem that additional information from tools connecting to the digital public space (for example GPS directionality) might be possible, and that we could use this to assist with our mental and other processes. This might function in the same way that some cognition currently appears to be offloaded to paper maps; when maps were used in the experimental tasks, constant checking of location against the map was observed, to reinforce knowledge of location. Digital maps move a step beyond what can be achieved by fixed paper maps because they can be centred on the individual rather than the location, on the person rather than the journey.

Can we 'sense' the higher dimension of digital public space, and instinctively 'travel' around it in the same way that the Northpaw device allows magnetic navigation? These are questions we cannot immediately answer, but the ease with which young people incorporate this technology into their lives suggests that neuroplasticity may be playing a role.

Digital public space: the next cultural revolution?

So far in this chapter we have discussed how proto-humans moved from social apes, to modern humans with elaborate cultural modes of living, and how the development of our larger brains may have been congruent with managing larger social groups. We have looked at tool usage and technology, and how language and storytelling developed with both of these: providing a means to think about the future, the past, and what others are thinking. This enabled a more effective way of managing social hierarchies and structuring how we live in large groups, and led to revolutions of cumulative technology. We have also touched upon how our thinking minds incorporate information from our physical senses and build this into our cognitive perception of the world. But what has this got to do with digital public space, and its design?

Previous revolutions mentioned earlier in this chapter involve specific technological breakthroughs which facilitated large groups living harmoniously and sharing information. The concepts of private and public are fundamentally linked to some information being shared, and some being restricted. Information may be selected for broadcast (such as the informational beadwork described above) or given to select chosen individuals. Writing in particular is an information technology which changed the nature of how information is shared and spread, and caused significant changes to society and culture as well as how individuals use their brains. The development of the printing press is also often considered a revolutionary technology because written text could be duplicated and distributed in ways never before considered. Because digital public space affords a new realm in which to share and distribute information and content, it may enable similarly large shifts in culture; a continuation of the ratchetting of technological revolutions built upon those of the past, with no way to return to an earlier status quo.

Stewart and Cohen (2009) refer to the repository of information outside of our individual brains (whether in oral history, written text or digital content) as part of a concept of 'extelligence'. They note that this is not simply a matter of 'keeping a record', but that individuals can access the information and also add to it and change it. Through this process, a cumulate external repository of culture and information is built that can in turn be accessed by others, scaffolding the development of humans as a social species. By this mechanism we link technology to our cognitive development, augmenting our intelligent minds. If digital public space is an expression of extelligence, we must consider what effects this 'upgraded' extelligence, which can be accessed and edited at significantly greater speeds, has on our individual behaviours. Andy Clark (1996) describes how the flow of information between our brains and the world is crucial to their function and our cognition. Rather than a 'filing cabinet' of fixed solutions, we scaffold our behaviour on external prompts that are constantly adapted. When this is linked in real time to the extelligence available via the digital public space, we may see, and are already seeing, adaptations to this resource.

Balsamo (2011) writes: 'The invention of novel devices, applications and tools necessarily involve the manifestation of an array of human practices; new languages; new body-based habits; new modes of interactivity; new forms of sociality; new forms of agency; new ways of knowing; new ways of living and dying'. In her book *Designing Culture* she explores the interaction between technology and culture, and digital public space is no exception to this.

> The social impact of imbedded computers may be analogous to two other technologies that have become ubiquitous. The first is writing, which is found everywhere from clothes labels to billboards. The second is electricity, which surges invisibly through the walls of every home, office and car. Writing and electricity become so commonplace, so unremarkable, that we forget their huge impact on everyday life.
>
> (Weiser and Brown, 1997)

If the digital public space is ubiquitous early enough, it may become embedded in our society as a new revolution. Is this already the case? Authors such as Josephine Green (2003) have spoken about the 'Knowledge Age' being a new societal shift comparable to the agricultural or industrial societal shifts. Perhaps if we are aware of a revolution in progress we can help shape the technology that is enabling it so that it is as effective as possible. As we have seen from discussions above of neuroplasticity and perception biases, we can take advantage of how our bodies already work in the development of the digital public space.

In Chapter 2 we have illustrated the evolution of the brain and how we developed language, narrative, tools and technology in a situated context, and how the digital technologies use these attributes and are in turn adopted by individuals and communities. We now turn to the complex relationship between the digital and the physical world, in particular the notion of the digital public space and the physical world. We consider what impact this has on modern culture and human behaviour.

Key points

- Our existence within a highly interconnected information network is a key feature of humanity, and forms part of the definition of 'culture'. This has developed as a consequence of our evolutionary history, which can be traced back through hominid ancestors via fossil and genetic records.

- Tool use and technological elaboration developed in tandem with increased brain size, but any link between them is poorly understood. A major shift in the archaeological record co-occurs with the emergence of anatomically modern humans, who created complex and symbolic artifacts.
- Some theorise that increase in brain size was related to living in larger social groups, to track social hierarchy and alleviate stresses arising from conflict: language is a critical factor for tracking social interactions in a large group.
- Theory of mind allows us to understand intentionality at different levels, and is tied to use of language. It may also allow flexibility of thinking that is important for the design and development of technology.
- Forward thinking and planning allows creation of higher-order tools (to create other tools).
- Technology is the result of cumulative culture and innovation, involving transmission and accumulation of ideas, improving on what came before: allowing innovation that no single individual could undertake alone.
- Language, as well as other shared behaviours such as laughter, singing and dancing, is an evolved behaviour which allows social interaction to occur.
- Much of our human behaviour reflects a propensity for narrative, storytelling and games, which are important for flexibility of action, rules of social living, and transfer of information via technology.
- Culture is made up of an accumulation of behaviours which are taught to members of a group and form the communal social behaviour expectations. Significant cultural changes due to ubiquitously embedded technologies may be called revolutions. After these it is almost impossible to return to a previous state.
- Integration of certain technologies of information into our behaviour and mental processes is possible due to neuroplasticity and learning ability.
- Digital technology may be 'changing our brains' but only in the same way all previous technology has: through neuroplasticity and distributed cognition allowing us to 'think' via external tools. In a modern environment we may give priority to different skillsets.
- We may interpret information in specific ways due to quirks of how our brains operate, based on historical evolutionary biases: brains and bodies must be considered as a fully integrated system, which can be extended into our technological augmentations.
- Digital public space may be the basis of a new technological revolution because of its function as a distributed cognitive engine for 'extelligence', the repository of total human information.

Notes

1 Fossils show that the hyoid bone, related to language use, resembled that of an ape rather than a human (Alemseged et al, 2006).

2 For an excellent detailed overview of the physiological basis for language, see *The Language Instinct* by Stephen Pinker (1995).

3 *The Domesticated Brain*, by Bruce Hood (2014), argues that this is due to self-domestication.

4 Interestingly, the gorilla was not seen by any untrained individuals given the same task, indicating that the radiologists may have been more perceptive of the detail of the lung scans.

References

Alemseged, Z., Spoor, F., Kimbel, W.H., Bobe, R., Geraads, D., Reed, D., and Wynn, J.G., 2006. A juvenile early hominin skeleton from Dikika, Ethiopia. *Nature*, *443*(7109), pp.296–301.

Balsamo, A., 2011. *Designing culture: the technological imagination at work*. Duke University Press, p.4.

Berger, L.R., Hawks, J., de Ruiter, D.J., Churchill, S.E., Schmid, P., Delezene, L.K., Kivell, T.L., Garvin, H.M., Williams, S.A., DeSilva, J.M. and Skinner, M.M., 2015. Homo naledi, a new species of the genus Homo from the Dinaledi Chamber, South Africa. *Elife*, *4*, p.e09560.

Botvinick, M. and Cohen, J., 1998. Rubber hands 'feel' touch that eyes see. *Nature*, *391*(6669), pp.756–756.

boyd, d., 2014. Is the Oculus Rift sexist? [online] *Quartz* Mar. 28. Available at: http://qz.com/192874/is-the-oculus-rift-designed-to-be-sexist/ [Accessed 26 September 2016].

Byrne, R. and Whiten, A., 1989. *Machiavellian intelligence: social expertise and the evolution of intellect in monkeys, apes, and humans*. Oxford Science Publications.

Carr, N., 2011. *The shallows: what the internet is doing to our brains*. WW Norton & Company.

Clark, A., 1996. *Being there: putting brain, body, and world together again*. MIT Press.

Cooray-Smith, J., 2014. *I left my watch at home & it's really, really bothering me. I feel really, profoundly uncomfortable. Like I've lost a finger.* [Twitter]. 5 November. Available at: https://twitter.com/thejimsmith [Accessed: 1 October 2015].

De Preester, H. and Tsakiris, M., 2009. Body-extension versus body-incorporation: is there a need for a body-model? *Phenomenology and the Cognitive Sciences*, *8*(3), pp.307–319.

Dean, L.G., Vale, G.L., Laland, K.N., Flynn, E. and Kendal, R.L., 2014. Human cumulative culture: a comparative perspective. *Biological Reviews*, *89*(2), pp.284–301.

Drew, T., Võ, M.L.H. and Wolfe, J.M., 2013. The invisible gorilla strikes again: sustained inattentional blindness in expert observers. *Psychological Science*, *24*(9), pp.1848–1853.

Dunbar, R., 2014. *Human evolution: a Pelican introduction*. Penguin UK.

Dunbar, R.I.M., 2004. Language, Music, and Laughter in Evolutionary Perspective. *In: Evolution of communication systems*, MIT Press, p.257.

Ehrsson, H.H., Spence, C. and Passingham, R.E., 2004. That's my hand! Activity in premotor cortex reflects feeling of ownership of a limb. *Science*, *305*(5685), pp.875–877.

Fischer, G., 2000. Symmetry of ignorance, social creativity, and meta-design. *Knowledge Based Systems*, *13*(7), pp.527–537.

Gibson, K.R., 1993. Tool use, language and social behaviour in relationship to information processing capacities. *In:* Gibson, K. R. and Ingold, T., eds. *Tools, language and cognition in human evolution*. Cambridge University Press, pp.251–269.

Gillan, C., 2014. Why can't I stop? *New Scientist*, *223*(2980), pp.28–29.

Gray, W.D. and Fu, W.T., 2004. Soft constraints in interactive behavior: the case of ignoring perfect knowledge in-the-world for imperfect knowledge in-the-head. *Cognitive Science*, *28*(3), pp.359–382.

Green, J., 2003. Thinking the future. *In:* Aarts, E. and Marzano, S., eds. *The new everyday: views on ambient intelligence*. 010 Publishers.

Hawkins, J. and Blakeslee, S., 2007. *On intelligence*. Macmillan.

Henshilwood, C.S., 2007. Fully symbolic Sapiens behaviour: innovation in the middle stone age at Blombos cave, South Africa. *In:* Stringer, C. and Mellars, P., eds. *Rethinking the human revolution: new behavioural and biological perspectives on the origin and dispersal of modern humans*. Cambridge University Press, pp.123–132.

Hood, B., 2014. *The domesticated brain: a Pelican introduction*. Penguin UK.

Hulit, L.M., Fahey, K.R. and Howard, M.R., 1993. *Born to talk: an introduction to speech and language development*. New York: Merrill.

Hunt, G.R., 1996. Manufacture and use of hook-tools by New Caledonian crows. *Nature*, *379*(6562), pp.249–251.

Jacobs, N. and Huck, J., 2017. Can we give ourselves extra senses? *In:* Heywood, I., ed. *Sensory arts and design*. Bloomsbury.

Krützen, M., Mann, J., Heithaus, M.R., Connor, R.C., Bejder, L. and Sherwin, W.B., 2005. Cultural transmission of tool use in bottlenose dolphins. *Proceedings of the National Academy of Sciences of the United States of America, 102*(25), pp.8939–8943.

Kuhn, S. and Stiner, M.C., 2007. Cognitive perspectives on modern human origins. *In*: Mellars, P. ed. *Rethinking the human revolution: new behavioural and biological perspectives on the origin and dispersal of modern humans*. McDonald Institute for Archaeological Research.

LeGuin, U., 2004. A rant about "Technology". [online] Available at: www.ursulakleguin.com/Note-Technology.html [Accessed 26 September 2016].

Malafouris, L., 2004. The cognitive basis of material engagement: where brain, body and culture conflate. *In:* DeMarrais, E., Gosden, C., and Renfrew, C., eds. *Rethinking materiality: the engagement of mind with the material world*. Cambridge: McDonald Institute for Archaeological Research, pp.53–61.

Maravita, A. and Iriki, A., 2004. Tools for the body (schema). *Trends in Cognitive Sciences, 8*(2), pp.79–86.

Mithen, S., 1996. *The prehistory of the mind: a search for the origin of art, religion and science.* Thames and Hudson.

Mullins, J., 2011. Squishybots: soft, bendy and smarter than ever. *New Scientist, 212*(2838), pp.48–51.

Pinker, S., 1995. *The language instinct: the new science of language and mind* (Vol. 7529). Penguin UK.

Reich, D., Green, R.E., Kircher, M., Krause, J., Patterson, N., Durand, E.Y., Viola, B., Briggs, A.W., Stenzel, U., Johnson, P.L. and Maricic, T., 2010. Genetic history of an archaic hominin group from Denisova Cave in Siberia. *Nature, 468*(7327), pp.1053–1060.

Shaw-Williams, K., 2014. The social trackways theory of the evolution of human cognition. *Biological Theory, 9*(1), pp.16–26.

Simons, D.J. and Chabris, C.F., 1999. Gorillas in our midst: Sustained inattentional blindness for dynamic events. *Perception, 28*(9), pp.1059–1074.

Smart, P. 2010. Cognition and the web. *1st ITA Workshop on Network-Enabled Cognition: The Contribution of Social and Technological Networks to Human Cognition*, USA.

Sposito, A., Bolognini, N., Vallar, G. and Maravita, A., 2012. Extension of perceived arm length following tool-use: clues to plasticity of body metrics. *Neuropsychologia, 50*(9), pp.2187–2194.

Sterling, B., 2005. *Shaping things.* Mediaworks Pamphlets, MIT Press, p.9.

Stewart, I., and Cohen, J., 2009. *Figments of reality: the evolution of the curious mind*, revised ed. Cambridge University Press.

Stout, D., Toth, N., Schick, K. and Chaminade, T., 2008. Neural correlates of Early Stone Age toolmaking: technology, language and cognition in human evolution. *Philosophical Transactions of the Royal Society of London B: Biological Sciences, 363*(1499), pp.1939–1949.

Strack, F., Martin, L.L. and Stepper, S., 1988. Inhibiting and facilitating conditions of the human smile: a nonobtrusive test of the facial feedback hypothesis. *Journal of Personality and Social Psychology, 54*(5), p.768.

Suddendorf, T. and Corballis, M.C., 2007. The evolution of foresight: What is mental time travel, and is it unique to humans? *Behavioral and Brain Sciences, 30*(03), pp.299–313.

Teng, S. and Whitney, D., 2011. The acuity of echolocation: spatial resolution in the sighted compared to expert performance. *Journal of Visual Impairment & Blindness, 105*(1), p.20.

Thompson, C., 2013. *Smarter than you think: how technology is changing our minds for the better.* Penguin.

Tomasello, M., 1999. The human adaptation for culture. *Annual Review of Anthropology*, pp.509–529.

Tomasello, M., Carpenter, M., Call, J., Behne, T. and Moll, H., 2005. Understanding and sharing intentions: the origins of cultural cognition. *Behavioral and Brain Sciences, 28*(05), pp.675–691.

Toups, M.A., Kitchen, A., Light, J.E. and Reed, D.L., 2011. Origin of clothing lice indicates early clothing use by anatomically modern humans in Africa. *Molecular Biology and Evolution, 28*(1), pp.29–32.

Tsakiris, M., Costantini, M. and Haggard, P., 2008. The role of the right temporo-parietal junction in maintaining a coherent sense of one's body. *Neuropsychologia*, *46*(12), pp.3014–3018.

Tsakiris, M. and Haggard, P., 2005. The rubber hand illusion revisited: visuotactile integration and self-attribution. *Journal of Experimental Psychology: Human Perception and Performance*, *31*(1), p.80.

Varga, S., 2011. Winnicott, symbolic play, and other minds. *Philosophical Psychology*, *24*(5), pp.625–637.

Waldron, S.M., Patrick, J., Morgan, P.L. and King, S., 2007. Influencing cognitive strategy by manipulating information access. *The Computer Journal*, *50*(6), pp.694–702.

Wayman, E., 2012. When did the human mind evolve to what it is today? [online] *Smithsonian Magazine.* Available at: www.smithsonianmag.com/science-nature/when-did-the-human-mind-evolve-to-what-it-is-today-140507905/ [Accessed 26 September 2016].

Weiser, M. and Brown, J.S., 1997. The coming age of calm technology. *In:* Denning, P.J. and Metcalfe R.M., eds. *Beyond calculation: the next fifty years.* Springer New York, pp.75–85.

Weizenbaum, J., 1976. *Computer power and human reason: from judgment to calculation.* W. H. Freeman & Co, p.20.

Wynn, T. and Coolidge, F.L., 2007. Did a small but significant enhancement in working memory capacity power the evolution of modern thinking. *In:* Mellars, P., Boyle, K., Bar-Yosef, O. and Stringer. C. eds. *Rethinking the human revolution.* McDonald Institute for Archaeological Research, pp.79–90.

Zhong, C.B. and Leonardelli, G.J., 2008. Cold and lonely: does social exclusion literally feel cold? *Psychological Science*, *19*(9), pp.838–842.

Section 2

What are the attributes and effects of digital public space?

3 The physicality of digital public space

This chapter and Chapter 4 will explore the impacts that digital technology, and digital public space in particular, have on modern culture and human behaviour. Many of the ways in which interactions take place in the digital public space are purely information based, which will be discussed in more detail in Chapter 4. But there are some aspects of digital public space which arise directly from how our minds have evolved in tandem with our physical bodies, and encroach more tangibly on how we experience things, and how we function in the world around us. Our evolutionary history has a direct impact on how this technology has developed, and we can explore how the experience of our bodies affects how we build and use digital public space, and how we can make it work more effectively.

The importance of physicality

As described in Chapter 2, our cognition and mental experience is heavily influenced by our physical experience of the world; how our bodies work and how they have evolved. This is in part because of the way in which we process sensory input; that some forms of information (such as that received by our eyes) may be prioritised and privileged over others. Inferences and extrapolations are made to turn piecemeal input into a coherent cognitive process, or at least the impression of one. However, another critical aspect, touched upon in Chapter 2, is how we use tools and physical objects external to our bodies and minds to extend our cognitive experience. This not only refers to the 'extelligence' that we build to hold our collective knowledge, but outsourcing cognitive power to physicality. An example of embodied 'active cognition' can be seen in the way that advanced players of the computer game *Tetris* use external 'epistemic' actions to reduce inner computation effort as described by Kirsh and Maglio (1994). Physically rotating the game pieces appears to give information on their potential fit with geographical spaces more quickly than mentally computing the possibilities, and this is the strategy used by experienced players. For this to work, there must be high co-ordination between mental and physical action, creating a unified constructed cognitive system including both the mental work and physical manipulation of the pieces. This use of external action to boost our cognitive power is not unusual, and is in fact a fundamental part of how we think. As Andy Clark puts it:

> We use intelligence to structure our environment so that we can succeed with less intelligence. Our brains make the world smart so we can be dumb in peace! Or, to look at it another way, it is the human brain plus these chunks of external scaffolding that finally constitutes the smart, rational inference engine we call mind.
>
> (Clark, 1996)

Digital technologies are not isolated from our physical existence, and are in fact closely interlinked with the spaces and places in which we live our lives. They can therefore become part of our external cognition in the same way as other features of our environment, and the connected nature of digital technology may mean that our cognition is boosted in ways it never has been before. Even those digital technologies which do not initially appear to have a physical component can affect our physical behaviour, by changing the capabilities of the space in which we live.

In some cases though, the causality is reversed. Some digital technology is specifically constructed in the way that it is because of our physical experience of the world. This in turn is due to the ways that our bodies and brains function, which is determined by our evolutionary history as explored in Chapter 2. By first exploring how designers and developers have approached purely digital, virtual spaces, many of which are public, we can shed light on what happens when interactions occur between digital and physical public spaces.

Virtual worlds, simulations and games

As early as the 1940s, technology was being developed to achieve simulation of real world experiences: in the first instance, flight simulation (Hillis, 1999). The main objective of this was military, with the goal of providing training without risking either the planes or the lives of the pilots. If you make a mistake while learning to fly a real plane, the results can be catastrophic – but if you are able to train on a simulator first, the risk is much reduced. This is an ethos which has carried through to modern training for (among other things) space flights, but flight simulators have also been used by the entertainment industry to give an experience of what flying a plane or a spacecraft is like. A key aspect of high level simulations is that they engage multiple senses – you not only see the effect of your actions on screen, but feel drag and resistance on the joystick or controls, and feel the roll and yaw of the craft in motion. These mechanisms use technology to create a simulation of physical space and tangible experiences: virtual reality. Virtual worlds, in contrast to the simulation of a particular environment and scenario, simulate the nature of space itself and create a new 'world' which can be explored.

Digital archives often invite spatial metaphors, such as 'navigating' a file system or learning the 'geography' of where things are catalogued. These do not always correspond exactly to physicality, and space is used simply as a linguistic tool, because this is how we are used to finding non-digital entries in an archive. However, early fictional versions of digital shared spaces, before such things were technologically possible, used actual geographic analogues to navigate. These are often referenced when talking about the conception of the internet or digital public spaces. In the novel *Snow Crash* for example, Neal Stephenson describes how his characters can visit shops and buildings which have been built along 'The Street', which circumnavigates the spherical, featureless virtual planet that is the 'Metaverse', travelling by foot or using the 'monorail' (Stephenson, 1992).

In 2001, Dodge and Kitchin discussed the mapping and spatialisation of 'cyberspace', suggesting that increasingly space-based representations would come to dominate how the online world was used, and giving a range of examples of spatial systems for information management and retrieval. Tangible navigation, based on replicating the feeling of traversing a space, seems to have been anticipated as the natural extension of existing virtual spaces and categorisation systems. It is interesting to note that the modern web has instead relied increasingly on search-based systems that take you directly to what you want rather than having to navigate the connections or route yourself.

Although the main structure of what we call the 'internet' today is not navigated with such a direct experience of presence, there has been a long history of the development of such spaces, particularly in games. Many games use maps and spatial locatedness as an underlying component of their function. Simon Garfield (2012) notes that one of the earliest computer games *Spacewar!* developed in 1961 at MIT, truly came into its own when a map of the (real) night sky was added as a background. 'Thereafter almost all screen games required some sort of map for effective play… the maps sometimes come in the box with the game, but more often they *are* the game, and the cartographical interpretation of the landscape is the ultimate challenge.' Modern games can be extensive and highly detailed in their mapping of imagined or real spaces. Some of these are what are known as 'unstructured games', where the creators have built a coherent world which can be explored by players. The extremely popular *Grand Theft Auto* series lets you drive around a city that even includes detailed street maps and an in-game 'satellite' navigation system which directs you to the address you want. However, the difference between map-based individual games and the metaverse, is that the latter could be visited and changed by many different people at once, who could meet and converse there. The virtual city was a *shared* space, and one that persisted, remaining the same (or at least, updated by other people) each time it was visited. This is a feature which exists in collaborative games and shared worlds.

Shared virtual spaces may sound like a new phenomenon, but in some ways they are not. In a sense, you could argue that the earliest form of virtual worlds was in shared storytelling. When stories are told, a rich sense of their setting and location is built up in the minds of those telling and listening to the story. Many people can contribute to what the world 'looks' like and how characters might interact in it, and writing a story down can give it persistence. However, the codification of shared storytelling worlds was facilitated by digital technology which allowed many people to share in a coherent persistent representation that was updated in real time by their actions. In the 1980s, MUDs or 'multi user dungeons' were extremely popular. Also called multi user domains, the word dungeon was used because these text-based collaborative games were often based on traditional dungeons and dragons style fantasy games. The games allowed users to participate in text-based collaborative play to explore a 'dungeon'. The surroundings were described, and players could choose where to go or how to interact with the people and objects they found, by typing commands. Downey (2012) calls these 'first generation' virtual worlds, noting that they often had a small number of players. He describes how the successors to these, such as the pioneering game *Habitat* by Lucasfilm, added graphical aspects to the virtual worlds, giving users avatars that they could control and see move around the space.

Many modern games are a direct descendant of MUDs. Popular 'Massively Multiplayer Online Roleplaying Games' (MMORPGs) allow large numbers of players to simultaneously explore a virtual world and, usually, complete activities or 'quests' within it. These are, according to Downey, 'third generation' virtual worlds, and an important distinction is their size. The world's most-subscribed MMORPG is *World of Warcraft*. At the end of 2014, the game had over 10 million subscribers[1]. This is a game that takes up hundreds of hours of people's lives.

Games can also attempt to duplicate specific physical spaces directly, and allow their manipulation as well as exploration. The popular game *Minecraft* does not try to emulate real life with photorealistic graphics, but consists of a world made of 'blocks' which can be arranged into complex configurations by the players. A critical aspect is that 'worlds' built by players are open and, if their creator desires, can be visited by any other player in the game. Although the game involves traditional 'game' activities such as gathering resources and fighting opponents, there are modes in which creation and building is paramount, and many versions of real world locations have been painstakingly constructed. The Danish government has recreated the entire country of Denmark within the game, including all buildings (BBC News, 2014). The Creative Exchange 'Blockanomics' project aims to take this simulation further, and has built a virtual version of the city of Carlisle which, as well as recreating the buildings, enables players in the space to vote on issues that they think are critical to the running of the city: both the virtual version and the real. The project, which involves city councillors, aims to use children's existing engagement with the game to raise awareness of civic issues, and act as a teaching aid.

Downey (2012) describes virtual worlds as: 'persistent, multidimentional, graphical environments consisting of open communities in which people can establish a sense of presence, learn, socialize, collaborate with others and express themselves'. In the examples above, persistence allows these worlds to be returned to at different times maintaining a perpetual existence. This persistence is one of the key features identifying something as a virtual world as opposed to some other form of interaction. The space remains in existence even when an individual is not using it, in fact would persist even if nobody was using it. The location of people and things and the ownership of objects is maintained and builds a coherent world that can be 'visited' and relied upon to have its own temporally stable existence. In this way, these digital spaces simulate the real environment and allow users to circumvent limitations of their daily life while replicating the 'feel' of being in the same geographic space as other people. Schwartz (2006) talks about how simulating physical aspects of the world and the environment can add to the perception that it is a 'real' space with persistent existence. Features such as ambient noise of habitation, and the development of specific 'attractions' and distinct features of different geographic regions within the space, can encourage a feeling of tourism and sightseeing to these fully developed worlds.

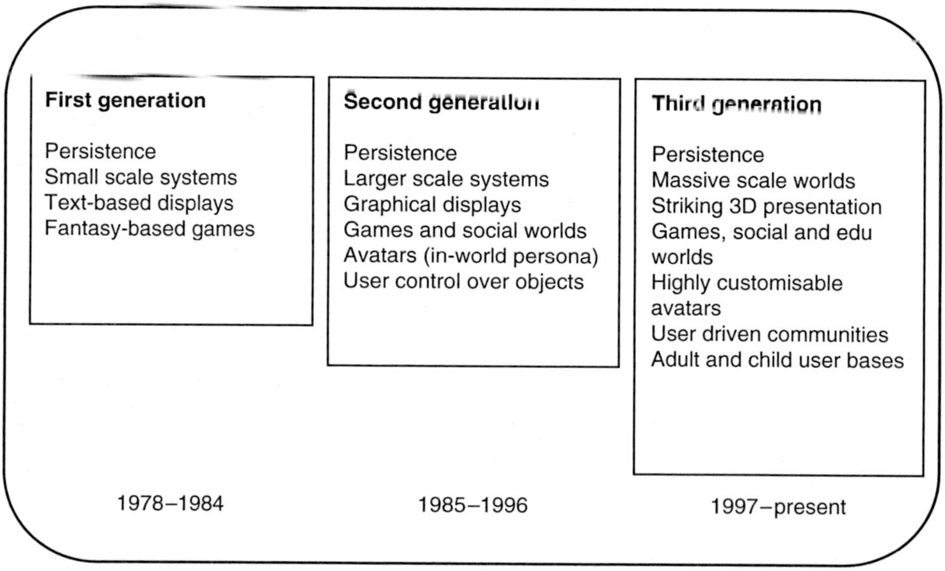

Figure 3.1 Summary of common traits found in virtual worlds during three generations of history
Source: adapted from Downey (2012)

Along with persistence, Thimm (2012) notes that interactivity and physicality are key. Interactivity does not just refer to communication between two parties, but that multiple people can participate at the same time, and the actions of one person affect the experience of others. Physicality identifies that we are talking about analogues of geographic spaces rather than other online social communication methods such as social media platforms. There is a first person experience of a physical environment, represented in some digital form. Having a spatial aspect to a virtual world offers a different type of flexibility in the way in which it is used by the participants. For example, the dimensionality of the world can give it specific affordances and allow experiences that map more closely to those which people are used to from the 'real world'. In a three dimensional representation that includes walls, and other barriers to line of sight, surprises can be found around corners. From a first person perspective in a three dimensional landscape, 'there is a "near" and a "far", an "above" and "below", and a "behind"' (Shaw & Warf, 2009). These experiences all feel 'natural' because they conform to cognitive and experiential systems that we use regularly in daily life, which have developed as part of our evolutionary history as described in Chapter 2. Dodge and Kitchin (2001) describe similarly how the processes of learning and understanding in geographic space may get applied to navigation in virtual space, for example learning landmarks to guide our path and concretise things in our mind.

Some virtual shared spaces are not traditional games in the sense of having objectives that the players are meant to achieve, but rather can be explored and used as a venue for social interaction. The most well known of these virtual worlds is *Second Life*. There are no concrete goals or tasks set by the creators; it is a 'life simulation' or a 'virtual social world' rather than a game. Bakioglu (2009) describes it as 'a relatively open-source multi-user virtual environment (MUVE) which comes into existence through the performative activities of its users'.

Such worlds allow users to present themselves as they wish, creating an avatar (a virtual version of oneself) that looks however they would like. Limitations of distance are also lifted, since *Second Life* allows instant 'teleportation' between regions – anybody can travel instantly large distances. Of course, being a digital space also means that individuals physically located large distances apart can be co-located in the virtual space.

A large variety of different activities 'take place' within *Second Life*, including but not limited to: showcases for large organisations, academic instruction and conferences, commercial businesses selling objects and fashion for avatars, museums, and purely social spaces (Ikegami & Hut, 2008). One key innovation of *Second Life* was that rather than charging a monthly fee for access, the creators at Linden Labs sell virtual space – plots or 'grids' that can be built upon by the people who buy them. *Second Life* allows people to create digital goods, and Linden Labs protects the intellectual property of the creators (Thimm, 2012). In this way, it encourages a functioning economy around the buying and selling of such virtual objects. This includes a virtual currency, the Linden Dollar. The ability to transfer money between 'real' and virtual economies is an important step in the actualisation of such worlds, and a topic that we will return to later.

While *Second Life* is a public space in the sense of being an egalitarian, open meeting space, it is not public in the sense of being publicly owned: such spaces are created and maintained by companies, who hold legal rights of access and ownership (the details of which are a topic of much academic discussion). Blizzard (the creator of *World of Warcraft*) can and has, for example, taken legal action against individuals who use third party software breaking the 'rules' of the world (Spooner, 2014). However, this is no different to the situation in physical 'public' spaces such as shopping centres and private parks. These are not public property, but spaces where the public gather and meet. There is more or less free access within the parameters of 'polite behaviour' laid out by the owners and societally agreed upon, and this can similarly be the case in virtual spaces.

Spatial and tangible interactions via digital media

When virtual spaces first rose in popularity, it was suggested that they might revolutionise social interaction and even work practices. Perhaps rather than travelling between meetings physically, or commuting to one's place of work, it would be easier to meet 'online' in a neutral digital space which can be accessed by anyone from anywhere. The complexities of the social interaction possible in such spaces led some (for example Benedikt, 2008a) to speculate that the field of architecture should turn its gaze to planning and managing these virtual spaces, with the new career 'cyberspace architect' emerging. It is arguable as to whether this has happened. Information architecture is certainly a critical area in digital public space, but normally more concerned with conceptual architecture of spaces that do not always map to physical analogues.

Despite this, there is a specific role that virtual worlds fill which is not always catered to by other digital spaces without an equivalent physicality. Various authors have described how these virtual worlds enable a feeling of 'togetherness' and co-presence that is absent from many other digital forms of communication, even video calling and other technologies where you can see the other person. Ikegami and Hut (2008) suggest that this is because the participants are free to 'explore the totality of their joint space', including walking around and coming across new areas. Their spatial nature, that the individuals are in some sense 'projecting themselves' into a virtual space, is important for this sense of co-presence. As described in Chapter 2, we are highly attuned via evolutionary selection to read social cues from others – be it via language, body position or how we observe interaction with others. The more closely virtual simulations of this experience map to our cognitive expectations, the more 'natural' the experience will feel and the more individuals will be inclined to use it for various purposes. Ikegami and Hut also note the social dimension of these spaces, that meeting in such a context will 'flatten age differences and social backgrounds'. In this sense it is very much a 'public' space, in that all who access are equal not just in their capabilities, but also in their right to be recognised as an equal individual based on their actions and interactions rather than their location, culture or social status 'outside' the virtual world.

In 2007, large media hype around the possibilities inherent in *Second Life* increased registered users significantly, and according to figures from Linden Labs, in 2014 there were around one million regular users. However, despite the fact that large numbers of people used and indeed still use *Second Life*, it never became the ubiquitous medium for business and socialising that a lot of people at the time predicted, although virtual world games such as *Minecraft* and *World of Warcraft* continue to be popular. Estimated worldwide figures for individuals engaging with virtual worlds are 20-30 million, with an average of 22 hours engagement each week (Spooner, 2014). But ubiquitous uptake for non-play purposes has not occurred, and this may be due to a variety of reasons; social, cognitive and technological.

While there is a significant extent to which people do 'connect' with their avatars, some of the issues that people experienced with early iterations of these spaces may have been that they did not fulfil the requirements of the senses necessary to 'feel' as if you were in a different place with different people. It is in fact relatively easy to relate yourself externally to a representative that you control, and to bind the control of that representation to your own senses. Some players even insist that a necessary part of being a 'good player' is to project your own body boundaries into the game so that 'they feel the joystick as an unconscious extension of the hand' (Hayles, 2002). But there is a greater threshold for naturalistic interpersonal interactions.

In Chapter 2, the embodied aspects of our experience and consciousness were discussed, and the way in which the brain is able to 'extend' the self into prosthetics and tools that act on behalf of the physical body. The rubber hand illusion tricks your brain into believing that a fake hand belongs to itself, simply because sensation and observed interaction link up. So it makes sense that when you control an external version of yourself, it is perceived as part of 'you', especially if it is controlled directly by your actions and you experience feedback on things that happen to the avatar. There is evidence that this is the case, and this embodiment phenomenon exists with avatars, where the player 'extends' their experience to various degrees into the avatar. Shaw and Warf (2009) describe how the use of three dimensions is an important aspect in facilitating this extension of the self, noting that 'on-screen events appear more "immanent" to the player [in three dimensions] than do the abstract representations found in two-dimensional games. This immanence is at its most pronounced when the game adopts a first-person perspective within a three-dimensional spatiality'.

It has even been found that the appearance of your avatar can map back across to how you perceive yourself and your own body, extending the experience in the other direction. Yee et al (2009) named this the Proteus Effect, finding that users who were given taller avatars showed more aggressive behaviours when negotiating than those who were given shorter avatars. This is similar to cognitive effects on embodiment in the physical sphere, such as the fact that changing the sound of your own footsteps can make you feel lighter or heavier (Burns, 2014), and that wearing different clothes can affect your perception of self to the extent that wearing a superhero top can make you rate yourself as stronger (Pine, 2014). This again relates to cognitive effects generated by how our brains take signals from the world to build an internal 'reality' that encompasses ourselves and our relationship with the world.

However, there are still limitations with the amount of embodiment and seamless interaction possible with systems such as *Second Life*. Much of this is because, at the moment, there is limited sensory feedback from virtual spaces. As stated previously, the whole-body experience is important for how we perceive the world, therefore if the match-up between our bodily experience and the actions of the avatar is not perfect, or we are only using one or two of our senses, the experience will be much less powerful. We might also experience problems if we project onto the avatars in a way that they are not equipped to handle; for example, reading body language as hostile or over-friendly when the sophistication is not good enough for the user to have intended this. As Donath (2014) comments: 'increasingly realistic avatars, with highly detailed faces and bodies, set up increasing expectations of responsiveness and subtlety that they are not equipped to fulfil'.

One way to trick the senses into a more 'realistic' experience is to co-ordinate input. In 1965, Sutherland wrote about an 'ultimate display' that would give the impression of being completely surrounded by 3D objects. This creates 'total immersion' and has been attempted by various technologies over the years, particularly for cinematic experiences. Large screens such as 'cinerama' and the large IMAX screens have attempted to maximise the amount of peripheral vision included in the experience. Cinema has also experimented with 3D, with various levels of success across the years as different methods were tried. Currently 3D cinema is available widely, but there is still not total acceptance that this will come to dominate the medium. Other senses have also been included in cinematic immersion, from seats which vibrate along with the motion of the film, to surround-sound, to 'aroma-rama' which spread appropriate smells to the audience. But all of these are presented passively to the audience who cannot interact with or affect what they are presented with, beyond what the film-maker intended that they experience. We have not yet perfected the 'ultimate display', though it has been presented in science fiction; an example which is probably familiar to most people is the 'holodeck' in Star Trek. This allows the crew of the *Starship Enterprise* to experience any setting they choose, and interact in a tactile sense with the people and objects they encounter there. While this is well beyond current technological capabilities, there is still progress being made in the technology of virtual interaction and simulation of experience.

High Fidelity[2] is a new virtual world in development by Philip Rosedale, the creator of *Second Life*. The key difference in this system is that it tracks the facial expressions and mannerisms of the users in real time – with extremely low latency – and maps them on to the avatar, giving a much more realistic interaction experience. The creators hope that this will finally enable the virtual world interactional experience to be 'as good as or better than the real world' (Farber, 2013). The system captures hand movements as well as facial expressions and even mouth movements, and has a much lower lag even than video chats such as Skype (with only a 100-millisecond delay for the action to be performed by the avatar) and therefore seems much more realistic. Jeremy Bailenson, director of Stanford University's Virtual Human Interaction commented that 'I used the system to interact with a person in real time and it felt like he was in the room with me' (Murphy, 2014).

Oculus Rift is another well publicised technology that hopes to integrate the physical experience more fully, initially in terms of virtual reality but which could be transferred to virtual world interaction. Based on headset technology which gives the viewer a '360 degree' experience of viewing the virtual world, the key feature of Oculus Rift that may lift it above previous virtual reality technology is full positional tracking. This enables the images in the headset to move in sync with the movements of the user. Another important aspect currently in development will bring in data about the user's body and produce replicas of their limbs in the virtual space, in the positions they would expect them to be. This will allow the brain to equate the limbs to those of the user's own body. Echoing the 'rubber hand' experiments, this will be magnified by the perfection of haptic feedback technology which gives touch sensation from the virtual reality systems to your body using full body suits and motion capture. This kind of technology is currently in development from companies such as Perception, who in 2014 ran a successful crowdfunding campaign for an extremely lightweight, low cost motion capture system that is compatible with systems such as Oculus Rift[3]. Haptic systems will in theory enable you to have a replica of your body and motion in the virtual world, and have a sense of touch that you can actually feel.

If these technologies are successfully integrated, it could be envisaged that rather than get on an airplane to attend a meeting, one could use a headset and be instantly 'transported' to a neutral space with full physical experience of being there: with the capability to shake hands, handle and feel virtual objects, and see every nuance of facial expression in your colleagues. If this becomes the case, the ownership, governance and public/private divide of such spaces will be important to consider, especially if they are linked and persistent, creating virtual 'real estate' that might be publicly accessible in some circumstances.

This type of remote working might not be restricted to virtual spaces. With tactile experiential feedback, possibilities are opened for the anticipated endpoint of a suite of technologies called telepresence, which allow extension of one's self to a remote space. The word was coined by Patrick Gunkel, and first popularised by Marvin Minksy who spoke about 'future instruments that will feel and work so much like our own hands that we won't notice any significant difference' (Minksy, 1980). In 1992, Steuer defined it as the 'experience of presence in an environment by means of a communication medium'. With communication technology now interspersed liberally through our entire lives, this could be interpreted as almost any means by which we express ourselves without being in the location of the recipient. However true telepresence, distinct even from 'video calling' technologies such as Skype and Google Hangouts which are now endemic within businesses, allows one to fully participate as if you were in the remote space. Although video calling technologies have revolutionised collaboration between geographically distanced partners, and allowed home working to flourish, the critical feature they lack is direct sensory feedback in real time. This gives a sense that you 'inhabit' the remotely operated device. When taking part in a video call you are restricted in your field of vision, do not have autonomy of movement, and usually only your head and shoulders are visible: this may reduce a lot of interpersonal information gained from body language. Both parties must also agree to initiate the interaction. True telepresence aims to actually give a physical presence, in one form or another, to the person who is attending remotely. One way of doing this is to give a physical 'body', in the form of a robot, to the person who is not present. Telepresence of this sort is not a virtual world, but a virtual intrusion onto the real world. Users of telepresence robots, such as the Beam and VGo, have reported that they felt much more agency and more connected to the space they are 'visiting'. Seth Stevenson reports how when using such a robot, people were forced to 'make space' for him:

> When I attended a… meeting in robot form… people were forced to move around the robot to take their seats. When I spun the robot slightly to its right to observe someone speaking at the far corner of the room, everyone could see where my attention was focused. When I piped up, all eyes turned to my screen, instead of vaguely staring at the speakerphone in the middle of the table as they'd have done if I'd phoned in.
>
> (Stevenson, 2014)

Similarly, he mentioned being able to arrive at the meeting when he chose, early or late if desired, rather than at the whim of the person controlling a video conferencing connection. This embodiment within the space rather than a device is what would bring telepresence into the realm of digital public space, because your representation could move autonomously within spaces that are considered public, via digital access. As this technology matures, there could be many implications for being able to transmit your experience to a physical avatar. From holiday-makers who could hire a 'body' to visit remote places, to people with physical disabilities who could leave their wheelchairs and participate fully in activities in spaces which were previously not accessible to them. There could also be significant improvements in productivity if rather than dialling in for a Skype call, you could come together with people in distant places for a fully immersive meeting where collaborative workshop activities and an equality of presence could be achieved.

Other types of work could also be affected by this kind of telepresence. Specialist fields may benefit from the ability to work at a distance not just via Skype calls and video conferencing, but in more tangible ways. It is now possible for a doctor to examine, diagnose and treat a patient who may be many hundreds of miles away, using a variety of technologies, and even perform surgeries (Marescaux et al, 2001). This kind of medical technology is important both for isolated areas where the nearest doctor may be unable to travel there quickly, and for the developing world where transport costs may prohibit access to traditional treatment. This may be critical for the development of other new technologies; astronauts on a mission to Mars may not be able to take a surgeon with them, but could have access to remotely operated surgical procedures that might become necessary on longer missions (though time delays in communication in such circumstances could be an additional hurdle).

Some have suggested that remote accessing of robots could be used to outsource menial tasks such as housework, which are still extremely difficult to automate. Services such as Amazon's Mechanical Turk currently allow work such as translation or image identification (which is difficult for computers but easy for humans) to be done by workers in developing countries, who work at much lower rates of pay. If telepresence might allow workers to control a robot that can perform housework, this might undercut prices and lead to a robotic worker revolution. There would be economic and social implications for such a move, but also, potentially, privacy ones. Early pilot studies of such a system revealed anxiety in those who did not want an anonymous presence in their household, since it was impossible to connect with the user or users of the robots in any meaningful way. Systems are therefore in development for ways in which the tasks could be carried out while also using filters and other technological solutions to obscure personal details such as private documents, or indeed any object which does not need to be identified to carry out the task at hand (Harris, 2014).

Digital/physical hybrid spaces

We have talked about virtual spaces that replicate physical ones, and about projecting digitally into remote physical spaces. Inhabiting virtual worlds such as those described in the previous section separates your experience from the 'real world'. Purely digital spaces, or even remotely accessed physical ones, must by necessity include the notion of moving from one space to another; hence phrases such as logging on, jacking in and so on. But increasingly, there is blending between the physical and digital experience of the world: spaces are imbued with digital technology and contain virtual aspects, but also have physical presence and tangible impact. As the virtual spaces are built from physical correlates and owe much to the geographic nature of how we interact with environments, so these new interconnected spaces must take into account the physicality of our daily lives. By augmenting either our environment or the physicality of digital experience, we can change the nature of our experience without switching to a 'different' reality.

As with other areas of new technology, experimentation with this new hybridisation of the virtual and physical has had early exploration in the realm of games. An example of a game which took the traditional notion of consoles and extended it into the physical world is the Nintendo Wii. The handheld controller for the Wii not only functions as a traditional gaming pad, but also includes accelerometers which, along with a sensor, can correlate the movements and rhythms of the player with the gaming world on screen. This allows physical interaction with the game world – when a player leans back, left or forward, this is represented in the actions of the avatar they control. Still more sophisticated use of physical interactions with gaming is the Kinect by Microsoft, which includes cameras that pick up bodily movement of players within its field of vision. In this way, movements can be transferred directly into the virtual world. The power of the Kinect system is such that it is being rapidly adapted for a myriad of uses outside gaming, such as in healthcare applications where it can read gait information and, for example, be utilised to monitor whether elderly people are likely to be at risk of falls (Stone & Skubic, 2013).

While the examples above bring physical interaction to screen-based activities, augmented reality seeks to enhance the real world with a digital overlay of information. Daniel Aliaga described early iterations of this technology, suggesting that:

> There are… situations in which the user might want to remain in the real world and instead of completely replacing the real world with a virtual world, might wish to merge the virtual world with the real world. An augmented reality system… could allow an architect to make actual-size modifications to an existing building, a homeowner to decorate a real house, or children to design and build virtual toys that could be used simultaneously with real toys.
>
> (Aliaga, 1997)

Aliaga's system used a head-mounted display to project information into a visual display for the user. Steve Mann, considered the 'father of wearable computing', has been wearing such a head-mounted display system for several decades, and has integrated aspects of its use into his everyday life, altering his view of the world and recording what he sees (Rheingold, 2002). Some integrated experiences using in-vision displays are now in production. Heads up displays used in military aviation (and in some modern consumer cars) provide overlays of information on the windscreen, and medical overlays allow physicians to be guided during surgery.[4] However, the majority of modern mainstream augmented reality technology has, until recently, tended to use device-based technology rather than head-mounted displays. Screen-based devices such as mobile phones and iPads, already widely owned and mobile, can be used as 'windows' into a digital overlay that adds data and information to the physical space. This might, among the myriad of uses, allow architects to view proposed buildings in their future setting, replace posters in your home with art from a museum (Coulton et al, 2015) or let game players view a hidden world that can be interacted with via their mobiles such as *Ingress* or *Pokemon Go*. This is however not a fully integrated experience. People do not generally experience the world through screens, and walk around with a mobile phone in front of them as a constant viewing portal – regardless of the fact that if you walk down a street you will see many people 'glued' to their phones!

Several large companies do currently appear to be pursuing integrated experience augmented reality using headsets, and although it is hard to tell with these projects when the products will come to market, it appears as if we are much closer than ever before to science fiction-like scenarios of ubiquitous digital overlays; at least if the hype is to be believed. One of the first of these to achieve a high public profile was Google Glass, which was released as a limited prototype in 2013. The system consisted of a device built into a pair of glasses, which displayed a 'floating' screen that projects an image into the eye of the wearer. Other features included voice activation and a touchpad on the arm of the frame. The Google Glass system received mixed critical reviews and some public ridicule. Many reviewers noted that wearing the headsets made them feel very awkward and self-conscious (Gibbs, 2014), and that the devices were limited in their capabilities. Although the prototypes could be purchased by 'explorers', they were still development sets and thus only included a small range of applications. Many of those testing the device noted the potential of the technology but that a 'killer app' had yet to be found.

While the device offered intriguing possibilities for use in public spaces, their use in such contexts also caused controversy over privacy concerns, since the headsets included a video recorder. Many articles were written in the popular press about the potential consequences of a device that could be recording any interaction at any time, and the devices were banned from several public environments such as hospitals, casinos and concerts (Gray, 2013). Responses from Google that they did not think the privacy concerns were justified simply reinforced the perception that the company had not thought through fully all the implications of the technology (Beschizza, 2015). The trial was halted in early 2015, with development of Google Glass moved to a different department of Google, and comments that they were focusing on future versions rather than maintaining the technology in its current form. Shortly after this, two new potential technologies entered the headlines – Magic Leap (also backed by Google) and Microsoft's Hologoggles, which includes gesture recognition enabling users to control, via hand gestures picked up by the system, the virtual objects they see. The potential in this technology, if it can be realised, is indicated by the fact that organisations such as NASA are already working on prototype implementations, with hopes to use it in Mars 2020 rover missions (Kelion, 2015).

These technologies often focus on vision as the sense which is augmented by technology, but the more senses are incorporated, the more integrated the experience is. Aliaga (1997) highlighted sound and tactility as desirable to create an integrated experience. Technology to incorporate non-visual sensory feedback is also progressing, for example Ultrahaptics, developed at the University of Bristol, which adds a sense of touch.[5] This uses an array of tiny speakers which emit high-frequency sound waves to exert pressure on the user's skin, creating the sensation of touching an invisible, floating object. Augmented soundscapes are also the subject of recent interest, leading some to speculate that the release of Apple's AirPod wireless headphone system is a precursor to them moving into this space.

A fundamental aspect of these technologies relating to their function in digital public space is their mobility. Provision of constant additional sensory information includes public spaces, so for example you might experience new content when walking down a street or using a park. This augmentation of the physical public space with digital content creates a new hybrid reality.

Many of the ways in which digital has been incorporated into public space has taken place in cities. This is, to quote Weise et al (2012), because 'cities are places which feature a high density in technological infrastructure and social activity and thus function as today's testing grounds for novel ubiquitous computing technologies'. Cities are by their very nature a way in which technology of mass habitation has been refined in order to allow large amounts of people to live together, pool resources and support community living.

One particularly common type of digital/physical hybrid space implementation is the near ubiquity of digital display screens that are becoming interwoven with the fabric of urban spaces. These range from digital advertising hoardings to public transportation information, and exist both inside publicly used buildings (such as arts centres and bus stations) and in open public space; for example BBC 'Big Screens'[6] and other large television screens now commonly seen in public plazas. When these displays are broadcasting fixed information, it is questionable how different this presentation is from that in non-digital forms such as static billboards. One difference is that the digital format allows the information available visually (as opposed to auditory broadcasts, such as radio or tannoy) to be updated remotely and much more rapidly. However, this is a matter of degree rather than a fundamental paradigm change afforded by the digital medium.

The real difference comes with digital displays which are reactive as well as simply delivering standard information: they may include interactive touch screens (such as service directories in retail centres), be accessible to the public using mobile phone connections, or incorporate sensors which enable them to display different information depending on whether someone is standing in front of the screen or even whether they are actually looking at it or not (Müller et al, 2010). Balsamo (2011) describes these activities as 'public interactives'; digitally mediated interactions which take place in public space. She describes the possibility of these as a 'new genre of infrastructure' and their importance to the creation of a future 'public culture'. These interactive displays do represent a significant change to the affordances of the environment, because they represent a portal to digital interactivity and information tangibly represented in the physical world. The mobile devices that we carry also serve this purpose, but large public devices are shared and often placed with some commercial or civic agenda, so the implications for the space itself are more complex.

This opportunity for significant change must be mediated by the fact that oversaturation of public spaces by digital information screens may reduce the attention paid to them. As described in Chapter 2, the attention which we give items in our visual field is not equally allocated, and although we may see things in our environment, we don't always perceive them fully unless they become the focus of our attention. Some research (such as the TILO project which investigated interactive screens in gallery spaces – see Case study 3.1) has shown that people in public spaces do not perceive screens in the environment unless they need to obtain specific information from them. They may not take in messages displayed even if those messages are aimed directly at them. This phenomenon is known as display blindness (Müller et al, 2009). While various tricks and techniques such as moving or flashing images may be used to pull the visual system in and increase the amount of attention given, they can also be extremely irritating and a form of visual 'pollution'. It is interesting to consider how this pollution might start to encroach upon physical spaces at the point where hybrid spaces become a canvas for advertisers even more than they already are (for example the myriad of advertising hoardings, many of which are themselves digital).

Case study 3.1: Awakening

Many urban spaces are a hybrid of physical and digital, threaded through with information conveyed both digitally and by other means – large screens surround us, which can be displaying anything from automated announcements and advertisements to entertainment or education. You might think that this information overload is something new which is changing the way that we interact with our environment. But busy urban spaces have always been jam-packed with information. Think of a busy marketplace in an agricultural society – sights, smells, movement and colour coming from all directions. Such an environment is overwhelming for someone who is not accustomed to it. But there are many evolutionary adaptations which allow us to make sense of such confusion, and we may now be using similar systems to navigate a world of digital information.

The Awakening project was a 'mini project' designed as an intervention on the TILO system for interactive hybrid screens for arts organisations. We wanted to explore how people's perceptions of the space could be changed by screens that interacted with them directly and personally. This is something that has often been seen in science fiction, and in fact the concept of aware digital environments is one that is so pervasive, many people do not realise how much of a challenging prospect it is with current technology. There are many examples of current technologies that pick up information about people inhabiting a space, but these have varying degrees of intelligence and integration. It is for example possible to tell whether people using a space are male or female, their approximate age and so on. What is currently a very hard problem is identifying individuals, and then following them around as they travel within a space. While this is not impossible, the computational power required means that it is extremely difficult to do this in a crowd situation for a large number of people; especially with an ever-changing group such as would be the case in any space accessible to the public.

We wanted to explore the potential of such technology without having to actually implement this extremely difficult technical challenge. So the Awakenings project mimicked the 'awakening' of the building by allowing it to respond directly to the people visiting – not in fact by technological wizardry, but the interventions of the researchers. Two types of targeted messages were displayed on the screen. The first were written based on careful observation and ethnography, and were designed to be responsive to the kinds of activities which are common in the FACT gallery in Liverpool in which the screens were deployed. The second set of messages were written live and in real time by our researchers who were discreetly observing visitors to the gallery space.

What we observed was fascinating. The majority of people passing the screen, though they may have been looking at it, did not appear to observe the text that was displayed. This was the case even when the message was directly addressing them with identifying personal information such as their appearance or behaviour at that time. The image of an eye was chosen to attempt to attract the attention of the people passing the screen, but because there were lots of other moving images at other times, it did not appear that this made a significant difference to the attention level. What did however attract significant attention from those passing the screen was their own image displayed back to them. A different intervention created a 'mirror' in which bursts of activity at a slightly delayed rate were displayed on the screen, which usually included those still in front of it. This allowed them to see themselves, and the result was noticeable – when seeing themselves on the screen, people were far more likely to stop and pay attention.

This is an example of perception bias – although we think that we see everything that goes on around us, a large part of what we 'see' is filled in by our visual system. We selectively attend to those parts that we are focused on. In a world increasingly covered in digital screens, and constant bombardment of marketing and advertising messages that are pushed at us without our request, it may be the case that we are selectively ignoring messages on large urban screens unless we specifically want something from them. For example in a train station or airport, you will stay attentively by the screen if you are waiting to see what gate your flight is at or which platform your train is leaving from. However, you are very unlikely to look at the screen if you already know this information or it is irrelevant to you. Therefore, if there was an important message about something else which happened to be displayed on the screen, would you see it? Perhaps not.

The other aspect of this project, when people did pay attention to the screen, was the concept of giving the building a personality. The messages that we conveyed to people were designed to illustrate the 'soul' of the building. Although this intervention was associated with an exhibition on science fiction, and we could have been said to be simulating artificial intelligence, there were also questions here about the sensing capabilities of the 'internet of things'. We questioned what this might mean for people in terms of what information they provide just by virtue of walking around. If sensors become ubiquitous, then perhaps it will be the case that the building will know your every need and be able to cater for it. Of course that means that this information is also available to companies who want to be able to benefit from knowing things about you – for example, if you always put the heating on when the weather drops below a certain temperature, you may get information from people who wish to sell you better insulation for your house.

Use of digital display screens does bring digital information into the physical space, in that it is available for us to experience using our senses in the environment. But there are ways that we can more tangibly change our experience of the space, that bring the digital out from the screens and integrate it into our physical landscape. Yet again, a rich area for experimentation has been the arena of games and play, since experiences that are playful can push the boundaries and encourage people to take part. Described by de Souza e Silva (2006) as 'the first hybrid-reality game', *Botfighters* by It's Alive of Sweden followed the model of traditional 'first person shooter' games but required players to move through the physical city. Interactions between players depended on their physical locations and distance from each other. These kinds of activities, which fundamentally incorporate the physical environment, are known as 'pervasive games', which co-exist in both digital and physical realms. Pervasive games, when interacting with mundane experiences, may provide benefits to those playing. *Cart-load-o-fun* is such a game that was developed by Exertion Games Lab (Toprak et al, 2013). In this game, passengers on public transport (in this case a tram) must co-operate in order to play the game, moving a character displayed on the tram to collect gems and avoid monsters. The game provides a means to interact with fellow passengers and relieve the potential boredom of travel.

Pervasive games fall under several different categories. Some, like the *Cart-load-o-fun* game and *Botfighers*, involve extensive movement in physical geography. This may enhance existing activity, such as the *Zombies Run!* game which provides incentive for exercise by turning your run into a thrilling chase from monsters. The augmented reality *Traces* app created by Ripple Inc, allows you to leave a digital 'present' for your friends (such as a video or song) visible as an augmented reality 'bubble', which can only be collected by going to a particular physical location at a particular time (Marks, 2014b). Others on the surface appear to be purely digital, but incorporate 'real world' location data and perhaps even activities of the player to affect the online game play. An example of this is *Cold Sun* which was developed as part of the Creative Exchange programme, which incorporates real-world weather data in the game play and aims to highlight the difference between weather and climate to educate about climate change. Still more are on a smaller scale but involve specific 'phygital' objects with which physical interaction affects the gameplay. These integrated phygital games are growing in popularity and are being produced by major toy manufacturers such as Disney Infinity and Skylanders.

This blurring of the boundary between virtual and 'real' does not just apply with regard to how and when we access digital content. It can also apply to ways in which our behaviour in the 'real' world is affected by the digital public space. To take a prosaic example, there were previously time and location constraints for many digital or media activities. For example, to watch a television programme, you had to be in front of a television set at a particular time and date. To read the news, you had to purchase a newspaper from a vendor once the news had been collated and printed for the day. Now streaming services and 'catch up' systems allow television programmes to be viewed at any time – and mobile devices and the linkage of profiles to many different screens means that the location in which they can be watched and in which news can be read is much more flexible as well. Although these technologies affect the affordances of particular physical locations, there are more fundamental ways in which the overlay of digital public space can interact with and enhance the physical spaces in which we live. This equally applies on the mass level of public discourse as well as for individuals – note the ways in which connected technology has influenced mass participation and protest, and that this can be expressed in the physical world such as by gathering 'flash mobs' of people very quickly together to a physical location to carry out some action, be it artistic or political.

Pervasive computing and the internet of things – how our objects interact with us

Based on the technological innovations described so far, we might say that we already live in an environment in which the physical experience of living is augmented by digital technology and information. But many people are working on technology to take this a step further by digitally connecting not just the people within an environment, but also objects. Although we are already surrounded in public areas by digital displays, these are mostly 'dumb' and do not interact with each other; simply transmitting information between different sources or reacting to specific sensor input. Pervasive computing aims to make our surroundings 'intelligent' by embedding in the environment or in objects innumerable sensors which collect information, and may undertake responses and actions based on what they think we 'need'. In this way, our objects will be connected and smart. This connection of digitally enabled objects is the backbone of the movement known as the 'internet of things'.

Objects have been digitally coded and embedded in larger processes for a long time. An example of this is barcodes, by which objects can be made machine readable. By using an appropriate scanner, any object with a barcode can be immediately recognised and categorised by a computer. Although these are passive objects, by giving them a digitally readable identification they can be made to interact with digital systems. However, the 'internet of things' envisages a world where this is taken significantly further. The miniaturisation and 'always on' capabilities of modern systems can enable ubiquitous connectivity, potentially for all manufactured objects.

One technology which is proving critical for this is RFID (radio frequency identification) tagging. Using small chips that can be attached to or embedded in an object, the key feature of RFID tags is that they transmit information proactively either under their own power, or when activated by a reader (depending on the variety). It is also possible for this information to be rewritten and changed, unlike a barcode which is static. Even a QR code, which can hold more information than a barcode, has fixed content. QR codes are often used to encode links to web URLs; while the content at that address may change, the URL itself cannot be amended. By contrast, the content of an RFID tag can potentially be updated and rewritten, and also can be read from a distance without having to be actively scanned. This gives items with embedded RFID tags agency to 'push' information about themselves out to the wider world and to other devices. The potentials of this technology are enormous: every object we buy could become part of our intelligent environment, with its own agency and ability to communicate what it 'wants' and 'needs'.

Figure 3.2 Barcode

Figure 3.3 QR code

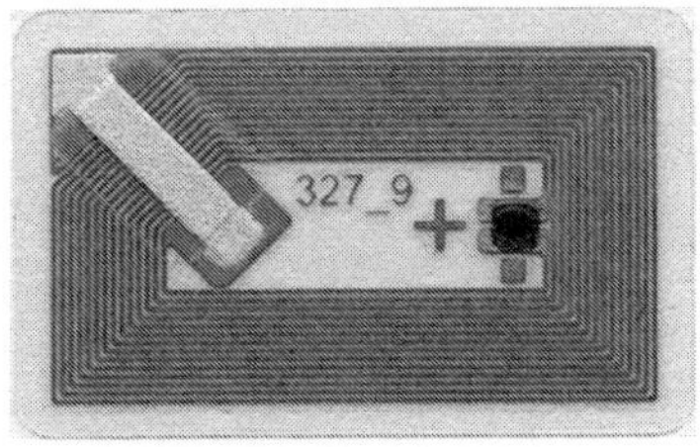

Figure 3.4 RFID tag

This could, for example, transform the way that retail supply chains are managed (Dodge & Kitchin, 2009). Mass production, transport of goods and the complex logistics systems which manage this are only possible with modern technological solutions. These can track and co-ordinate huge quantities of stock, which may be perishable or utterly reliant on immediate consumer demand. Up until now, although tracking systems such as barcodes and even RFID have become a staple part of the manufacturing and delivery process, they have not typically been used domestically. Once the produce gets out of the supermarket and to your house, you no longer have any use for the barcode. But this may be changing. Appliances within your home could use the information in an RFID tag to gain information about how they should interact with a product. 'Smart' packaging using RFID might tell your microwave how long your ready-meal should be cooked for. Instructions in clothing could tell your washing machine the correct temperate and spin cycle. Your fridge could identify when food is approaching the sell-by date, and email you a recipe to use a particular product up, or send a reminder to your shopping list that more needs to be bought. This 'invisible' exchange of unique ID data could happen in the background without you even being aware of it, with software which enables items you own to work for your benefit without explicit human intervention.

RFID is only one of many technologies which exist to link objects together. The internet of things is now a huge area of interest for researchers, companies who want to take advantage of the new technology, and also policy makers who are interested in how these vast changes will impact on wider society. Several different categorisation schemes have been developed which try and order both the existing devices, and potential varieties that might not be feasible with today's technology but are on the near horizon.

Dodge and Kitchin (2009) classify objects with embedded software code into two categories: peripherally coded objects, and 'codejects'. It is this latter that are particularly interesting, because they are items in which the software capabilities are integral to their function. An example of a peripherally coded object is an oven with a digital timer. The oven will still cook food if the timer breaks, therefore still serving its primary function in the absence of the code. However, codejects are useless if the code fails. Dodge and Kitchin further break down these codejects into three categories: hard codejects in which the functions are fixed and cannot be adapted by the user; unitary codejects which can be programmed to different behaviours; and logjects which have 'awareness' and keep logs of their activities which can be shared across networks. It is these logjects which are the backbone of the internet of things.

In his book *Shaping Things* (2005), Bruce Sterling includes speculative classes of objects in his categorisation, and classifies the role of objects in terms of their human-object relationships. As well as the perhaps more familiar categories of 'artifact', 'machine', and 'product', he includes 'gizmo' (a user-alterable and programmable multi-functional object) and the neologisms 'spime' (a tracked, networked object with extensive informational support) and 'biot' (an entity that is both an object and person that provides data to the network). The RFID tagged objects described above are logjects and spimes, and this category is most closely associated with the internet of things[7].

Many people when hearing about the internet of things, think only of objects that are privately owned – examples such as we give above; your toaster talking to your phone, or your fridge talking to the supermarket to tell you that you need to buy more milk. It is certainly true that technologies of the home and workplace are ones that have been a focus of pervasive technology and digital automation for several decades. In 1991, Mark Weiser famously wrote about the possibilities of ubiquitous computing by describing the experimental Xerox Palo Alto research centre. 'Doors open only to the right badge wearer, rooms greet people by name, telephone calls can be automatically forwarded to wherever the recipient may be, receptionists actually know where people are, computer terminals retrieve the preferences of whoever is sitting at them, and appointment diaries write themselves.' Weiser speculates about presentation rooms in which the light, sound levels and text size on presentations adapts itself based on the number of people in the room. None of these processes alone use complex technologies but the world that Weiser writes about, where this type of connected information is ubiquitous, is still a way off; in part perhaps because we have not yet reached the tipping point of uptake.

Rapoport (2013) wrote that 'smart homes promise users an intelligent environment that learns and adapts to their preferences and needs while ensconcing the inimitable virtues embedded in the concept of "home"'. Existing 'smart homes' can control the environment carefully, attending to many variables set by the owners, and enabling adaptation of conditions based on different needs. Smart homes act as 'intelligent' agents working away in the background, almost invisibly, to provide a service in the best interests of the inhabitants. Such services might include setting the correct temperature via air conditioning, controlling the lights, monitoring (and even reducing) energy consumption, automatic recording of television programmes likely to be of interest, and even monitoring the health of inhabitants. For example, some systems can monitor the footsteps and movements of elderly people, and are able to notify the health or emergency services if they have a fall and injure themselves.[8] Homes, and buildings more generally, might even respond to information coming from outside, or beyond their immediate surroundings; for example the weather outside, or who might be travelling from their place of work and likely to return to the house. In the work sphere, intelligent spaces might shift resources depending on who is working in the building and how they need to flexibly work together, giving personalised information based on who they are and their ideal collaborators (Myerson & Ross, 2003).

But beyond private objects and spaces such as homes, and semi-private spaces such as workplaces, there are also profound implications for public spaces. By digitally connecting objects in physical public space, these areas can be transformed into hybrid spaces. A simple example of this would be public street furniture that can transmit information. This might be in a functional capacity, such as streetlights which brighten or dim depending on information about who is standing under them, and can also record and transmit information about unexpected activity (Rutkin, 2014a). It might help interaction of the public with the space, for example giving pertinent information to visitors such as park closing times. The information might even be less pertinent and more light-hearted, as in the case of the Bristol-based art project Hello Lamp Post,[9] which gave residents and visitors the opportunity to engage with the objects in the city. Using pre-existing maintenance codes which exist on the majority of street furniture (such as lamp posts, post boxes and parking meters) people could communicate with the objects by sending text messages addressed to their specific code. In this way, it was possible to start a 'conversation' with the objects by answering questions they sent and seeing what others had replied. You might struggle to see how this is useful, but consider, for example, if similar technology was implemented across a city and helped co-ordinate how many people were waiting at a particular bus stop, or when dustbins were full and needed emptying. If public space is a collection of objects in a space that is accessible to anyone, then providing a small amount of 'intelligence' to those objects, and connecting them up, can potentially improve the use of the space for those who visit and make maintenance easier. If, for example, benches in a public park transmit information about how many people sit on them, it could help plan which benches to replace. But this connected information does not just have to apply to publicly owned objects, it can also be used to make the information generated by our private objects public in a way that, theoretically, would improve our lives generally and allow companies who make such objects to provide us with a better service.

Many of us already carry such smart objects with us at all times. Smartphones are almost ubiquitous, carried everywhere, and include sophisticated sensing components such as pedometers, heartrate monitors, gyroscopes and light sensors. This means that we are perfectly set up to be the recipients of 'anticipatory mobile computing' (Pejovic & Musolesi, 2015) in which mobile phones are able to sense the environment and your behaviour, and act in a way that it thinks you want it to: for example, knowing that when you are at home you do not want to receive telephone calls from your boss. This context awareness can be combined with predictions of anticipated needs and responses to provide 'intelligent reasoning' and act as a personal assistant that knows what you need before you ask for it. This might even extend to being able to 'see' what you are doing, for example in the case of the 'sense-see' system which extends the capabilities of the built-in camera to give phones 'peripheral vision'. The system can read hand gestures and know when you are leaving the room, and remind you to take your mobile with you (Yang et al, 2013).

This awareness and reaction to context does not need to be restricted to individual devices, but can also be built into technology used by multiple people, including broadcast media. Paul Coulton of Lancaster University has worked with the BBC to build prototypes exploring the possibilities of reactive environments that could be linked to television broadcasts to provide sensory responses – for example, releasing smells associated with the programme you are watching, or causing electronically augmented toys to activate when particular things happen in the programme. This work uses responses generated by pre-existing cues which are currently used for things such as subtitles and audio descriptions for the visually impaired. The Perceptive Media project (Gradinar et al, 2015) takes this a step further by adapting the broadcast itself to the surrounding context, such as the specific location of listeners and viewers, the current weather conditions, or how many people are currently watching in the room.

By linking this type of responsive awareness to connected objects, a more immersive experience can be achieved. Similarly, it may be possible to target reactive environments to individual people in the room, with images that are polarised to appear only from certain positions, or appear in augmented reality. Technology is under development that can even deliver sounds to particular individuals only (Marks, 2014a), so you can choose whether or not to listen to content as well as view it on ubiquitous screens. This could allow objects to guide the visually impaired around a smart home, or allow audio messages to be left that could be delivered when the person they are for arrives at a particular spot.

Another way to incorporate many of these sensing technologies more fully into daily life is to embed them in devices that are worn on the body: wearable technology. These devices are soaring in popularity and include 'smart watches', and 'quantified self' monitoring devices such as the Fitbit and Jawbone. The field of wearables has been in existence since the 1990s, but the development of smartphones significantly widened the potential. This was due to advances in technological miniaturisation, increased power of the computer systems involved, and the fact that consumers are now more comfortable with digital technology that they carry around at all times. Many modern wearables connect to mobile phone operating systems, thus requiring less computational power in the devices themselves. Much has been made of the health and fitness potential of wearables that track personal data and give feedback on, for example, the amount of exercise that people do. However, whether or not this has a positive impact on the health of individuals depends on their own personal motivations, and also on them being able to make sense of the data received (Tsekleves, 2014).

Smart textiles may provide a way to incorporate smart devices into our clothes and accessories. Several clothing based devices have been demonstrated as proof of concept prototypes, but these have to date not been commercialised. This may be because of a lack of manufacturing infrastructure to produce them on a commercial scale, or current lack of acceptance and demand for their domestic usage. There is much potential in such devices, not least because they would be able to fully take advantage of engaging multiple senses and be part of our embodied experience of the world. To use these would not need significant departure from the way in which we utilised clothes as a technology for several hundred thousand years. Marion Verbücken (2003) highlights the importance of making new points of contact between 'mixed reality' environments, and our senses, so that they are accessible and can fit in with our daily experience of the world.

In describing the potentiality of the internet of things and connected devices, we have talked about allowing software to take over and carry out tasks on our behalf. This in theory will make all of our lives easier; by offloading some menial tasks onto objects, we no longer have to undertake them ourselves. This drive to remove some of the work we do onto technological objects is one that has been a major driver throughout history, whether it is a bucket that lets us carry more water than our own hands, a bicycle which lets us travel greater distances without tiring, or an electronic telephone directory which stores our most-used contacts. The internet of things is in some ways simply an extension of this. With the internet of things, we each own objects which communicate and make our lives easier. But a more ambitious scenario is that all objects and devices talk to each other and combine with data which we generate (both via devices such as smartphones, and more generally captured through, for example, CCTV cameras and public sensors) into a linked information system of pervasive computing. True anticipatory computing would turn our environment into one in which machines become 'invisible' or so embedded within our lives that we no longer have to be aware of them. This scenario, with information shared between multiple objects and individuals and extended into shared spaces, would create a new type of digital public space where our public spaces were digitally enabled to react to our needs both individually and communally. Aarts and Marzano (2013) describe anticipatory computing of this type as the 'tripartite relationship' of man, machine and environment, and list five primary features which are characteristic of it: the technology must be a) physically embedded (perhaps hidden within the surroundings), b) context aware and responsive to environmental changes, c) personalised and tailored to needs of the user, d) adjustable according to input, and e) anticipatory of human desires.

But if we 'trust' these smart objects to make decisions for us and take on significant tasks within our life, there may potentially be greater risks to consider. For one thing, if we place too much reliance on objects embedded in our homes and our lives, we risk being unable to cope if for whatever reason they suddenly fail – if your front door is programmed to only open for you and your family and be impregnable to anyone else, what happens if the power fails and it will not let you out? Or what if you fall and are injured, and people trying to lend assistance are unable to access the house to help you? Codejects that cannot perform their tasks without digital functionality become reliant on that functionality being available at all times.

We also must trust that the decisions being made for us are correct and in our best interests. Many people will be familiar with stories of drivers who blindly follow the instructions of their GPS satellite navigation systems and end up in situations they would never normally have entered of their own volition, such as driving into a river. This is already offloading some of the cognitive decision making on to the satnav, but what if the car itself made the driving decisions? Self-driving cars are beginning to be rolled out by companies such as Google, but although they may be significantly safer than relying on human judgement and reaction times, there remains the issue of responsibility if the system does fail. Who is responsible should the car go to the wrong place, or even worse, end up in an accident? And what about moral decisions: given the decision between hitting and killing a child, or swerving and killing the passengers, which action should a self-driving car choose?

Privacy is also a potential issue; how do we ensure that information collected is only used in ways we want it to be? When we add connected things to our homes or networks, we assume that the only people who will be able to manage how these objects will act will be ourselves. It is however possible to imagine scenarios where your smart home could be 'hacked' and take actions that you do not want – such as the one described above where the doors are locked and you cannot leave. This privacy issue also extends to the information that is 'remembered' by logjects. A commercially sold item may track every aspect of its lifecycle journey, from production to disposal, but what if we do not want those items to be forever associated with us, or give indications as to how we use them? Recently introduced 'right to be forgotten' rules as mentioned in Chapter 1 (and discussed further in Chapter 6) mean that individuals can ask for outdated information which is misrepresentative to be excluded from search results about them. Perhaps a similar right to be forgotten may need to be applied to certain objects if they are tagged and cannot be disassociated from us. Will this drive us away from built in obsolescence, as we are loath to dispose of items that can reveal how wasteful we are? Such issues of privacy and security are becoming more and more pertinent as the digital public space begins to affect all areas of our lives, and this will be explored in more depth in subsequent chapters.

Finally, we must not forget the technical considerations which will be necessary to achieve the lofty goals spoken about by advocates of the internet of things. Achieving this will require the storage, transfer and sorting of vast amounts of data. Although processing power is ever increasing in line with Moore's Law (which states that the performance of integrated circuits doubles every 18 months) we may be approaching hard limits which will be insurmountable without significant breakthroughs in computer technology (such as quantum computing).

If these breakthroughs are achieved however, digital technology that receives, transmits and has its own power source might be packaged small enough to be delivered in 'smart dust' or 'digital paint'. It is crucial that before this is deployed we have already considered the implications of creating fully distributed 'ambient computing' that surrounds us at every moment. Already sensors less than $0.1cm^3$ can transmit data about temperature, humidity and other environmental conditions. Such devices must be self-powering, or obtain power from external sources during their lifetime, because the wide distribution and their location in hard to reach places would mean that replacing batteries would be impossible. We would therefore be setting these computers 'loose' in the environment. Science fiction writers have already begun to consider the potential implications of such an intelligent environment, such as the setting described in the short story 'A Swarm of Living Robjects Around Us' by Adam Roberts (2014), which was written in consultation with an expert in distributed computing. These kinds of fictional explorations of potential are important, as will be explored further in Chapter 8.

Digitally enhanced systems in physical space

In this chapter we have explored the ways in which our physical experience of the world influences digital versions that we create, and also how digital augmentation can add extra dimensions to our physical experience. In many of the examples above, analogies are made with existing uses of space, or reference made to sensory or factual information delivered via digital technology. However, there are more systemic ways in which digital technology can affect our experience of space, and the uses we make of it.

In Chapter 2, the nature of revolutions was discussed, quoting Weizenbaum (1976) who talked about technologies which cannot be removed from societal structures once they are embedded. Maps are a technology mentioned by Weizenbaum as revolutionary, and their use has changed our uses and perceptions of space. They are also still an evolving technology, with digital innovations. Maps are ancient, with pictorial representations of geography going back at least 13,500 years (Utrilla et al, 2009). Cartographers throughout history used maps to define the shape of the world as they saw it, to describe journeys, or to act as a guide. Many famous maps give us historical impressions not only of how people perceived the world they live in, but also cultural information on how people lived, their religious beliefs and their goals and wishes.

However, GPS technology, with satellites that can locate you accurately anywhere on the planet, means that a map no longer has to be giving information about a particular space or journey but can be centred on the individual, the journey of that particular moment. Equipped with a smartphone and Google Maps, it has become common practice not to consult a map before making a journey but simply 'on the fly', to check mid-journey only the parts of the route that you are uncertain about – even if this is most of it. The knowledge which we are required to retain about our location in the world and its relation to other spaces has changed fundamentally and is changing our interaction with travel. Some have bemoaned this as a step backwards in individual knowledge and awareness of space. Simon Garfield suggests:

> When we're looking at maps on our dashboard or on phones as we walk, we tend not to look around or up so much. It is now entirely possible to travel many hundreds of miles – to the other end of the country, perhaps, or even a continent – without having the faintest clue about how we got there. A victory for sat nav, a loss for geography, history, navigation, maps, human communication and the sense of being connected to the world all around us.
>
> (Garfield, 2012)

He also quotes Norman Dennison of the A-Z map company, creators of the London A-Z map, as saying: 'If we don't use [paper] maps we lose the idea of where we're really going. Where London lies in relation to Bristol or Newcastle is getting lost to the youngsters – you just enter the postcode now'.

However digital maps can also be used for new purposes which were not possible with the paper versions. Collaborative mapping uses digital map technology to collect real-time information from vast numbers of people. This creates powerful mapping tools that can be used for all sorts of things: for example the Ushahidi platform, created in the wake of the 2007 Kenyan crisis, collected eyewitness reports of violence during the political crisis (Thompson, 2013). Patrick Meier, Ushahidi's former head of crisis mapping, describes the power that such maps have. 'Having a real-time map is almost as good as having your own helicopter. A live map provides immediate situational awareness, a bird's-eye view – and thus an additional perspective on events unfolding in time and space' (Meier, 2015).

With this power, maps as representations of a space are no longer fixed. They do not just provide the information known by and seen as vital to the original map-maker, but create more abstract virtual versions of places, which can provide constantly updated information that is relevant at the time you are looking at it. They can function almost like the magical 'marauders map' in J.K. Rowling's *Harry Potter* novels, which shows the current location of all individuals within the grounds of Hogwarts School of Witchcraft and Wizardry.

This incorporation of digital geolocation technology into the structure of our society runs deep, and may justifiably be called a revolution by Weizenbaum's criteria above, in the sense that disrupting the embedded technology may cause chaos and upheaval. As Garfield writes:

> GPS is now such a significant part of our lives that the effect of failure would be catastrophic… the loss of GPS would now affect all emergency services, all systems of traffic control including shipping and flight navigation, and all communications bar semaphore. It would affect the ability to keep accurate time and predict earthquakes. It would set the guidance and interception of ballistic missiles to haywire. What would begin with gridlock at road intersections would very rapidly turn the world dark, and then off. Everything would stop.
>
> (Garfield, 2012)

Our perceptions of space that we travel in, often perceived as public, is shaped by the lens of the maps we use to navigate it and our ability to identify exactly where we are at any given time. How workplaces function is a similarly good example of systemic spatial practices which have historical precedents of major cultural change and are now also being affected by digital technology. The majority of people spend a large part of their daily life involved in the practices of work, and technology has influenced the form which this takes.

Robert Owen was the founder of the eight-hour work movement in the UK, in 1817 coining the famous idiom 'Eight hours labour, eight hours recreation, eight hours rest'. The movement campaigned for the rights of workers in the midst of the industrial revolution to achieve a work-life balance in which they had better quality of life (Myerson & Gee, 2014). The introduction of factories as part of industrialisation meant that it was common for workers to travel to factories and undertake extremely long shifts, often doing repeated tasks for many hours. Jobs such as this still exist, though labour laws in developed countries mean that there are limits to how long any individual can be 'clocked in'. However, in the last 100 years a new paradigm of work has arisen, in which a cluster of new technologies such as the lightbulb, elevator, telephone and typewriter revolutionised the world of work, with many workers housed within dedicated office buildings.

Digital technology is bringing a new paradigm shift to the world of work. Digital communications technology has meant that it is increasingly easy for work to be distributed, and for the line between work time and leisure time to be blurred. The concept of the 'nomadic office' is now on the rise, with individuals being able to undertake work from other places than the traditional office.

> Work stretches across a continuum of locations, from home and high street to a transport hub and serviced club – wherever people need to be. The predictable working day is being replaced by a 24/7 culture in which people are connected to the organization and its clients from wherever they are.
>
> (Myerson & Ross, 2003)

In their essay on 'Redefining working life' (quoted above), Jeremy Myerson and Philip Ross write about the changes in the workplace in the 21st century. They list digital technologies and mobile devices, along with ubiquitous technology, as factors that force organisations to rethink what their buildings are for. Do people really need a fixed workplace when the work they do relies solely on the technology they carry with them at all times? There are no longer clear fixed physical boundaries between 'work' and 'leisure' for a lot of people. This means that the uses we put space to can change dramatically. If you can do work from your home, does your home also become your office?

The line also blurs between which space is 'private' and which is 'public', at least as far as your employers are concerned. Most companies have full access to emails sent by their employees, even when they are not on the premises. We must nowadays question whether it is proper for business to use information from social media when considering candidates for jobs, or for them to dictate proper behaviour for their employees on social media even on 'personal' accounts. Many companies are starting to encourage employees to wear health tracking devices – nominally to encourage good health and well-being practices, but which effectively allow monitoring of activity. Some of this digital tracking data is already being analysed by employers; for example the Bank of America, who in 2009 asked employees at a call centre in Rhode Island to wear sensors made by Sociometric Solutions to record movements, interactions and tone of voice (Rutkin, 2014b). The data provided insights into how the call centre worked and led to changes in the office structure to encourage chatting – since workers who were more social were found to be more productive. There are however privacy concerns with these sorts of interventions. Could an employer insist on such wearables being used by their staff? What about outside of strict working hours, for example to monitor who might be getting enough sleep to do their job effectively? These sorts of questions, which impact our right to workers' privacy in the digital space, will be explored further in Chapter 6.

The emergence of new digital practices is just one example of digital public space as a facet of working culture, and work space itself is but one aspect of our physical environment. However, these kind of revolutionary changes are being seen across all spheres of modern life. Sometimes extending the boundaries of physical spaces into a digital realm can affect our perceptions of the physical space; for example, there are implications for statehood and nationality in the fact that information is not restricted by space. Estonia, for example, now offers e-residency. Individuals who wish to take advantage of this can open a bank account in the country or start a business. They may also be able to have a say in law-making: although e-residents are not currently citizens and do not have voting rights in elections, the country has one of the most advanced internet voting systems allowing citizens to vote from wherever they are worldwide (in the last election, from 98 different countries). This concept is extending the boundaries of what is to some extent a purely geographic construct (a country or region) by allowing 'entry' to it on a virtual level.

De Souza e Silva (2006) describes how the proliferation of mobile devices creates 'hybrid spaces' by creating a dynamic relationship with online space; 'embedding it in outdoor, everyday activities'. She describes how this constant connectivity creates new types of social environments. 'Without the traditional distinction between physical and digital space, a hybrid space occurs when one no longer needs to go out of physical space to get in touch with digital environments'. As described above, the boundary between what is digital and what is physical becomes blurred, and no 'logging on' distinction needs to be made between the two.

In considering the internet of things, Paul Coulton (2015) similarly suggests 'we should in the near future no longer be considering the Internet as a space we visit but rather as the place we live in'. This notion of living *in* the internet, or more specifically the digital public space, rethinks entirely the previous iteration of digital spaces as ones that are apart and virtual. In this new hybrid space, we will encounter all the experiences of interaction that previously were limited to purely physical aspects; both good and bad.[10]

Hybrid spaces and the internet of things make this linkage between the physical and the digital, but rely on pervasive information being available to our devices, from the 'cloud' of linked public data. The information that is contained and exchanged in digitally connected objects is now surrounding us at all times, creating a digital public space of information – that imbues our material world but is not made of the same substance. This digital information space will be explored in Chapter 4.

Key points

- Our experience and cognition is influenced heavily by our physical environment and senses: embodied cognition describes how we use aspects of the world around us, including digital technology as part of our mental processes.
- Simulations replicate physical experience in a digital space. When these are persistent and shared they become virtual worlds.
- Providing simulations of physical experience may provide a more 'natural' experience and assist learning, because of our familiarity with concrete spatial reality.
- Virtual worlds can become spaces for virtual 'togetherness' and facilitate social interaction at a distance; these owned spaces may act as public fora.
- The nature of experience in virtual spaces depends on qualities of avatars and how immersive they are.
- Telepresence, and ultimately, embodiment, could change the nature of how we consider presence in physical space.
- The world is increasingly not divided into digital and physical spaces, but a hybrid space with no clear division between them.
- This may take the form of augmented reality (on handheld devices or head-mounted displays), digital information embedded within public display systems, or use of data to enhance use of space such as in pervasive games.
- Access to digitally stored information may also change our perceptions and use of physical space, allowing access to information in unprecedented ways.
- Pervasive computing and the internet of things bring 'intelligence' to our objects and systems, allowing data transfer and communication objects that we surround ourselves with.
- This may apply to objects we own or use individually in our homes and workplaces, but also public, shared spaces and objects.
- Anticipatory ubiquitous computing, if technically feasible, would embed machines invisibly in the fabric of our modern world, bringing many benefits but also challenges of trust, security and over-reliance.

- Digital enhancement of our space and experience of it, with GPS and mapping, and new forms of work, are already showing how culture and society is radically shifting due to this technology.

Notes

1 Figures taken from http://www.statista.com/statistics/276601/number-of-world-of-warcraft-subscribers-by-quarter/
2 https://highfidelity.io/
3 www.kickstarter.com/projects/1663270989/project-perception-neuron
4 This has been used, for example, in breast cancer surgery (Sato et al, 1998) and endoscopic surgery (Nakamoto et al, 2012) and many other domains.
5 http://ultrahaptics.com/
6 www.capitalnetworks.com/clients/case-study/bbc-big-screen
7 Biots, on the other hand, are associated with the quantified-self movement which will be discussed in subsequent chapters.
8 For an overview of healthcare applications of home based sensor systems, see Acampora et al, 2013.
9 www.hellolamppost.co.uk/
10 For an example of the latter, young people may now be using digital media to undertake self-harm, see Robinson, 2013.

References

Aarts, E. and Marzano, S., 2013. *The new everyday: visions of ambient intelligence*. 010 Publishers.

Acampora, G., Cook, D.J., Rashidi, P. and Vasilakos, A.V., 2013. A survey on ambient intelligence in healthcare. *Proceedings of the IEEE*, *101*(12), pp.2470–2494.

Aliaga, D.G., 1997. Virtual objects in the real world. *Communications of the ACM*, *40*(3), pp.49–54.

Bakioglu, B.S., 2009. Spectacular interventions of Second Life: Goon culture, griefing, and disruption in virtual spaces. *Journal for Virtual Worlds Research*, *1*(3), p.5.

Balsamo, A., 2011. *Designing culture: the technological imagination at work*. Duke University Press.

BBC News, 2014. *Minecraft: all of Denmark virtually recreated*. [online] Available at: www.bbc.co.uk/news/technology-27155859 [Accessed 27 September 2016].

Benedikt, M.L., 2008a. *Human needs and how architecture addresses them*. Austin: University of Texas Press.

Beschizza, R. 2015. Google Glass chief "amazed" by privacy issues that helped kill his project. [online] Boing Boing. Available at: http://boingboing.net/2015/03/19/google-glass-chief-amazed.html [Accessed 27 September 2016].

Burns, C., 2014. Hear yourself happy. *New Scientist*, *224*(2996), pp.40–43.

Clark, A., 1996. *Being there: putting brain, body, and world together again*. MIT Press, p.180.

Coulton, P., 2015. Playful and gameful design for the internet of things. *In:* Nijholt, A., ed. *More Playful User Interfaces*. Springer Singapore, pp.151–173.

Coulton, P., Murphy, E. and Smith, R., 2015. Live at Lica: collection access via augmented reality. *Digital R&D Fund for the Arts Research & Development Report*.

de Souza e Silva, A., 2006. From cyber to hybrid mobile technologies as interfaces of hybrid spaces. *Space and Culture*, *9*(3), pp.261–278.

Dodge, M. and Kitchin, R., 2001. *Mapping cyberspace*. London: Routledge.

Dodge, M. and Kitchin, R., 2009. Software, objects, and home space. *Environment and Planning A*, *41*(6), pp.1344–1365.

Donath, J., 2014. *The social machine: designs for living online*. MIT Press, p.260.

Downey, S., 2012. Visualizing a taxonomy for virtual worlds. *Journal of Educational Multimedia and Hypermedia*, *21*(1), p.53.

Farber, D., 2013. Philip Rosedale's second life with high fidelity. [online] Cnet. Available at: www.cnet.com/news/philip-rosedales-second-life-with-high-fidelity/ [Accessed 27 September 2016].

Garfield, S. 2012. *On the map: why the world looks the way it does.* Profile Books, p.289.

Gibbs, S., 2014. Google Glass review: useful – but overpriced and socially awkward. [online] *The Guardian.* Available at: www.theguardian.com/technology/2014/dec/03/google-glass-review-curiously-useful-overpriced-socially-awkward [Accessed 27 September 2016].

Gradinar, A., Burnett, D., Coulton, P., Forrester, I., Watkins, M., Scutt, T. and Murphy, E., 2015, September. Perceptive media – adaptive storytelling for digital broadcast. *In:* Abascal J., Barbosa S., Fetter M., Gross T., Palanque P. and Winckler M., eds. *Human–Computer Interaction,* Springer International Publishing, pp. 586-589.

Gray, R. 2013. The places where Google Glass is banned. [online] *The Telegraph.* Available at: www.telegraph.co.uk/technology/google/10494231/The-places-where-Google-Glass-is-banned.html [Accessed 27 September 2016].

Harris, M., 2014. Are you looking at me? *New Scientist, 224*(2998), p.21.

Hayles, N.K., 2002. Flesh and metal: reconfiguring the mindbody in virtual environments. *Configurations, 10*(2), pp.297–320.

Hillis, K., 1999. *Digital sensations: space, identity, and embodiment in virtual reality.* University of Minnesota Press.

Ikegami, E. and Hut, P., 2008. Avatars are for real: virtual communities and public spheres. *Journal of Virtual Worlds Research, 1*(1), pp.1–19.

Kelion, L., 2015. *Windows 10 to get 'holographic' headset and Cortana.* [online] BBC News. Available at: www.bbc.co.uk/news/technology-30924022 [Accessed 27 September 2016].

Kirsh, D. and Maglio, P., 1994. On distinguishing epistemic from pragmatic action. *Cognitive Science, 18*(4), pp.513–549.

Marescaux, J., Leroy, J., Gagner, M., Rubino, F., Mutter, D., Vix, M., Butner, S.E. and Smith, M.K., 2001. Transatlantic robot-assisted telesurgery. *Nature, 413*(6854), pp.379–380.

Marks, P., 2014a. Beams of sound immerse you in music others can't hear. *New Scientist,* 2954

Marks, P., 2014b. Messaging app lets you leave secrets on street corners. *New Scientist,* 2981

Meier, P., 2015. *Digital humanitarians: how big data is changing the face of humanitarian response.* CRC Press, p.157.

Minsky, M., 1980. Telepresence. *OMNI magazine,* June 1980. Available at: http://web.media.mit.edu/~minsky/papers/Telepresence.html

Müller, J., Wilmsmann, D., Exeler, J., Buzeck, M., Schmidt, A., Jay, T. and Krüger, A., 2009, May. Display blindness: the effect of expectations on attention towards digital signage. *In: International Conference on Pervasive Computing,* Springer Berlin Heidelberg, pp.1–8.

Müller, J., Alt, F., Michelis, D. and Schmidt, A., 2010, October. Requirements and design space for interactive public displays. *In: Proceedings of the 18th ACM International Conference on Multimedia.* ACM, pp.1285–1294.

Murphy, S., 2014. New virtual world put the real you on screen. *New Scientist, 222*(2966), pp.19–20.

Myerson, J. and Gee, E., eds. 2014. *Time and motion: redefining working life.* Liverpool: Liverpool University Press.

Myerson, J. and Ross, P., 2003. *21st century office.* Laurence King Publishing, p.10.

Nakamoto, M., Ukimura, O., Faber, K. and Gill, I.S., 2012. Current progress on augmented reality visualization in endoscopic surgery. *Current Opinion in Urology, 22*(2), pp.121–126.

Pejovic, V. and Musolesi, M., 2015. Anticipatory mobile computing: a survey of the state of the art and research challenges. *ACM Computing Surveys (CSUR), 47*(3), p.47.

Pine, K.J., 2014. *Mind what you wear: the psychology of fashion.* Kindle edn. [ebook].

Rapoport, M., 2013. Being a body or having one: automated domestic technologies and corporeality. *AI & Society, 28*(2), pp.209–218.

Rheingold, H., 2002. *Smart mobs: the next social revolution.* New York: Basic Books.

Roberts, A., 2014. A swarm of living robjects around us. *In*: Amos, M., Page, R., eds. *Beta Life.* Comma Press.

Robinson, H., 2013. How many teenagers are using Ask.fm to self-harm? [online] *New Statesman.* Available at: www.newstatesman.com/sci-tech/2013/08/how-many-teenagers-are-using-askfm-self-harm [Accessed 28 September 2016].

Rutkin, A., 2014a. Bright lights, smart city. *New Scientist, 223*(2981), p.17.

Rutkin, A., 2014b. Off the clock, on the record. *New Scientist, 224*(2991), pp.22–23.

Sato, Y., Nakamoto, M., Tamaki, Y., Sasama, T., Sakita, I., Nakajima, Y., Monden, M. and Tamura, S., 1998. Image guidance of breast cancer surgery using 3-D ultrasound images and augmented reality visualization. *IEEE Transactions on Medical Imaging, 17*(5), pp.681–693.

Schwartz, L., 2006. Fantasy, realism, and the other in recent video games. *Space and Culture, 9*(3), pp.313–325.

Shaw, I.G.R. and Warf, B., 2009. Worlds of affect: virtual geographies of video games. *Environment and Planning A, 41*(6), pp.1332–1343.

Spooner, M.A., 2014. It's not a game anymore, or is it: virtual worlds, virtual lives, and the modern (mis)statement of the virtual law imperative. *University of St. Thomas Law Journal, 10*, p.533.

Stephenson, N., 1992. *Snow crash.* Bantam Books.

Sterling, B., 2005. *Shaping things.* Mediaworks Pamphlets, MIT Press.

Steuer, J., 1992. Defining virtual reality: dimensions determining telepresence. *Journal of Communication, 42*(4), pp.73–93.

Stevenson, S. 2014. Wish I were there: the beam telepresence robot lets you be in two places at once. [online] *Slate*. Available at: www.slate.com/articles/technology/technology/2014/05/beam_pro_telepresence_robot_how_it_works_and_why_it_is_strangely_alluring.html [Accessed 27 September 2016].

Stone, E.E. and Skubic, M., 2013. Unobtrusive, continuous, in-home gait measurement using the Microsoft Kinect. *IEEE Transactions on Biomedical Engineering, 60*(10), pp.2925–2932.

Sutherland, I.E., 1965. The ultimate display. *Proceedings of the International Federation of Information Processing Congress, 2.*

Thimm, C., 2012. Virtual worlds: game or virtual society? *In*: Fromme, J. and Unger, A. eds. *Computer Games and New Media Cultures*. Springer Netherlands, pp.173–190.

Thompson, C., 2013. *Smarter than you think: how technology is changing our minds for the better.* Penguin.

Toprak, C., Platt, J., Ho, H.Y. and Mueller, F., 2013, April. Cart-load-o-fun: designing digital games for trams. *In: CHI'13 Extended Abstracts on Human Factors in Computing Systems*. ACM, pp.2877–2878.

Tsekleves, E., 2014. Wearable tech for Christmas? It probably won't help you get fit. [online] *The Guardian.* Available at: www.theguardian.com/media-network/2014/dec/01/wearable-technology-gadget-christmas-health-fitness [Accessed 28 September 2016].

Utrilla, P., Mazo, C., Sopena, M.C., Martínez-Bea, M. and Domingo, R., 2009. A palaeolithic map from 13,660 cal BP: engraved stone blocks from the Late Magdalenian in Abauntz Cave (Navarra, Spain). *Journal of Human Evolution, 57*(2), pp.99–111.

Verbücken, M., 2003. Towards a new sensoriality. *In:* Aarts, E. and Marzano, S., 2003. eds. *The new everyday: visions of ambient intelligence*, 010 publishers.

Weise, S., Hardy, J., Agarwal, P., Coulton, P., Friday, A. and Chiasson, M., 2012, September. Democratizing ubiquitous computing: a right for locality. *In: Proceedings of the 2012 ACM Conference on Ubiquitous Computing*. ACM, pp. 521-530.

Weiser, M., 1991. The computer for the 21st century. *Scientific American, 265*(3), pp.94–104.

Weizenbaum, J., 1976. *Computer power and human reason: from judgment to calculation*. W.H. Freeman & Co.

Yang, X.D., Hasan, K., Bruce, N. and Irani, P., 2013, October. Surround-see: enabling peripheral vision on smartphones during active use. *In*: *Proceedings of the 26th Annual ACM Symposium on User Interface Software and Technology*. ACM, pp.291–300.

Yee, N., Bailenson, J.N. and Ducheneaut, N., 2009. The Proteus effect: implications of transformed digital self-representation on online and offline behavior. *Communication Research, 36(2)*.

4 Inhabiting digital information space

In Chapter 3 we explored virtual worlds that recreate geographic spaces, digital enhancement of physical spaces, and how physical concerns influence the design of connected digital technology. However, we can also look beyond that, to new digital public spaces that are not directly comparable to physical ones, but still exist as publically accessible and shared realms. Unlike virtual worlds (which recreate spaces with physical properties in a virtual format) these may have no physical correlates at all, or might provide entirely new experiences, effects and phenomena in physical space. This chapter will discuss why it is so pervasive to talk about intangible digital experience using the language of physical space, and how discussion of 'inhabiting' the digital public space must also consider how these intangible digital information spaces affects our lives.

DPS as an intangible space

> With recorded images duplicated and transmitted everywhere at the speed of light it is simply a fact that we hardly need head-mounted displays and gloves, and technology of 'virtual reality' to experience the irrelevance of spatiotemporal distance, to understand what it means to dwell in a global sea of pure information and to come to believe implicitly, indeed pragmatically, that 'I plug (or tune or log) in, therefore I am'.
>
> (Benedikt, 2008b)

As we have seen, humans exist as a species which has evolved in a specific environment within physical space, highly dependent on knowledge of surroundings and social interactions with others. This knowledge is built from information received by the senses; constantly updated through exposure to the environment, and might include knowledge of how to acquire food, how to protect oneself against the elements, and information about one's peers. This latter is critical, because while living in groups provides advantages, it also creates stresses which must be offset by carefully managed social interaction.

There are several phases of access to knowledge:

Without communication, adding to knowledge is only possible by use of one's own senses – you can smell where the best food is, you can see others around you.

Communication with others means that resources can be shared and alliances can be made; a group can be warned when predators are sighted by one lookout, you can provide support to another on the basis they will do the same for you later, and can see who else you have supported. Communication is common across many living things, at various levels of complexity: from plants which send out chemical signals to others when being attacked, to monkeys with specific alert calls for different predators.

> *Fully developed language* however, allows detailed information to be transferred robustly from one individual to another. You do not have to see a river yourself to be told where it is and how to get there. You can even be passed information about ideas and concepts – things that did not exist in physical space but only between minds. Now to access knowledge you can find the person who has that information and receive a transfer into your own head.
>
> *Writing and representation* refine this process even further, and widen the scope of where information might be found. Not only can you gain information directly from your own senses, and second hand from those of other people, but also from concrete representations of what others thought important enough to make writing or pictures. Now the space of 'information that could be known' encompasses information objects, such as written documents and books. To access the information within these, they either must be public (such as signs, notices or personal adornment), you have to own them (as individual texts), or go to a place where they are stored and available to access, such as a library. The act of accessing information from these objects still has a physical component, and in a certain sense the information has a tangible existence associated with the object.

At the start of the computer age, transferring information into a digital format did not change this. Information stored in a computer could be manipulated much more rapidly, but it was still 'inside' the computer and physically located. Much of the language used to describe computer interfaces reinforces this: we talk about 'files' and a 'desktop'. The point at which this relationship between information and objects started to break down was when networked computation began to become the dominant form of digital interaction. Information and files could now be transferred easily, duplicated and shared between remote locations. Making a new copy of a book or document became effortless and meant that information could be made public in a way that eliminated the need to go to a specific physical location.

The experience now of using highly networked technology is that it may not feel like there is a physical location associated with the information at all. Wireless technology and cloud computing have broken this perceived link to a computer as a storage device – public information can be accessed from anywhere, at any time, and devices are simply used to pull out the relevant part when needed. Many people may not even fully comprehend that there is a physical component to data storage at all, though this is in fact the case: Google for example have vast data storage centres, and all information held 'on the internet' is in fact on an individual server located somewhere physically. Although wireless sends things 'through the air', the main connections of the underlying structure of the internet consist of huge undersea cables that can by physically disrupted (Holpuch, 2013). The illusion comes from the speed of transfer, the networked nature of servers and also the fact that much of the information is mirrored and copied and exists simultaneously in several places at once. From the point of view of the user however, the important thing is that computers are not seen as the repository of public information in the same way as an encyclopaedia might be – but that they are simply portals to an ephemeral miasma of digital content that exists 'out there'. This is the digital space, and increasingly individuals are blurring the line between the public and private aspects for the sake of convenience of access.

This kind of experience is not something which necessarily fits within the paradigms of situations our brain has evolved to handle. Our reaction to it is shaped by the analogies which we have to make in order to relate it to things that our bodies (as well as our brains) understand. As mentioned previously, it is far easier to catalogue information based on physical space and relate it to whole-body actions. It may therefore be true in a certain sense to say that there is a 'geography' of the digital public space, but that it is a constructed one, different for every individual. In Chapter 1, we described the conceptual space of networked digital objects and information as if it were a higher dimension – invisible and intangible but layered over our physical existence, with points where the two realms intersect.

This concept of 'higher dimensions' is one which is borrowed from physics and mathematics, and is worth considering more carefully in its relevance to our understanding of experience in digital public space. These ideas are not new, and in fact were explored in a very straightforward manner in a well-known book written in 1884 by Edwin Abbot. *Flatland* was not a textbook, but a satirical novella described in its subtitle as *A Romance of Many Dimensions*. Abbot was not just writing about mathematics but about social dynamics. In it, he tells the story of A. Square, who lives in a two dimensional world, but has a terrifying experience with a being from a higher dimensional existence: a sphere.

We can use an analogy inspired by *Flatland* to explain how higher dimensional objects would theoretically interact with three-dimensional space. Imagine a two dimensional being that lives, say, on the surface of a pond, like a pond-skater insect. This creature, travelling across the surface, knows nothing of the air above, or the water beneath; they can only see and interact with things on the surface layer, and can travel forward or backwards, side to side within that one plane. Now imagine a three-dimensional human puts their hand slowly into the pond. Even as the hand approaches the surface, the pond skater is entirely unaware of it, because nothing above the surface can be perceived. When the first finger touches the surface of the pond, the pond-skater becomes aware of it as appearing from nowhere. Three other fingers and a thumb break the surface, but are perceived as four different entities, unconnected; the connections between them are in the third dimension, above the surface, and thus impossible to comprehend. As the hand moves through and deeper into the water, the five 'finger' objects get larger and larger, until they merge when the palm of the hand is submerged.

In this analogy, although content held in the digital public space may seem to consist of many different objects in many different physical locations, it is all part of one higher-dimensional object that intersects there; the physical location in three dimensions from which this knowledge space is accessed is irrelevant to what it contains, and many different areas can access it at once. As the three-dimensional world exists 'above' and 'around' the two-dimensional plane of the observer, so the dimension of digital information space sits 'above' and 'around' our physical domain. The space expands and contracts dependent on the information, content, and infrastructure we build for it, and stretches beyond what we might be able to imagine or draw on a piece of paper into esoterica such as the 'space' created by algorithms which automate how telephone signals are transferred and resolved into single connections on a mobile.[1]

In effect, we have widened the world in which we live to include this new dimension. This process began with the collapse of space possible with remote communication, and is further the case with mobile technology: when you speak on a mobile phone while walking down the street, you are to some extent 'existing' in two worlds at once: remotely, and in the physical space. These may overlap in new and surprising ways. Jen Southern (2012) describes the phenomenon of 'comobility'; being mobile with others at a distance, enabled by technology which allows you not just to communicate with others, but spatially locate them and understand their movements in real time despite not being with them physically. With digital communication technology, our awareness may be in many different overlapping spaces simultaneously. Access to these may take many forms, from email alerts on our mobiles to wearable computers providing augmented reality overlays as described in Chapter 3. The implications of this are many and varied and this chapter will explore some of them in more detail.

Social constructions in digital space

Although this new dimension is not a spatial one, we still use spatial analogies and language when talking about it. Some of these, like the 'desktop' metaphor mentioned above, are simply used to better interpret user interfaces. But deeper uses of spatial terminology, seen consistently across digital interactions, reflect the way we socially construct space around us. Take for example the notion of boundaries and thresholds in geographic space. We put up physical or contextual 'barriers' (that might consist of brick walls, bead curtains or a change in pavement type) that define the space and inform those crossing between them that the use of each space is different and fits within a specific social context. Benjamin Koslowski (2014) has talked about the way the 'screen' of the theatre relates to the 'screen' of the digital space – in both circumstances we are taking on board social conventions of understanding that tell us things about the nature of the action we are viewing and experiencing, and place us within a particular context of space and meaning.

With this in mind, we can talk about being 'inhabitants' of the digital public space both literally and figuratively, and examine the effects that being 'in' such a space has on our behaviour, in the same way that moving from a nomadic lifestyle to permanent urban dwellings created changes in the way we live.

In 1926, in an interview by John B. Kennedy, Nikola Tesla described his vision of the future of wireless transmission over long distances.

> Wireless will achieve the closer contact through transmission of intelligence, transport of our bodies and materials and conveyance of energy… When wireless is perfectly applied the whole earth will be converted into a huge brain, which in fact it is, all things being particles of a real and rhythmic whole. We shall be able to communicate with one another instantly, irrespective of distance. Not only this, but through television and telephony we shall see and hear one another as perfectly as though we were face to face, despite intervening distances of thousands of miles; and the instruments through which we shall be able to do his will be amazingly simple compared with our present telephone. A man will be able to carry one in his vest pocket.
>
> (Nikola Tesla, quoted in Kennedy, 1926)

Tesla's prediction seems remarkably accurate from today's perspective. He was talking principally about wireless electricity, but it is wireless communication that is the major driver of cultural change and has created the new digital public space. To those who have grown up with mobile phones, the internet, and the world wide web, instant access to information and communication seems the natural state of things and one that it is hard to do without, despite the fact that it was inconceivable 30 years ago. This ability both to communicate over large distances and to access a vast constantly updated reference library has changed our daily interactions – from pub quizzes and discussions in the bar, to how we arrange meetings and transport.

Taking part in a modern pub quiz is a very different experience than it was 20 or even 10 years ago. 'All mobile phones away' is a common rule, and some quizzes have even changed the nature of their questions – from those that require knowledge of specific information, to those which require lateral thinking and the ability to make connections. Information is no longer a scarce resource – in fact it is the opposite. The digital public space, with information constantly uploaded and made available, gives individuals immediate access to vast amounts of knowledge almost instantly. Neville Brody (2012) compares this digital knowledge space to the growth in the availability of information and the spread of knowledge that immediately preceded the Renaissance.

The way that social meetings are now arranged is distinct from the pub quiz example in that it is the social expectation of connectivity which has changed, rather than the access to fixed information. In this case, the new ability given by digital connectivity is that of discovering (via mobile telephony or social media) exactly where your friends are at any given moment, and to communicate with them instantly should plans change. This means that rather than having to have fixed and definite arrangements for meeting (for example, 'We'll meet under the clock tower at 3pm.') arrangements can be much more fluid ('We'll meet in town later. I'll call when I get there.'). Rheingold (2002) talks of this in terms of a changing sense of time: flexibility allows the future to be less structured. While mobile telephony is not specifically an aspect of digital public space because it consists of private communication between individuals, it is part of a wider phenomenon of information density and connected information space. In a sense, there is a contraction of the space and time in which interactions happen, because information is constantly at hand regarding the location and requirements of your social groups. This can be updated on the fly, and therefore plans can always be changed even if people are not co-located, because information can be published and passed between groups or individuals at a distance. This has similarities to the way that the telegraph changed the way people perceived distance and the speed of information flow in the 19th century (Bargh & McKenna, 2004), but the broadcast nature of much of this information affects the nature of being 'in public' rather than point to point communication.

These two examples indicate that the digital public space is changing assumptions about what information and content is instantly available to an individual, and subsequently allowing changes to social interaction. It also demonstrates how interactions in the public space can now contain much more information because it is not constrained geographically or temporally. This constant connection to content has also been called ubiquitous media (Featherstone, 2009) since with the spread of mobile devices there is now available almost constant access to, and means to create, media content in a huge variety of forms; from blog posts to YouTube videos to surveillance footage to photographs on Facebook.

Some lifestyle and behavioural changes enabled by digital connectivity might be due entirely to the disconnection from specific times and locations, rather than the activities themselves being quicker or easier. For example, by allowing 24-hour access to financial transactions, internet banking removes the obligation to visit a bank during opening hours and, potentially, make an inconvenient journey to the physical location. This kind of time saving might significantly change both the ways in which tasks are structured throughout the day and the amount of time that is needed to be devoted to them. Such changes to how time is used can also be seen in what can be achieved when multiple spaces can be inhabited at once, digitally. For example, in the past, waiting at a bus stop or travelling might have resulted in 'dead time'. This may have been spent reading fixed content such as a book or newspaper, by interacting with others directly present, or simply unproductively. With connectivity to the digital information space, this time can now provide opportunities through our mobile devices to be a dynamic extension of our social life or workspace; our physical location no longer has to directly influence the type of activity we are engaged in. Andrews and Hartevelt describe consequences of this disconnection between physical location and connectedness:

> Networks allow us to easily book holidays or conduct financial transactions anywhere, at any time, and 'going on holiday' no longer implies isolation from those back home. This increased connectivity influences the way people, organizations and society

> organize themselves. Friends stay in touch across continents, family ties survive dispersion and cosmopolitan networks of friends replace local village communities.
>
> (Andrews & Hartevelt, 2003)

This effect on communities will be explored further below.

Ubiquitous access to information can subsequently have an impact on the way that we use physical spaces, and may lead to changes that can significantly affect people's lives. Many public or semi-public spaces such as parks, cafes and museums now have wifi coverage, or at least data coverage, enabling those there to be fully connected. Connectivity may be freely provided, on a subscription/pay for use service (such as BT Cloud hotspots which cover many public areas in the UK), or selective access such as the eduroam network (which allows anyone who works or studies at a university to access the internet in any academic institution across Europe). Mobile data access is almost ubiquitous in the UK, with 3G available for over 99% of the country in 2014 (Ofcom, 2014). This can change how people make use of public spaces. Such blanket coverage is now expected by those making use of the spaces, and it may be that such facilities enter the regulated category of utilities that are considered a public service.

Some may consider talk of restructuring culture extreme, and unlikely. But it is important to think of this in relation to previous major technological revolutions and the way that they have become embedded to the extent that they are no longer noticed by us. Most people in developed nations take the constant presence of electrical power for granted, noticing it only when it is not there during a power cut. Similarly, the development and subsequent ubiquity of writing allowed information content to fill the world around us; it is almost impossible in any area with permanent human settlement (with a very small number of exceptions) to avoid writing or symbolic written representation in one form or another: from books to product labels to signposts to name badges, writing surrounds us at every moment. Pervasive computing may be affecting a move beyond static writing and information to dynamic information flows surrounding us, described by Manuel Castells (2011) as 'the formation of a multimodal, multichannel system of digital communication that integrates all forms of media'. Floridi (2007) also describes this transition, suggesting that 'the threshold between online and offline will soon disappear, and that once there is no difference, we shall become not cyborgs but rather inforgs, that is, connected informational organisms'.

By examining more closely some specific ways in which this information overlay changes behaviour and experience, we can explore the impacts of the intangible nature of the digital public space. Memory, shared experience and community networks are all structures which are being affected by this aspect of digital space and will be explored below, as well as the way in which we inhabit space by creating digital identities for ourselves in the form of profiles.

Memory, recall and knowledge: cognitive augmentation

Augmenting our cognitive capabilities is not a new idea. In fact as we have established in Chapter 2, it is something that is already common with existing technologies. However, digital technology offers the opportunity to enhance our recall and memory to an extent not previously seen. This has been one of the intended goals since the early days of personal computation. For example, in 1945 Vannevar Bush described a theoretical system he called a 'memex' which would serve this purpose: 'A memex is a device in which an individual stores all his books, records and communications, and which is mechanized so that it may be consulted with exceeding speed and flexibility. It is an enlarged intimate supplement to his memory' (Bush, 1945). Though the details of its function are different, he might have been describing the modern world wide web.

In the example of the pub quiz, above, the format of the questions has had to be changed because simple information-based questions are no longer a straightforward way to differentiate teams. Whereas previously having learned certain facts and being able to recall them from memory was a specific skill, it is now easy to pull those facts from the internet. Functioning like the memex as an intimate memory supplement, the extended information network of digital public space becomes a new type of cognitive storage.

> I did a calculation in the mid nineties of the number of documents that the seated user at home could get to within a minute and found that over a period of four years the number increased by a factor of ten thousand. One order of magnitude came because the disks are bigger, and three orders of magnitude came because of the Internet. That's a huge change in a short time. If I want to know, for example, whether "foodchain" is spelled closed up, or if there's a space between food and chain, I can whip over to my workstation, type it both ways, find the number of people on earth who have used it each way, and know definitively where the majority usage is. I would never have known that before, and it happens in almost the same amount of time that it takes me to search my own memory. It's as if I have a strap-on cortex!
>
> (Stu Card, quoted in Moggridge, 2006)

This ability to 'recall' information on any topic has changed our experience of the world. If we watch a film, and recognise an actor but cannot place where from, it is common practice to immediately reference the 'internet movie database' (IMDB) and find out where we have seen them before. When travelling, we no longer need to buy paper maps for the place we are going; in fact there is no need to consult a map at all before leaving, because the map can be accessed at any time. Simon Garfield (2012) thusly describes the process of setting a destination part way through a modern car journey with a satnav: 'You set off in the car from a familiar place, and only when you get a little nervous do the satellites take your hand'.

In this way then, our knowledge can be fluidly augmented at any time, like Stu Card's 'strap-on cortex'. To those who are firmly embedded in digital information space, it is almost as if this vast store of data is accessible as an extension of their own minds; augmenting our memories in the same way that previous technology has allowed augmentation of our physical capabilities as discussed in Chapter 2. A clear example of this is described by Thad Starner whose thesis detailed his experience with prototype wearable computing over several years. He described an incident in class where he began to answer a question about something that was discussed previously, before failing mid-sentence because he was not able to retrieve his notes as he had anticipated.

> Volunteering, I said, 'We said the importance of deixes is… uh… uh… humph, whoops! Uh, I'll get back to you on that'.
>
> The class, most of whom were Media Laboratory graduate students familiar with wearable computing, began to laugh. I had not known the precise wording of the answer and had tried to retrieve my class notes on the topic. Having done this routinely in the past, I had expected to have the information in time to complete my sentence. Due to a complex series of mistaken keystrokes, I had failed so badly that I could not cover my error, much to everyone's amusement.
>
> (Thad Starner, quoted in Rheingold, 2002)

Starner was so accustomed to his external memory that he was inconvenienced when it unexpectedly failed, as if he had forgotten what he was going to say. Most people do not yet have the information quite so readily available that it is accessed mid-sentence, but failure to 'recall' due to digital content being unavailable is becoming a more frequent frustration for those who come to expect it.

While portals to access the information space might differ in form and function, the content is accessible extremely quickly and often retrieved using search functions, or software based on cataloguing systems that can be 'navigated' in a way that mimics physical space. This is not necessarily just coincidence or a lack of innovation. External recall is not a phenomenon entirely outside the realm of pre-digital human experience, so digital extension of this is just a new order of magnitude. It has for a very long time been the case that humans have consulted external sources of information. Memory aides, such as a 'knotted handkerchief' concrete the act of remembering something outside of the brain. In a sense, this is the function that writing serves, to extend our memories and thinking beyond ourselves onto the paper we write on. Hutchins (1995) points out that it is not a metaphor to call drawing a line on a piece of paper remembering, and erasing it forgetting. In a sense, when we erase it we remove it from our 'external' memory. The boundaries of our memory change, depending on other objects that we are using to do our thinking and remembering for us.

Mementos, souvenirs and treasured possessions, especially representative ones such as photographs and letters, are physical objects which represent memories and can call them to mind when needed. These 'intimate media'[2] are 'browsable' in that you can walk through a room, pick things up and have emotions triggered and memories recalled. They act as an extension of our sense of self and can evoke recall and associated feelings (Grayson & Shulman, 2000). Even abstract 'objects' with no physical form can act as external memory aides, particularly if they are part of the shared cultural landscape. For example, many traditional religious festivals have their origin in agricultural timekeeping, whereby the festival marked the appropriate time to remember to undertake certain actions as a community, such as planting or harvesting the grain.[3] Alternatively, mnemonics and attachment to external events such as dates can help remember lists, such as having specific days of the year associated with Catholic saints in order to be able to recall them all.

Digital implementations of this kind of externalisation of memory are very popular. Storage of information in a digital format is much more flexible and malleable than creation of physical objects, because it is so much easier to return to a digital object and make changes to it. In the same way that people may prefer to draw initially in pencil rather than permanent ink, Hillis (1999) describes how digital drawing (using a 'lightpen') made people less anxious and hesitant to draw and create something 'as though they are not committing in the same way as they do on paper'. A digital drawing can be changed or manipulated, copied or edited by other users without any wasted resources or permanent record of the 'imperfect' first try; unless that is what you want. Digital archives therefore, in their function as a collection of objects and information that can be personalised and include personal mementos, serves as an expansion of this natural tendency to use external objects as prompts for memories and emotion.

By archiving our 'possessions' digitally, we can store a lot more information in a searchable format. However, there may be losses in terms of how much we can 'browse' said possessions. This is in part to do with sheer volume, and also due to the fact that digital objects are not as easy to display in one's general field of visibility, and do not have the same tangibility as, say, photographs in frames or mementos on a shelf. This might be overcome as digital space comes to encompass the physical space as well as the information space; with some of the digital physical interaction explored at the end of Chapter 3. For example, physical objects might be created which hold digital information to enhance the memories associated with the object itself. Records that elicit memory traces through digital recordings might also be associated with physical objects,[4] combining the tangibility of objects with the high fidelity and data content of digital media.

Case study 4.1: Digital mementos

One consequence of digital connectivity is that information is collected and stored in ways that were not previously possible. These might range from digital interactions on services such as Twitter, to constantly monitored information about ourselves collected by wearable computers that measure our pace, gait and pulse, or even video footage of each experience. This data is often transient, and perhaps difficult to relate to our lives. What if these could be made a part of our experience, and our memories? Bettina Nissen has been exploring this as part of her work with Creative Exchange, designing tangible artifacts created out of data, digitally fabricated to create personalised objects that are unique, expressive and creative. For example, ISIS Arts in Newcastle worked with audiences at an art exhibition to create personalised souvenirs. As visitors left the installation, they were asked for their responses and reactions to what they had seen, which were represented in a unique artifact produced on a cutter-plotter, which they could take away with them as a souvenir of their visit.

Nissen explains how this provided a more concrete connection between the audiences, their experience of the art, and how they evaluated it: 'This novel approach deepened and extended visitors' interaction with the concepts of the artwork while it also provided the arts organisation with a novel perspective on how audiences can be engaged with the process of evaluation through more creative means. Audiences engaged more deeply with their experiences of events concepts or practices'.

Another piece of research resulted from working with people attending a technology conference, and produced individualised, 3D printed objects which 'translated' their Twitter usage during the conference into a physical object, a clip which they could display on their lapels or name badges. These material representations of digital interactions prompted discussion and reflection, and even led some people to change their behaviour in order to affect what was produced (Nissen & Bowers, 2015).

By taking home these individualised, personal mementos that they had a part in creating, the conference attendees or exhibition visitors now own an object that is not only a tangible, physical reminder which can elicit memories of the event and their experiences, but also contains unique traces of what happened on the day, which support and enhance the memories. An important aspect of this work was that the data translation objects were created with the participation of the people whose data it was. The participants themselves were therefore able to invest meaning in the objects. Because the process of translation was part of their experience, meaning and value is added to the items and souvenirs, and the data traces become a tangible part of externalised memories and mementos.

If we are able to extend our memories in this way with external digital resources, which increasingly have large capacities and high fidelity, might be we able to get to a stage where we 'remember' everything and have perfect recall? Current movements such as quantified self and lifelogging are based on continuously recording everything that happens to us. If we can access this as part of our memory, will this mean we never forget anything? Additionally, if these 'memories' are shared and part of a generally accessible digital public space, might we 'remember' things that happened to someone else and are part of a shared cultural memory rather than an individual one? The possible implications of these questions are explored further in Chapter 8.

But digital augmentation of what goes on in our heads may extend beyond memory capabilities. In Chapter 2, we discussed the neuroplasticity of the brain and the fact that it can adapt quickly to additional inputs to the senses. The brain's sense of self extends out to items which are used to gather sensory information, such as tools. Like James Cooray-Smith's tweet about his watch (see p. 43), many people describe feelings of having 'a limb missing' when they are apart from their smartphone. This may indicate that digital extensions of memory and other cognitive processes might function in a way which resembles physical extensions of our sense of self. Might the flow of information, constantly accessed from these devices, become something which is integrated into the infrastructure of our brains, the 'strap-on cortex' as coined by Stu Card?

Andy Clark and David Chalmers (1998) refer to the incorporation of external objects into our mental processes as extended cognition. In his book *Natural-Born Cyborgs* (2003), Clark expands this theory, and suggests that this is something that has already been happening throughout the history of our species, and continues to happen as technology improves. He argues that cognition does not just take place in our heads, in our 'skin-bag' bodies, but in the interaction between our brains and the non-biological resources which we structure around ourselves. The key distinction that Clark makes between humans and other species is the plasticity and learning ability of our brains, and how they adapt to external cognitive aids. It is this that has allowed our vast intellectual development. Again, this is not something new with digital technology, but is part of the way we have always used external aids to thought: such as moving physical scrabble tiles around to help find words.

High level technological augmentation of the brain sounds like a positive thing – an improvement on what we are able to achieve with biology. Sergey Brin for example, seems to see this as an ideal state: 'Certainly if you had all the world's information directly attached to your brain, or an artificial brain that was smarter than your brain, you'd be better off' (Newsweek, 2004).

There may however be downsides to this 'sea of information'. It has been hypothesised that because we no longer have the necessity of holding information in our brains, we end up storing less of it. This is because we know that we will be able to access it by other means whenever we need it; the concept of transactive memory as referenced in Chapter 2. To quote David Brooks:

> I had thought that the magic of the information age was that it allowed us to know more, but then I realized the magic of the information age is that it allows us to know less. It provides us with external cognitive servants – silicon memory systems, collaborative online filters, consumer preference algorithms and networked knowledge. We can burden these servants and liberate ourselves.
>
> (Brooks, 2007)

Some people are concerned that by offloading memory capacity to computers in this way, we will no longer exercise our own mental capacity, to negative effect. Carr (2011) suggests that there may be trade-offs in letting technology do the work for us – that we might be 'getting dimmer'. By automating cognitive processes, he suggests, we may be getting 'worse' at them. There is also the risk that users of these technologies mimic previous mechanisms of cognition in a way that is subtly inferior without us really noticing. For example, research has shown (Evans, 2008) that there is a correlation between the advent of online academic publishing, and a decrease in the number and breadth of articles cited. Evans suggests that poor indexing of print journals meant that researchers had to browse more widely and thus were exposed to a broader sphere of work than the much faster consensus of 'important findings' that can arise using online searching.

Coming to rely on technology to 'think' and 'recall' for us could also be dangerous if the technology fails. For example, it is increasingly common to rely on digital public maps and GPS data, rather than planning routes in advance using paper maps. It may even be the case that routes are not planned in advance at all, because people 'know' that they can 'always' find their location and plan a new route. However, GPS coverage is not total, and can fail with little warning, leaving lost those who were relying only on Google Maps to find their way. A similar phenomenon has been reported multiple times in the media,[5] of individuals who follow the guidance provided by satellite navigation systems while driving, and end up in perilous situations because the directions given are inaccurate. These incidents can only occur when users follow the route directions blindly rather than applying 'common sense'. If we become used to the 'just in time' mind-set, that Balsamo (2011) describes as people being 'confident that when they need to know something they'll know where to find it' then a removal of this external knowledge due to technological or other restrictions may be devastating. This is particularly worrying if the 'space' that such information is held in is not owned by individuals, or even collectively by public bodies, but within open spaces owned by companies. If your 'memory' is stored in the 'cloud', on a server owned by a large company, is it still *your* memory, and reliably holding what you want it to? These issues of ownership will be explored further in Chapter 5.

Extelligence explosion: cognitive surplus

When talking of cognitive augmentation we can consider not just the enhancement of individual brains (through access to a larger 'memory' or processing power), but benefits on a larger scale through the combination of high computing power and vast amounts of data collected. In effect, these computers are acting as 'brains' on behalf of entire societies, undertaking cognition in ways that are not possible for humans.

Modern computing power enables the volume of data being produced and handled daily to be massively larger. This extends into many different spheres: from the 55 million Facebook posts made each day and 300 hours of video uploaded to YouTube every minute, to the experiments in the Large Hadron Collider which produce and record vast swathes of data about particle collisions. In the past there would be nothing that could be done with such data because the computing power simply did not exist to cope with it. Huge amounts of information are also collected from the many sensors and recorders that are liberally spread throughout both public spaces and private homes. The 'Nest' thermostat for example, uses sensors to automatically set the heating schedule for a private house based on temperature, humidity, weather, and who is in the house: calculating schedules based on previous and expected activity. Google purchased Nest, and therefore now has access to heating and, potentially, house usage information for everyone who uses it. This could be used, for example, by electricity companies to calculate demand on a wider scale.

Because we can capture and compute with this information, the extelligence of our species is increasing exponentially. Individual human minds may not have the capacity to understand or cope with the volume, but we can use technology to tease out the critical parts and bring it back in a way that is relevant to what we need. Public opening of this data might create a digital collective data space; a shared realm of publically collected data that can be used for good or ill. This might be as straightforward as data mining to understand social and digital networks more clearly. But there are also emergent properties of large data sets.

New types of knowledge can be created simply by contributing and co-ordinating large amounts of (human) cognitive power to certain problems. Having access to large numbers of Twitter followers means that you can quickly gain answers to complex questions (for example, 'What is the best restaurant for vegans within walking distance of the Natural History Museum?' or 'Does anyone remember the colour of a character from a particular television introduction sequence?'). This phenomenon has been referred to as the Twitter 'hive mind'.[6] Clay Shirky (2010) has talked about the 'cognitive surplus', and how, because many people have small amounts of spare time, vast benefits are being gained through collaborative creation activities such as Ushahidi maps[7] and Wikipedia edits. 'Frequently Asked Questions' lists, which had notable popularity in information-sharing Usenet groups in the 90s, are often constructed out of questions, and sometimes answers, supplied by a diverse group; honing down what is the most commonly sought-after information. This builds a collective knowledge base, and may function as a form of transactive memory with those who you have never even met. Rheingold (2002) calls this collective knowledge production and sharing 'smart mobs', and notes that by allowing cooperation and acting in concert in ways that were not previously possible, collective action is enabled.

Some activities are more effectively carried out by humans than computers, and again the ability to gather this input in small amounts from many people can be leveraged to large impacts. This is the idea behind crowdsourced activities such as *Foldit*. This game requires players to 'solve puzzles for science' and allows the public to contribute to predicting the structure of protein folding; a task very important for drug development and medical science but difficult to automate. As the connected population gets ever larger, particularly those playing games, it may be that this gets leveraged to more crowdsourced tasks. In his 2011 novel *Reamde*, Neal Stephenson imagines a MMORPG in which the players are undertaking airport security tasks by looking for 'goblins' coming through a particular area. The 'goblins' are created by the real-time algorithms of the game taking note of particular suspicious activities, thus players provide much higher levels of attention than a single security guard watching monitors could provide, and higher accuracy than a computer alone because of human pattern matching skill. This kind of utilisation of game players' abilities for real-world tasks is now actually being implemented, for example by plans to incorporate disaster response tasks into *World of Warcraft* (Hodson, 2013).

The potential in gathering and utilising large amounts of information forms part of the concept referred to as 'big data', which in turn contributes to the notion of 'smart cities'. A modern city is a technologically governed space. For example, transport systems are becoming more and more digitally controlled and automated: from mass transit timetabling systems to the lights that govern traffic flow. This allows cities to be larger and more self-regulating, and react more quickly to changes. Some cities, such as Chicago, are taking this one step further. In 2014 a pilot scheme was trialled which used predictive models based on a wide range of data from sources such as home inspection records and census data. These were able to identify buildings that are likely to experience problems such as lead poisoning in children (Rutkin, 2014c). Similar systems are being used in other cities, with the dream being to create truly 'smart cities' that have an interwoven fabric of data, drawn from a myriad of sources, that can predict and react to changing circumstances and create a better environment for its inhabitants. Such smart cities might be considered in a sense themselves cognisant organisms that can take 'decisions' based on what they sense. By connecting resources and needs, and using algorithmic processing that can account for more variables than human planners, cities that are more sustainable and environmentally sound can be designed.

There are however some people who worry that this digital integration in cities can lead to security risks. Ghena et al (2014) found significant security flaws in the traffic signalling system used commonly across the US and demonstrated how an attacker could disrupt the system at an intersection without any direct physical access to the lights. They explain how general traffic chaos could be caused by such an attack, or more subtle effects such as giving a vehicle access to 'green lights' for the entirety of its journey. With increased networking, this kind of attack becomes more feasible on a wider scale than if aspects of city infrastructure are individually managed.

Manipulation of large connected data sets may also inconvenience us as individuals if the priorities of those controlling the systems do not match our own. The sea of information in which we exist does not just include that which we seek out ourselves, but more which might be thrust upon us. We are already constantly being bombarded by unsolicited information – mostly by advertisers – which is designed to get us to pay attention to it. Even if information that we are provided with is relevant and interesting to us, there is a danger that there is too much of it: information overload.

> [T]he powerful tools for discovering, filtering and distributing information developed by companies like Google ensure that we are forever inundated by information of immediate interest to us – and in quantities well beyond what our brains can handle.
>
> (Carr, 2011)

When dealing with such vast amounts of data, either on a city-wide level or an individual level of one's data archives, the issue becomes not the remembering of correct information, but sorting and filtering it in such a way that we can retrieve what we need quickly. The alternative to this is handing over control of decision making to the data collection itself, by means of algorithms. These problems become even more difficult when you consider that new content is being added and updated all the time due to the collaborative constructional nature of these public data collections, and that by using these digital public spaces as your 'brain', you are in fact sharing your cognitive space with all other users.

Shared experience space

Information in the digital public space is not static. The public digital information space we have been describing is not a collection of discrete facts, nor are the contents curated or uploaded by a single author. A critical part of why this new digital dimension of information is powerful is the ability to co-create and upload by all participants. The most obvious example of this is Wikipedia, the encyclopaedia which is made up of knowledge to which anyone can contribute. However, the interactive nature of digital public space is more pervasive and fundamental than this. In Stu Card's example above, the value comes from knowing how other people are using the word 'foodchain'. This is possible because all of these multitudinous usages have a permanence that comes from the online medium they are situated in, but might evolve over time. Each time the word or words are used, it is stored for retrieval by anyone, for any purpose; even those undreamt of by the people creating the original content. Online information is both constantly changing, never complete, and yet a historical record of interaction; Paul Booth uses the word 'intratextuality' to describe meaning that forms within the content of such an ever-updating record (Booth, 2010).

Benkler (2006) suggests that the ability to independently create and discover information, as well as the ability to fluidly move between collaborative endeavours, might enhance individual autonomy, giving individuals 'a significantly greater role in authoring their own lives'. However, he also gives an argument that the chaotic nature of production can reduce the capacity to communicate effectively: 'When everyone can speak, the central point of failure becomes the capacity to be heard – who listens to whom, and how that question is decided'. This is a factor in the 'Babel' effect; that the democratisation of production creates an unmanageable cacophony, and may even reduce the overall quality of the content, since there are no restrictions or curation of what can be uploaded. However, the quality concern may simply be matter of volume and an expression of what is generally known as 'Sturgeon's Law' – 90% of everything is trash. The issue again comes back to discerning the quality from the 'trash'.

We have gone through a paradigm shift, from scarcity to abundance of information. However, another shift is from information as something static, 'facts' that can be referenced in a printed encyclopaedia, to something that we are a part of; fluid extelligence that is growing exponentially. Henry Jenkins (2006) talks of 'consumption communities' where collaborative contribution is an intrinsic part of shared knowledge: 'consumption has become a collective process… none of us can know everything; each of us knows something; and we can put the pieces together if we pool our resources and combine our skills'. Balsamo (2011) goes even further, remarking on the onus that is put on consumers to also put things back for others: 'Just as we harvest bits and fragments of information from various media flows, so too are we called to actively contribute to the information streams we fish in'.

Not only can digital content be added to, but it can be changed; remixed and reimagined. Once something is uploaded to digital public space, it in some sense becomes public property. Although digitally available content may still be owned and subject to copyright provisions, in real terms it can be shared, copied and distributed almost infinitely; and also adapted and changed.

Some researchers have focused on the digital age as the time of the *prosumer*; those who do not just produce or consume, but an amalgamation of both (Ritzer & Jurgenson, 2010). This drive to produce as well as consume is relevant to a large part of interaction with the digital public space; from community journalism, to mashup videos on YouTube, to collaboratively written 'fanfiction'. Digital public space enables uploading of 'your' content to be viewed by anyone, a modern equivalent of the Roman 'Acta Diurna' by which news and announcements were presented daily in public places to be viewed by all.

This has arguably also facilitated a democratisation of information, the largest since the invention of the printing press changed books from a luxury item that only the very richest could afford, to something that was mass produced. The digital public space effects a similar level of change: it has meant that *creation* of distributable material, something that previously needed support from industries such as publishing, film, or music, can now be undertaken by an individual with limited technical experience and financial resource. There are also consequences in terms of the speed at which collaborative creative production can take place. By being able to send content and communicate almost instantly, discussion, research, and creativity can take place over greater distances and faster timescales than ever before.[8]

Rapid communication also provides a greater opportunity for dialogue between 'professional' creators, and consumers. Examples of this include instant consumer feedback via social media and other avenues of consumer marketing, and also in the creative industry; the digitisation of 'fandom'. While there have been organised collections of those appreciating an entertainment commodity for a very long time (for example the outpouring of public grief and the wearing of black armbands when Sherlock Holmes was 'killed off' by Sir Arthur Conan Doyle), there are now often direct avenues of feedback and discussions for fans which travel and develop at much faster speeds. The author Diane Duane, who frequently uses the internet to interact with her fans, described the difference in these interactions in the last 10 to 20 years as 'a sea change'.

> Suddenly, the people you wanted most to touch directly are within range, yet at the same time the people who have it most in their power to completely misunderstand what you're doing are also in range. And they can get at you as easily and quickly as you can get at them. So you learn a certain level of discretion as quickly as you can.
>
> (Duane, in Anglo-Filles, 2014)

As Duane suggests, closeness between professional creators and consumers can backfire and lead to outcry when producers are not as reactive to the desires of fans as the fans think they should be.

The facilitation of collaborative production and the speed of communication all serve to homogenise information available to communities, and create a 'shared experience space' that is richer and more diverse than it would be otherwise. By sharing experiences in 'real time' and being able to upload shareable content, individuals become a cohesive group and form strong community ties (Cook, 2015). This may also be related to why there is still value in experiencing things 'together' which some forms of digital broadcasting encourage, although there is now starting to be a reduction in time-specific necessity (for example TV on demand negating the need for broadcast at a specific time). Examples of this encouragement include 'second screen' activities which tie social media and internet content to a specific live television broadcast, allowing communities to interact because they are all experiencing the content at the same time. This link is acknowledged by the National Theatre, who live-stream their theatre productions to cinemas across the world. Although this results in a filmed version of the play, there is, as described in Chapter 1, a focus on the 'experience' of being in the theatre, and the productions are only generally screened either live or within a very short period following the theatrical run. They are not made available on DVD for home viewing, with the intention that there is still the communal experience of 'going to the theatre' as an intrinsic part of the product which they produce.

This connection between experience and interaction, both on and offline, can be affected by perceptions of boundaries as described at the start of this chapter. The Chattr project (developed with the Creative Exchange) attempted to replicate some of the features of a digital space in the physical world. It did this by creating a café area (with facilities better than those elsewhere in the festival space) which could only be accessed by signing an agreement that all conversation could and would be recorded and shared online. The privacy aspects of this were interesting and will be discussed later; however, what is relevant here is how people perceived the spread of their conversation via the online sphere. In the first iteration of the intervention, tweets containing the transcripts of conversations were shared online and paper transcripts were displayed in a different part of the gallery space, away from the café. However, in the second iteration, a screen displayed the transcriptions within the Chattr space. Rather than the display increasing the sensitivity of participants to the fact that their conversation was shared publicly, it was observed that they thought it spread *less* widely, perhaps because they associated the display screen with their physical location and did not consider that it was also part of the larger digital sphere. People's speech indicated different ways in which the perception of the digital context changed their actual behaviour (Salinas et al, 2016).

The link between digital collective experience and 'real-world' activities can be seen even more clearly in the use of digital media to influence and inform activism and action which is effected in physical space. There are many well-known examples of how activist groups or political movements have used digital spaces to coordinate and share information for collaborative action in the real world, often because such action is unsanctioned or illegal. The 'Arab Spring' was well publicised as a movement which did not necessarily originate in social media but was facilitated by it: those involved were made aware of where demonstrations and actions were due to take place by using digital platforms. The power of social media may be changing the nature of political activism:

> It strikes me that social media embodies the connection between action and expression. For example, you can tweet that you are going to a demonstration. The hashtag connects you to others, acts as an expression of your opinion and a call to action, and builds solidarity. It is democratic, efficient and endlessly variable. It is personal but increases social capital for the movement.
>
> (Beckett, 2011)

This shared experience, whether taking place purely in the digital space or reaching back into the 'real world', depends on having people to share it with. This may consist of an extra channel through which to interact with those you see regularly in person, or may connect people who live remotely and would not otherwise have a way to interact at all. It may even consist of creating a platonic 'other person' with whom to communicate; the nonspecific addressee to whom we direct social media posts and blogs. They are constructed from our general impression of the readership; be that our friends, a particular social group or the wider world. This internalisation and reconstruction of the audience is considered by some to be novel to new media.[9]

These changed perspectives on how we interact and share experience with those closest to us may affect the ways in which we manage our social lives and communities, which as discussed are closely related to the development of human cognition and culture.

Digital friendship

One of the key arguments in Chapter 2 was that social relations played an important part in the development of human cognition, through evolutionary selection pressure on brain size to manage social group increase. Robin Dunbar, cited extensively in that chapter, is known for his theories of social group size. Based on an analysis of group sizes in different primates, he suggested that intelligence and brain size are directly related to the demands of group living, and that brain size predicts a maximum for the number of relationships that can be maintained by an individual.

Plotting neocortex size against group size for different species, Dunbar found a strong correlation that predicted a 'group size' for humans of around 150. This has become known as Dunbar's number (Dunbar, 2010), and there is much evidence that this really is a figure which has real-world relevance – popularly described as 'the number of friends we have.' Examples which back this up include census data for hunter-gatherer populations with an average group size of 153 (Aiello & Dunbar 1993), the Hutterite communities in the USA and Canada, the size of 18th century and Neolithic villages (Dunbar 1993), and a study which looked at the number of Christmas cards sent by households and found a mean network size of 153.5 (Hill & Dunbar, 2003).

Why is the number of friends and contacts important? Because of the amount of 'social capital' that it can garner you; in other words the people that you can expect to gain support from. Social capital describes how much help you can expect to get from members of your network. Before language, one theory goes, this social capital would be built up by reciprocal grooming. Language significantly decreases the time required for building these connections, because while grooming is one on one, language allows conversations between up to four people at once (Dunbar et al, 1995).

But what might happen when some of the cognitive load involved in relationship tracking is offloaded to digital platforms? It could be hypothesised that social media provides a new venue for reciprocal 'grooming' and sharing of social information, able to handle a greater number of simultaneous transactions even than language. Many researchers have sought to examine these new forms of social interaction and whether they might affect 'group size', and push us beyond Dunbar's number. Evidence suggests that some changes may be occurring, but their nature is not yet clear. For example, surveys of Americans' closest networks (the people they confide in and discuss important information with) seem to show that these became smaller (not larger as one might expect) over the 20 years from 1984 to 2004 (McPherson et al, 2006). More recent work[10] has however disputed this.

Gonçalves et al (2011) used data from Twitter to model networks of conversations and found that Twitter users were able to maintain a maximum of 100-200 'stable relationships' (as characterised by a certain level of reciprocated interaction). They used this to justify Dunbar's number and maintain that the hard upper limit on social interactions still exists. Dunbar (2012) has himself suggested that more time spent on online relationships takes time away from 'real' ones. But many studies have shown that the more people use the internet, the more friends they have and the more social contact they have with those friends (Wang & Wellman, 2010). Kanai et al (2011) found that the number of Facebook friends was larger than real-world social networks, but was significantly correlated with them; people with more 'real-world' contacts were more likely to have a greater number of Facebook friends. A clue to untangle why this might be, and why conflicting results might be seen, can be gained from their analysis of brain activity carried out alongside the work. They found that there were specific brain areas associated with online social network size, and (though there was correlation between the two) these were different ones than those that are associated with real-world social network size. The areas which correlated with online social network size included those associated with being sensitive to other people's intentions, and memory capacity; particularly for pair association such as between names and faces.

Another way of looking at the differences digital technology makes to relationship networks might be to look at the *way* friends interact online and offline, rather than the volume.

Recent work has shown that online social media use, in particular Facebook, can contribute to 'relationship maintenance' and strengthen ties, bringing people closer together (Burke & Kraut, 2014). Work by Lai and Katz (2012) similarly found that mobile telephones were mainly used to maintain existing social networks rather than create new ties, but that using them, particularly with text messages, solidified the social group. Social media services may provide similar network maintenance, but have the advantage of making it easier to do this with multiple people at once.

Enclosed systems like Facebook can be used for maintaining existing social capital; preserving links with people that you otherwise might fall out of contact with even as acquaintances. The original social media networks highlighted this aspect, for example Friends Reunited whose selling point was finding those you had lost touch with, and Facebook which promoted itself in maintaining networks of college friends. More 'open' networks like Twitter and Tumblr where sharing outside of closed friendship groups is encouraged, might be more suited to a different form of social capital utilisation. It is possible that by enabling sharing more widely, social resources can easily be gathered even from outside one's direct network, leveraging 'friends of friends' (Ellison & Vitak, 2015).

This 'strength of ties' is an important concept, as is how we track them. Dunbar's number refers to the number of 'friends' that we have. But this uses quite a specific definition of friendship – it is not people that we know, or people that we like, but rather those that we might expect to do us a favour if we asked. There are different divisions of social connections; Zhou et al (2005) suggest a geometric progression of different categories of human social organisation, and it appears that these layers might exist at around 5, 15, 50, 150 (and maybe 500 and 1500), each inclusive of the previous level and representing a different level of intimacy.

De Ruiter et al (2011) also point out that the cognitive load of social interaction does not only include our friends – it is also important to keep track of negative relationships; those who we might want to avoid. While Dunbar's number is clearly important, it may not represent the total sum of our social interaction network. Research based on, for example, the number of Facebook friends or Twitter connections that people have, may not take into account that an individual's distributed network of social connections may be spread across several different platforms and systems, as well as offline. It is now easier for people to belong to several different networks and communities at once, which may or may not overlap and might therefore increase the overall number of social connections maintained. There is also a difference between networks of friends and networks of relations – the latter not needing as much maintenance to remain close. Additionally, culture and rituals can bond a much larger group through means other than direct person to person interaction, by introducing common mind-sets and recognisable 'in group' identifiers. These encourage co-operation and kinship even to those you might not have ever met before but recognise as belonging to your 'group' by how they dress or act.

What kind of support can digital connections provide? It is sometimes suggested that social media is less effective at relationship building because it is usually using only one sensory modality, often text. Things like video calling or telephone conversations are more 'satisfying' and therefore useful for building 'strong, emotionally intense relationships' (Dunbar, 2012). However, we may need to re-evaluate meaningfulness in the digital information age. Social links in digital public space might indeed be less rich and possibly weaker, but are more widely distributed, and we should consider whether this is a different, yet equally valuable system for social capital.

In his 2012 paper, Dunbar suggested that 'Merely being part of a Twitter-created crowd does not constitute having more friends, no matter how intoxicating it may seem at the time' and that 'quality relationships require focussed time, not the anonymized chatter of a broadcast medium such as Twitter'. But just because it is not a 'quality relationship' does not mean the link is valueless. Granovetter described in 1973 the concept of weak ties, the links you might have with those who are not your friends but are contacts in your network. He tested how people got jobs, and found it was more likely to be from weak ties. But a key finding was that it was also short chains of ties that helped; rather than a message being passed through a chain of increasingly distant contacts, direct information came from the people with a weak tie to the person asking, or one contact away. Having more weak ties therefore was beneficial because it gave a broader reach to a wider set of information.

This ability to gain assistance from those you do not have close ties with can be useful for reaching a wider range of possible contacts, all with their own sources of potential information. But it can also be useful for general data gathering in terms of factual information such as train times, weather or answers to obscure questions. Asking an interesting question on Twitter might not just get answers from those who follow you, but lead them to 'retweet' it to their wider networks and potentially allow you to get answers from a much wider range of people. Additionally, there is increased visibility of communication, and ties that were previously hidden might now be visible to other connections in the network. Hampton et al (2011) suggested that 'this enables flows from person-to-person and from tie-to-tie, which are likely to increase access to social capital'. Judith Donath hypothesises that in this way 'super networks' with vast numbers of weak ties can expand users' range of information sources. She argues that this is potentially more relevant for a contemporary mobile society, where people travel frequently and meet a large variety of people in a variety of different contexts. 'The key role of technology in supporting social super-networks is to help us turn weaker ties into richer relationships by helping us discover more about each other' (Donath, 2014).

If we can find reliable support via our weak networks (such as a trusted babysitter) we may not need strong ties as much. This ability to reach a wide number of people with whom you might only have a tangential relationship is therefore a key factor of digital social network services. We may not be able to keep track of any more 'strong' relationships because of cognitive constraints, but in the same way that cognitive augmentation can happen with digital memory stores, we may be able to maintain a much greater number of persistent weak ties across our lifetime that can convey other benefits.

Social effects of the digital public space are difficult to measure because the landscape is changing so fast; the connected digital information space of today is very different to that of five years ago, or even a year ago, and will be different again in five years' time. Many of the studies quoted above were carried out when the internet was new, and people were still learning how to use it or indeed inventing new ways to use it. It is therefore difficult to say in detail what long-term effect digital public space will have on interpersonal relationships other than to suggest that it will have an effect. It changes the way in which we communicate with and maintain links to other people, and this is a fundamental component of our lives, and one of the driving forces behind human evolutionary behaviour. While we cannot therefore make predictions about long term effects, we can look more broadly on trends in larger community interactions.

Communities, interpersonal networks and ambient intimacy

The word 'community' is used frequently in discussions of the internet; from how existing communities use it to communicate and cement ties, to how new communities of interest are created and maintained. A community can be defined simply as a group with shared interests or locality, but there are more abstract qualities that make something a community rather than just a collection of individuals. Some of these are to do with culture, with shared attitudes and territory, as described below:

> Every mature and healthy community shares a few basic principles, irrespective of culture, scale or enabling media. Community members need mechanisms for representing themselves (identity) and for building up their reputation (reward and recognition). A community needs its own territory within which people can adopt specific habits and rituals. Community members need a way to express their shared values and keep their collective memory. This collective intelligence is the main embodiment of the community's culture, which is in turn the foundation of that community.
>
> (Andrews & Hartevelt, 2003)

The digital public space provides many new opportunities for communities. Unlike traditional communities, geographic proximity is not a factor: it is easy to 'meet' and share 'space' with people from all over the world who hold shared values and interests. The 'territory' within which rituals and habits are built can be purely virtual: a bulletin board, a chat room, a Facebook group. This is facilitated if the spaces used to congregate are public, and can be accessed by individuals from many different places.

As danah boyd (2007) puts it: 'Networked publics serve many of the same functions as other types of publics – they allow people to gather for social, cultural, and civic purposes and they help people connect with a world beyond their close friends and family'. Digital public space can help maintain and nurture existing communities, but also seems to be a very effective mechanism to build new ones; by eliminating the effects of geographic distance, it becomes much easier to find people with similar interests as well as to discuss, share and develop these relationships. But boyd also notes that they allow restructuring of people's engagement with the 'public space' of networked publics, and with each other, giving new dynamics to the interaction.

Josephine Green agrees that these digital communities are providing alternative forms of belonging and support, given the decline of traditional 'forms of belonging, such local communities, church, family and nation state'. She suggests that we can exist within multiple different groups at the same time, allowing us to have 'fluid' identities. 'These emerging fluid networks mean that the personal and the social is in constant state of becoming' (Green, 2003).

What effect are virtual communities and relationships maintained via digital public space having on our lives, and 'real world' interactions? This is a topic which is still under much debate, with strong advocates on both sides. Some research, such as that of Haythornthwaite in 2001, suggested that the time people spent on the internet was taken from time previously devoted to other activities. Surveys and studies carried out around this time indicated that time spent online displaces face to face social activity, notably with family members.[11] But it is not as clear cut as this: while some studies found that internet use reduced the level of social activities taken part in, others found that it increased; for example, work by Wellman et al (2001) found that internet communication with friends and family was in *addition* to telephone contact, leading to a greater level of interaction overall. Note also that in 2001 'going online' necessarily meant sitting in front of a fixed terminal, whereas nowadays mobile internet use means it can be interwoven with other activities.

More recent research[12] has found that offline and online social interaction affect each other, and DiMaggio et al (2003) conclude that 'research suggests that the Internet sustains the bonds of community by complementing, not replacing, other channels of interaction'.

This 'reinforcement model' suggests that the internet facilitates the creation of social capital by making information flow more efficiently, and enables other activities such as shopping, communicating with existing friends, and seeking out information (for example to enable travel). Studies over longer time periods (for example Crang et al, 2007) have found that rather than decreasing socialisation, there is some evidence that it might be increased. Rather than 'using the internet' being an activity that replaces others, it may be an enabling process, and thus it becomes difficult to separate out as an 'activity' that consumes a certain amount of time. While there is still evidence on both sides, it appears that internet use does not have a purely negative effect on 'real world' relationships.

Why might this be? Digital technology makes flow of information more efficient. And it may not be any specific content in this information that is critical, but rather its flow within a group. Earlier we described how a significant factor in the development of our intelligence and the drive of technology was interaction between individuals and managing the stresses of group living. Digital public space gives an extra dimension to these interactions, and provides different affordances than face to face contact; for example, having access to a conversational history, obscuring factors such as race and gender that might otherwise impact interaction, and allowing asynchrony of conversation (Donath, 2014). This affects not just relationships between individuals, but the construction and composition of groups and communities. Because digital public space contracts time and space and allows multiple simultaneous interactions, you can 'broadcast' information about yourself and others, and be provided with constant updates and information about social cues. This can provide a constant awareness of our contemporaries.

This constant connection to peers can be maintained in many different ways. Historically, similar effects might have been seen in small, connected communities that spent much of their time together. Through networks of information sharing and gossip, community members were kept constantly updated on what everyone was up to: the 'village' model. New technology has allowed this awareness to spread out geographically to people who are more separated in space, something that may have been lost through the move towards urban living. In this way the 'village' model is replicated by enabling individuals to share context and intimate details of daily life, but on a larger and more diverse scale, perhaps larger than 'at any time in human history' (Hampton et al, 2011).

Blogger Leisa Reichelt calls this phenomenon 'ambient intimacy': 'being able to keep in touch with people with a level of regularity and intimacy that you wouldn't usually have access to, because time and space conspire to make it impossible' (Reichelt, 2007). It replicates the feeling of being in the same physical location as contemporaries. Information and experience can be shared with a group as it occurs, and solidifies the sense of shared experience, building a community. In this context a community refers, as described above, to shared group ethos and mental models. Community members can understand unique language and jargon and feel comfortable without having to explain themselves all the time; sharing reference points, expected standards of behaviour, and accepted codes of conduct.

Text messages have been shown to help with this ambient awareness: those who are remote can share thoughts and remain in each other's awareness by sending these short messages back and forth. This phenomenon was described by Pasi Mäenpäa in reference to the use of mobile phones by Finnish teens, who sent communication that appeared to have minimal informational content:

> Such chatting hardly resembles real exchange of information or even intercourse, as much as merely sharing one's life with others in real time. It is a question of living in the same

> rhythm or wave with one's closest friends, the feeling of a continuously shared life. The repetitive communications by phone are not merely an exchange of information; they also open another world of experience beside, or instead of, the one inhabited at the moment.
> (Mäenpää, 2001)

Manuel Castells (2011), on talking about mobile telephony, described this 'perpetual connectivity' as the key feature of wireless communication rather than mobility. Relationships that might otherwise lose their closeness can be 'pulsed' by short messages, 'to maintain an almost constant, low level of communication' (Cass et al, 2003).

Social media services, which allow people to upload personal messages and 'status updates' to a communal space, are perfectly built to facilitate this exchange of short, constant information pulses and therefore ambient intimacy. In an interview by Clive Thompson on this topic, founder of Facebook Mark Zuckerberg spoke about his decision to include a 'news feed', which collates information from those listed by users as friends, and presents it as a push-based updated system rather than pull-based. Zuckerberg described how initially, there was resistance against this use of the technology (Thompson, 2008). Users could not see the point, or why they would want to be constantly sent information. But once they started using it, it quickly became the central feature. Rather than having to go and seek out information on particular friends, Facebook presents you with all (or at least a chosen selection) of the information. This transfer to push-based services means that you can dip in and out of an information flow which is delivered to you without effort.

To fully utilise and get the benefits of ambient intimacy requires a certain level of initial effort and engagement with it. Like reading and writing, there are 'start up costs' to integrating this new technology into the way we live our lives; in the case of Twitter for example, there are two aspects to this: entering the 'flow' of information, and constructing broadcast information to contribute to the flow that is contextually relevant for your 'followers'. This latter may seem awkward at first, as the information being shared may not seem important. Approaches taken to this can be different – carefully crafting each individual status update, or simply posting everything that comes into your head. But the social space that you are entering into and audience you are addressing is important, at least if you wish to have the intended response from those reading it, whatever that might be. This requires internalisation of other people and 'theory of mind' at potentially higher levels, creating, as described above, a generic 'reader' who is an amalgamation of all existing and potential recipients. Sometimes people can even forget that those who exist outside of the parameters of this 'ideal reader' will see the contributions, leading to conflicts where those who were not intended to see the social flow, for example potential employers, are able to access it. Possible consequences of this will be explored further in Chapter 6.

Any single message on an ambient intimacy platform can seem meaningless, trivial or banal. 'No message is the single-most-important message' says Marc Davis, chief scientist at Yahoo and former professor of information science at the University of California at Berkeley (also quoted by Thompson, 2008). Facebook has an obvious appeal to new users because 'news' of your friends is something that is sought after, but news is not all that is presented. Twitter seems more initially alienating because of its open premise; you can follow content from almost anyone, with no obligation for reciprocity. Who should you 'follow' and why would you be interested in what they have to say? This may be why new users often follow celebrities, whose banality may appear more 'interesting'. We expect intimacy only with those who reciprocate it, but digital public space makes one-sided intimacy possible, or at least the illusion of it, a phenomenon known as parasocial interaction. We can become immersed in the day to day private details of someone we have never met, and feel closer to them than to people with whom we spend every day in physical proximity. It is when a certain base level of interaction is reached that the stream becomes useful, and this can apply to people we know, or people we do not:

> This is the paradox of ambient awareness. Each little update — each individual bit of social information – is insignificant on its own, even supremely mundane. But taken together, over time, the little snippets coalesce into a surprisingly sophisticated portrait of your friends' and family members' lives, like thousands of dots making a pointillist painting. This was never before possible, because in the real world, no friend would bother to call you up and detail the sandwiches she was eating. The ambient information becomes like "a type of E.S.P.," as [Ben] Haley described it to me, an invisible dimension floating over everyday life.
>
> (Thompson, 2008)

One potential drawback of immersing oneself in the stream of ambient information is that once you are connected in to it, disconnecting can feel very disconcerting. Elizabeth Minkel (2014) has described how this disconnection can be generated even by simply being in a different timezone to the majority of those in your 'stream' – the constant flow is temporally tethered, and moves at a very fast pace. More than ever, things that are not recent are no longer worth bothering with, and the period of 'recency' is becoming shorter and shorter: take for example Snapchat, which deletes photographs shortly after they are shared. Becoming out of synch with the information flow of your peers can feel very lonely if you know that communication is happening but you are not directly part of it. Another issue is the risk of information overload; critical to the stream being useful is the relevance of the information. Effective filtering must be set up so that the information that is being fed to you is a representational slice of that which is relevant and interesting to you personally; the general volume is often so overwhelmingly large that you cannot hope to even keep up with everything interesting, let alone the whole stream, especially if it is 'contaminated' with things that are not relevant.

However, it may not be so easy to know which things are going to be of use to us. It may not be things that align with our interests or existing opinions. Additionally, when maintaining digital communities, it is important to bear in mind that the technology may not always behave exactly as it appears to. It may be that what we are fed via mediated technological publics is not actually representative of the reality, as will be explored further in Chapter 7.

There may also be potential downsides of the very open and visible nature of ambient intimacy and densely connected community networks. Connections and interactions that once were transient or hidden might now be visible on a wider stage, and while that helps in extending networks, it can lead to its own problems. Binder et al (2009) suggest that interpersonal communications being visible might have unintended negative consequences and lead to the destabilisation of networks. They give an example of messages appropriate for one particular relationship, for example talking to a friend, which might end up being shared in less appropriate contexts such as visible to a family member on Facebook. This can not only lead to embarrassment but can affect the overall structure of relationships, if for example the family member then takes a dislike to the friend because of the nature of the conversation.

This is a result of the fact that geographic spatial boundaries that normally separate different groups are not present online. 'Users may have to actively uphold structural (offline) features of their networks that are ignored by technology if they want to avoid social clashes' (Binder et al, 2009). This is compounded by the fact that the messages persist and can be accessed later when they were composed as fleeting, of the moment communications, as well as the fact that the audience may be wider than initially perceived. A final factor may be that you do not necessarily wish to present the same 'self' and behaviour to different overlapping communities to which you belong. Binder et al found that Facebook users with more diverse networks experienced more tension. If you want to present different faces to different groups, it may cause problems if this is conflated online in single profile. However, many digital platforms encourage this very behaviour, by purporting to represent your 'real' self.

Profiles: the individual in the digital public space

In order to 'exist' in the digital public space and interact with others, it is necessary to virtualise information about yourself. In the analogy we were using before, it is necessary to create a higher dimensional version of yourself that can exist and traverse the digital public space, and represent you to others who might encounter you there. In most cases, this entails owning a digital profile.

Creating a profile is a critical aspect of your digital existence. It carves out an individualised 'space' and 'body' within the larger context and allows representation of your personality across different aspects. If this profile is created by you, it offers an opportunity to create a distinction between your 'real world' person and the digital version, for careful curation of the identity and information that is presented to the world. This might be explicit in the case of the appearance of an avatar and factual information such as birth dates presented, or more subtle, such as which other profiles are linked with it and the language and content of associated online interactions.

The digital profile can represent an idealised version of one's 'real' identity, or can be subverted to create pseudonymity, or multiple personality facets. Sometimes, multiple profiles exist for different audiences. In other cases, pseudonymity is necessary to prevent negative attention to the originator; either because they are acting in a way that is not socially acceptable and do not want to suffer the repercussions, or because they are putting themselves in danger by making themselves part of the online community and speaking publicly.

When the internet was new, there was a general culture of distrust around putting accurate information in one's profile. It was common practice in the early nineties to choose an anonymising 'handle' or 'nick' with which to refer to oneself. In using one's real name, it was felt that you might be inviting unwanted attention from 'internet axe murderers' to your 'real life'; there was a need for separation between the two domains. However, with the ubiquity, and almost necessity of having a digital profile, there is now a move more towards representing oneself accurately in the online space. This may not be a representation of one's whole existence, being a 'work' profile or a 'private' one for consumption by friends, but the implication is that it accurately represents a facet of our true personality. This extension of our 'real selves' into the digital public space may be an indication that users feel more comfortable treating this public digital space as another part of everyday existence, rather than a game or unreal state in which a costume or disguise might be donned. Alternatively, it might simply reflect the fact that having an online profile is more socially acceptable, and there is no shame (and therefore no need to hide identity) associated with having a profile online and spending time in digital interaction.

Indeed, companies such as Google and Facebook have the aim of bringing the whole world into the digital space, and now actively discourage the use of pseudonyms. There is of course an ulterior motive to this: these companies do not charge customers, but instead commoditise the information held about them. By using 'real' names, this information can be effectively networked and the companies can create their own profile, which may contain much more information than the 'customer' suspects might be known. Work by Kosinski et al (2013) found that even if not given directly, private information such as political leaning, ethnic origin and sexuality can be inferred with high levels of accuracy simply from the 'likes' given on Facebook. Such data mining and analysis could be used to deliver targeted advertising.

A more benevolent justification for the use of real names is the maintenance of responsibility for things that are posted online, as a prevention tactic against those who would make use of their online 'persona' to post abuse. The creation of an anonymous persona may lead to diminished feelings of responsibility and towards the violation of social norms through actions such as bullying and abuse. Aaron Ben-Ze'ev speculates that this is because the anonymity of online interactions reduces the shame attached to such poor behaviour; the activities are associated with the online persona and not the person who created it (Ben-Ze'ev, 2003). Much has been written about the worrying prevalence of cyberstalking, online abuse and hate speech, a large proportion of which is directed at women. Amanda Hess points out that while the abusers can hide behind a blanket of anonymity and online identities that are separated from their own, the victims are often operating professionally online and in contexts highly connected to their offline activities. She quotes Nathan Jurgenson, a social media sociologist at the University of Maryland, who points out that, 'It's a lot easier for the person who made the threat – and the person who is investigating the threat – to believe that what's happening on the Internet isn't real' (Hess, 2014). Kent Norman suggests that the absence of face to face cues can cause communication to become 'disembodied', as if talking to oneself, and recommends that to combat this, 'cyberspace designers convey a greater sense that one is in a public space' in order to emphasise publicness and thus personal responsibility (Norman, 2008).

But anonymity does not just have negative consequences. It can also lead to greater intimacy because of the freedom to not be identified with the statements: witness the popularity of the online art project 'PostSecret'[13] which displays confessional 'postcards' that contributors have created containing personal statements of feelings and emotions. The value of the website is not just that it is voyeuristic but that comfort is gained from knowing others may be experiencing similar issues that may never usually be spoken of publicly.

Insistence on 'real' names is also an issue for certain groups for whom a name is a more fluid entity, or one that needs protecting. Examples of groups who have had issues with these policies include transgender people whose birth names do not match those they currently use,[14] those of non-western origins whose names are not recognised as 'real' by the organisations in question, and sufferers of abuse who use a different name online so they cannot be found by their abusers. There are also many people who are identified by nicknames and user-names so much that linking their online profile to their 'real' name would lead to their friends and family being unable to find said profile. Even in the non-digital sphere, we construct our own selves, and digital spaces give a freedom to do this more fluidly, showing different facets and aspects of our self at different times and in different spaces. Rigid policing of the nature of the self cuts off the many opportunities this brings for inhabitants of the new digital public space. Bakardjieva and Gaden (2011) observe that many users already employ 'inventive tactics' to compartmentalise their communication. By doing this they can avoid being restrained by the push of social media platforms to flatten their networks and homogenise their profile to be applicable to all their contacts; communicating via alternative platforms where necessary, so that not all of their lives and relationships are taken over.

Beyond this, it must also be understood that our footprint in the digital public space does not just consist of the information that we put there ourselves. Our digital profiles are not just created by us, but may be affected and added to by other people. We shed data everywhere we go, that is created in spite of ourselves, that streams behind us and may be picked up by many different people. This may include traces of our physical experience that are picked up by sensors in the environment – such as the CCTV cameras that litter our cities. Alternatively, they might be data trails that we are complicit in; from supermarket loyalty cards and website cookies, to our smartphones. Every transaction is logged, every GPS co-ordinate check saved (unless convoluted steps are taken to avoid it, and even this is sometimes not possible). We give implicit (or explicit) permission for this data to be collected, we even encourage it sometimes in terms of technology like wearable devices that give some of the information back for our use. But we seldom consider what it might be used for further down the line, and how it might be connected with other information. Sometimes this quantification of the self is even more closely linked to our health and wellbeing, for example those who have medical monitoring such as linked pacemakers or diabetic insulin pumps. These provide significant amounts of data which can be used by doctors, and by patients to help understand and manage their body function – but only if they are able to understand the data that is gathered. This may need interpretation by medical professionals or another visualisation to make it understandable. Generated data only has value if it is interpreted correctly, and may have different uses for different people, therefore the data on 'yourself' might be accessible to you, but that does not mean you understand it or how it might be used by others. Your digital public space footprint might be more visible to other people than it is to yourself, and may influence how you are perceived by others.

All of this contributes to our digital profile that exists independently and can often be accessed without our knowledge. There are many concerns of privacy related to the data that we create as we go about our modern lives, and these might have far reaching implications from restrictions of civil liberty to long-distance murder via medical devices. These issues will be explored further in Chapter 6. This all sounds very sinister but it can also be used to improve our wellbeing through the use of 'big data' as described above, for example by cross referencing location and demographics with health data to provide better tailored medical care, or delivery of individualised content that anticipates need.

Ultimately, even if the digital public space acts as a geographically analogous dimension, it is one where we have vested interests in certain areas. This is inevitable, as we have always been more interested in things that relate to us than to others. This can be seen by returning to the history of maps and how they relate to our sense of place in the world. Andrew Garfield relates the story of the *Mappa Mundi*, the medieval map created around 1300 CE, probably in Hereford where it has been displayed since. On the map, which displays the extent of the known world at the time, the place where Hereford is marked is worn and damaged – from people touching the place where they were. He relates this to the way that people use modern digital maps such as Google Maps in exactly the same way:

> And where did people search for first? The very same place they had looked for when they viewed the Mappa Mundi at the end of the thirteenth century, the place where they lived. 'Always,' Brian McClendon told me. "And every new version, people go and say how does my town or my house look." This is a part of human nature – the desire to know where we fit within the grander scheme of things. But it is also emblematic of the new form of cartography that Google and its digital counterparts represent: Me-mapping, the placing of the user at the instant centre of everything.
>
> (Garfield, 2012)

This shows how even with almost limitless amounts of information, we relate it directly to ourselves and pay special attention to certain parts, and pick what we are interested in. The consequences of this selection, and how external factors can bias our view of the world, are explored in the following chapters.

Key points

- As cultural humans we depend on accessing knowledge of our surroundings from a variety of sources, which grow further from individual sensory experience as information technology improves and which can be shared more easily.
- Connected digital technology and wireless networking separates information experience from physical storage.
- Information space exists as a 'higher dimension' which touches all spaces and times, and can be accessed from anywhere and simultaneously.
- As inhabitants of this information space our daily interactions are changed; from arranging meetings to accessing knowledge and services to use of physical spaces.
- Being able to instantly access information is a form of cognitive augmentation, giving us an external 'memory'. This is not dissimilar to physical mementos and aide-memoires. However digital mementos (which might also be embedded as physical objects) can be more easily searched, updated and catalogued.
- Extended, embodied cognition may lead to trade-offs in how we store knowledge internally. Overreliance on digital cognition is something to be wary of.
- Combined computing power and large data sets can have emergent properties which can provide new opportunities and forms of knowledge.
- 'Big data' can have benefits such as collective problem solving and digital integration in cities, but could be overwhelming or used for non-altruistic purposes.
- Vast volumes of collectively shared information mean that collaborative contribution, remixing, reimagining and re-sharing are important parts of the digital information space: we are in the age of the prosumer. Barriers between producers and consumers of information content are being eroded.
- Connecting in real-time is still important to create a shared experience space, but perceptions of the barriers (or lack of them) between information in digital and 'real' space can be unclear.
- Shared experience allows us to connect with others.
- Social relations are a critical human experience. Digital platforms may affect our social networks of friends, and how and why we interact with other people.
- Digital platforms may allow 'weak tie' networks to flourish, which can provide different forms of support to traditional networks.
- Communities in digital public space are not bound by geography, and can provide alternative forms of belonging and support.
- It is possible to exist in several different communities, online or offline, at once; they appear to complement each other rather than compete.
- 'Ambient intimacy' is the constant awareness of others in your community afforded by social media and constant information updates on peers, solidifying the sense of shared experience and community. Being disconnected from this intimacy can be disconcerting, and it must be carefully curated.
- Having multiple networks and connections may destabilise relationships, especially if information travels between them in unintended ways.

- Creating profiles allows separation of different facets of digital life. Social norms of when it is acceptable to use real names, pseudonyms and anonymity are in shift, and all have benefits and risks.
- Digital profiles create footprints of our personal information that might be used in ways we do not intend.

Notes

1 See Dodge et al (2009) for further discussion of how software creates space.
2 Defined as such by Cass et al (2003).
3 For example, the 'counting of the Omer' in Jewish tradition, which marks the 40-day period between the barley and wheat harvests, and the two harvest festivals associated with these.
4 For early work on the implications of this for recall, see Kawamura et al (2003).
5 For example, a cab driver who drove into a river, as reported on BBC News in April 2008: http://news.bbc.co.uk/1/hi/england/norfolk/7362254.stm
6 There is however no guarantee that the answers given are unbiased (see Chapter 7).
7 See Chapter 3.
8 It is important to bear in mind that our view of communication times in the past can sometimes be distorted in ways we do not expect: in the 1890s for example, multiple posting times of up to six times a day in London meant letters travelled quickly, and postcards were popularly used for rapid updates similar to text messages or social media (Gillen, 2016).
9 Adrian Chan (2009) provides an interesting self-analysis of audience construction on Twitter.
10 For example Hampton et al (2009 and 2011).
11 Nie (2001) gives a good overview and discussion of these surveys.
12 DiMaggio et al (2014) cites several studies including Etzioni & Etzioni (1997), Rheingold (1993), Wellman & Gulia (1999).
13 www.postsecret.com
14 See the experience of Zoe Cat for example, an employee of Facebook (Cat, 2015).

References

Abbott, E.A., 1884. *Flatland; a romance of many dimensions, by A. Square.* Seeley & Co.

Aiello, L.C. and Dunbar, R.I., 1993. Neocortex size, group size, and the evolution of language. *Current Anthropology, 34*(2), pp.184–193.

Andrews, A. and Hartevelt, M., 2003. Community, memory and ambient intelligence. *In*: Aarts, E. and Marzano, S., eds. *The new everyday: views on ambient intelligence.* 010 publishers, p.185.

Anglo-Filles, 2014. Alina's best day ever [podcast]. *Anglo-Filles.27.* Available at: http://anglo-filles.madeoffail.net/episodes/anglofilles-episode-27-alinas-best-day-ever/ [Accessed 11 October 2016].

Bakardjieva, M. and Gaden, G., 2011. Web 2.0 technologies of the self. *Philosophy and Technology, 25(3).*

Balsamo, A., 2011. *Designing culture: the technological imagination at work.* Duke University Press, p.134.

Bargh, J.A. and McKenna, K.Y., 2004. The internet and social life. *Annual Review of Psychology, 55*, pp.573–590.

Beckett, C., 2011. *After Tunisia and Egypt: towards a new typology of media and networked political change.* [online] Polis. Available at: http://blogs.lse.ac.uk/polis/2011/02/11/after-tunisia-and-egypt-towards-a-new-typology-of-media-and-networked-political-change/ [Accessed 11 October 2016].

Benedikt, M.L., 2008. Cityspace, cyberspace, and the spatiology of information. *Journal for Virtual Worlds Research, 1*(1), p.4.

Benkler, Y., 2006. *The wealth of networks: how social production transforms markets and freedom.* Yale University Press.

Ben-Ze'ev, A., 2003. Privacy, emotional closeness, and openness in cyberspace. *Computers in Human Behavior*, *19*(4), pp.451–467.

Binder, J., Howes, A. and Sutcliffe, A., 2009, April. The problem of conflicting social spheres: effects of network structure on experienced tension in social network sites. *Proceedings of the SIGCHI Conference on Human Factors in Computing Systems*. ACM, pp. 965–974.

Booth, P., 2010. *Digital fandom: new media studies*. Vol. 68. Peter Lang Publishing.

boyd, d., 2007. Social network sites: public, private, or what. *Knowledge Tree*, *13*(1), pp.1–7.

Brody, N., 2012. Modelling the digital public space: the new renaissance. *In:* Hemment, D., Thompson, B., de Vicente, J.L. and Cooper, R., eds. *Digital Public Space.* FutureEverything Publications. Available from: http://futureeverything. org/publications/digital-public-spaces/ [Accessed 23rd September 2016].

Brooks, 2007. *The outsourced brain*. [online] *The New York Times*. Available at: www.nytimes. com/2007/10/26/opinion/26brooks.html [Accessed 5 October 2016].

Burke, M. and Kraut, R.E., 2014, April. Growing closer on Facebook: changes in tie strength through social network site use. *Proceedings of the SIGCHI Conference on Human Factors in Computing Systems*. ACM, pp.4187–4196.

Bush, V., 1945. As we may think. *The Atlantic Monthly 176*(1), pp.101–108.

Carr, N., 2011. *The shallows: what the internet is doing to our brains*. WW Norton & Company.

Cass, J., Goulden, L. and Koslov, S., 2003. Intimate media. Emotional needs and ambient intelligence. *In:* Aarts, E. and Marzano, S., eds. *The new everyday: views on ambient intelligence*. 010 Publishers.

Castells, M., 2011. *The rise of the network society: The information age: economy, society, and culture, volume 1*. John Wiley & Sons, p.1976.

Cat, Z., 2015. My name is only real enough to work at Facebook, not to use on the site. [online] Available at: https://medium.com/@zip/my-name-is-only-real-enough-to-work-at-facebook-not-to-use-on-the-site-c37daf3f4b03#.2b6wejs5c [Accessed 13 October 2016].

Chan, A., 2009. If you think twitter is weird, you're not alone. [online] *Gravity 7*. Available at: www. gravity7.com/blog/media/2009/06/if-you-think-twitter-is-weird-youre-not.html [Accessed 11 October 2016].

Clark, A., 2003. *Natural-born cyborgs: minds, technologies, and the future of human intelligence*. Oxford University Press.

Clark, A. and Chalmers, D., 1998. The extended mind. *Analysis*, *58*(1). Oxford University Press, pp.7–19.

Cook, T., 2015. Always keep (nerd)fighting: Durkheim's collective effervescence and fandoms as reformative social movements. *In: PCA/ACA National Conference 2015*. Available at: http://ncp.pcaaca.org/presentation/always-keep-nerdfighting-durkheim%E2%80%99s-collective-effervescence-and-fandoms-reformative-so [Accessed 26 March 2016].

Crang, M., Crosbie, T. and Graham, S., 2007. Technology, time–space, and the remediation of neighbourhood life. *Environment and Planning A*, *39*(10), pp.2405–2422.

de Ruiter, J., Weston, G. and Lyon, S.M., 2011. Dunbar's number: group size and brain physiology in humans reexamined. *American Anthropologist*, *113*(4), pp.557–568.

DiMaggio, P., Hargittai, E., Neuman, W.R. and Robinson, J.P., 2003. Social implications of the internet. *In:* Nissenbaum, H., Price, M.E., and Lang, P., eds. *Academy and the Internet,* p.48.

Dodge, M., Kitchin, R. and Zook, M., 2009. How does software make space? Exploring some geographical dimensions of pervasive computing and software studies. *Environment and Planning A*, *41*(6), pp.1283–1293.

Donath, J., 2014. *The social machine: designs for living online*. MIT Press.

Dunbar, R.I., 1993. Coevolution of neocortical size, group size and language in humans. *Behavioral and Brain Sciences*, *16*(04), pp.681–694.

Dunbar, R., 2010. *How many friends does one person need?: Dunbar's number and other evolutionary quirks*. Faber & Faber.

Dunbar, R.I., 2012. Social cognition on the internet: testing constraints on social network size. *Philosophical Transactions of the Royal Society of London B: Biological Sciences*, *367*(1599), pp.2192–2201.

Dunbar, R.I., Duncan, N.D.C. and Nettle, D., 1995. Size and structure of freely forming conversational groups. *Human Nature*, *6*(1), pp.67–78.

Ellison, N.B. and Vitak, J., 2015. Social network site affordances and their relationship to social capital processes. *The Handbook of the Psychology of Communication Technology*, pp.205–227.

Evans, J.A., 2008. Electronic publication and the narrowing of science and scholarship. *Science*, *321*(5887), pp.395–399.

Featherstone, M., 2009. Ubiquitous media: an introduction. *Theory, Culture & Society*, *26*(2–3), pp.1–22.

Floridi, L., 2007. A look into the future impact of ICT on our lives. *The Information Society*, *23*(1), pp.59–64.

Garfield, S., 2012. *On the map: why the world looks the way it does.* Profile Books.

Ghena, B., Beyer, W., Hillaker, A., Pevarnek, J. and Halderman, J.A., 2014. Green lights forever: analyzing the security of traffic infrastructure. *8th USENIX Workshop on Offensive Technologies (WOOT 14)*.

Gillen, J., 2016. How early picture postcards were the Edwardian equivalent of Instagram. [online] *The Conversation.* Available at: https://theconversation.com/how-early-picture-postcards-were-the-edwardian-equivalent-of-instagram-61870 [Accessed 11 October 2016].

Gonçalves, B., Perra, N. and Vespignani, A., 2011. Modeling users' activity on twitter networks: Validation of Dunbar's number. *PLOS One*, *6*(8), p.e22656.

Granovetter, M.S., 1973. The strength of weak ties. *American Journal of Sociology*, pp.1360–1380.

Grayson, K. and Shulman, D., 2000. Indexicality and the verification function of irreplaceable possessions: a semiotic analysis. *Journal of Consumer Research*, *27*(1), pp.17-30.

Green, J., 2003. Thinking the future. *In:* Aarts, E. and Marzano, S., eds. *The new everyday: views on ambient intelligence*. 010 Publishers.

Hampton, K.N., Lee, C.J. and Her, E.J., 2011. How new media affords network diversity: direct and mediated access to social capital through participation in local social settings. *New Media & Society*, p.1461444810390342.

Hampton, K.N., Sessions, L.F., Her, E.J. and Rainie, L., 2009. Social isolation and new technology. *Pew Internet & American Life Project*, *4*.

Hampton, K.N., Sessions, L.F. and Her, E.J., 2011. Core networks, social isolation, and new media: how internet and mobile phone use is related to network size and diversity. *Information, Communication & Society*, *14*(1), pp.130–155.

Haythornthwaite, C., 2001. Introduction: the internet in everyday life. *American Behavioral Scientist*, *45*(3), pp.363–382.

Hess, A., 2014. Why women aren't welcome on the internet. [online] *Pacific Standard.* Available at: https://psmag.com/why-women-aren-t-welcome-on-the-internet-aa21fdbc8d6#.e7k54g5wy [Accessed 13 October 2016].

Hill, R.A. and Dunbar, R.I., 2003. Social network size in humans. *Human Nature*, *14*(1), pp.53-72.

Hillis, K., 1999. *Digital sensations: space, identity, and embodiment in virtual reality*. University of Minnesota Press.

Hodson, H., 2013. Online gamers harnessed to help disaster response. *New Scientist*, *219*(2928), p.21.

Holpuch, A., 2013. Damaged undersea internet cable causes widespread service disruption. [online] *The Guardian.* Available at: www.theguardian.com/technology/2013/mar/28/damaged-undersea-cable-internet-disruption *[Accessed 29 September 2016].*

Hutchins, E., 1995. *Cognition in the wild.* MIT Press.

Jenkins, H., 2006. *Convergence culture: where old and new media collide*. NYU Press.

Kanai, R., Bahrami, B., Roylance, R. and Rees, G., 2011. Online social network size is reflected in human brain structure. *Proceedings of the Royal Society of London B*. p. rspb20111959.

Kawamura, T., Fukuhara, T., Takeda, H., Kono, Y. and Kidode, M., 2003, July. Ubiquitous memories: wearable interface for computational augmentation of human memory based on real world objects. *Proceedings of 4th International Conference on Cognitive Science (ICCS2003)*, pp. 273-278.

Kennedy, J.B., 1926. When woman is boss: an interview with Nikola Tesla. *Colliers, Seattle*.

Kosinski, M., Stillwell, D. and Graepel, T., 2013. Private traits and attributes are predictable from digital records of human behavior. *Proceedings of the National Academy of Sciences*, *110*(15), pp.5802–5805.

Koslowski, B., 2014. Screenscapes: of theatre, audiences and social media. *In*: Vicente, A. and Ferreira, H., eds. *Post-screen: device, medium and concept*. Faculdade de Belas-Artes da Universidade de Lisboa, 2014.

Lai, C.H. and Katz, J.E., 2012. Are we evolved to live with mobiles? An evolutionary view of mobile communication. *Periodica Polytechnica, Social and Management Sciences*, *20*(1), p.45.

Mäenpää, P., 2001. Mobile communication as a way of urban life. *In*: Mäenpäa, P., Warde, A. and Gronow, J., eds. *Ordinary consumption*. Routledge, p.111.

McPherson, M., Smith-Lovin, L. and Brashears, M.E., 2006. Social isolation in America: changes in core discussion networks over two decades. *American Sociological Review*, *71*(3), pp.353-375.

Minkel, E., 2014. ICYMI: the internet has ruined our conception of time. [online] *New Republic*. Available at: https://newrepublic.com/article/117886/icymi-internet-acronym-destroying-our-conception-time [Accessed 13 October 2016].

Moggridge, B., 2006. Adopting technology. *Designing Interactions*, pp.237-317.

Newsweek, 2004. All eyes on Google. [online] *Newsweek*. Available at: http://europe.newsweek.com/all-eyes-google-124041 [Accessed 5 October 2016].

Nie, N.H., 2001. Sociability, interpersonal relations, and the internet reconciling conflicting findings. *American Behavioral Scientist*, *45*(3), pp.420-435.

Nissen, B. and Bowers, J., 2015, April. Data-things: digital fabrication situated within participatory data translation activities. *Proceedings of the 33rd Annual ACM Conference on Human Factors in Computing Systems*. ACM, pp. 2467–2476.

Norman, K.L., 2008. *Cyberpsychology: An introduction to human-computer interaction*. Vol. 1. NY: Cambridge University Press, p.327.

Ofcom, 2014. *The communications market 2014*. Available at: www.ofcom.org.uk/__data/assets/pdf_file/0031/19498/2014_uk_cmr.pdf.

Reichelt, L., 2007. Ambient intimacy. [online] Available at: www.disambiguity.com/ambient-intimacy/ [Accessed 13 October 2016].

Rheingold, H., 2002. *Smart mobs: the next social revolution*. Basic Books.

Ritzer, G. and Jurgenson, N., 2010. Production, consumption, presumption: the nature of capitalism in the age of the digital 'prosumer'. *Journal of Consumer Culture*, *10*(1), pp.13-36.

Rutkin, A., 2014c. How data can save a city. *New Scientist*, *224*(2990), pp.24-25.

Salinas, L., Coulton, P. and Dunn, N., 2016. Using game design as a frame for evaluating experiences in hybrid digital/physical spaces. *Architecture and Culture*, *4*(1), pp.115-135.

Shirky, C. 2010. How cognitive surplus will change the world. [online] *TED Talks 2010*. Available at: www.ted.com/talks/clay_shirky_how_cognitive_surplus_will_change_the_world [Accessed 10 October 2016].

Southern, J., 2012. Comobility: how proximity and distance travel together in locative media. *Canadian Journal of Communication*, *37*(1), p.75.

Stephenson, N., 2011. *Reamde*. William Morrow & Co.

Thompson, C., 2008. Brave new world of digital intimacy. [online] *The New York Times Magazine*. Available at: www.nytimes.com/2008/09/07/magazine/07awareness-t.html?pagewanted=all&_r=1 [Accessed 13 October 2016].

Wang, H. and Wellman, B., 2010. Social connectivity in America: changes in adult friendship network size from 2002 to 2007. *American Behavioral Scientist*, *53*(8), pp.1148–1169.

Wellman, B., Haase, A.Q., Witte, J. and Hampton, K., 2001. Does the internet increase, decrease, or supplement social capital? Social networks, participation, and community commitment. *American Behavioral Scientist*,*45*(3), pp.436–455.

Zhou, W.X., Sornette, D., Hill, R.A. and Dunbar, R.I., 2005. Discrete hierarchical organization of social group sizes. *Proceedings of the Royal Society of London B: Biological Sciences*, *272*(1561), pp.439–444.

Section 3

What are the consequences of digital public space?

5 Transactions, payment, and ownership

So far we have illustrated the notion of different digital spaces: geographic, physical or abstract – and how we interact with and within them, tangibly and intellectually, through information and communication.

We are now moving through time and virtual space as *information entities*, as avatars (collections of constructed information), as creators and consumers of *information objects*. Spaces we inhabit may be constructed and designed in ways that represent physical geographies or containers (libraries, files) or may represent novel informational structures that do not have physical correlates. We are architects, explorers and consumers in this digital space. Our challenge is understanding how we operate in what is actually a 'public' space, as individuals.

Now that we have defined the nature of these digital spaces, we can begin to explore what it means to inhabit these worlds. In order to do this, we can look at some specific ways in which it might change behaviour. The first of these we will look at is in the nature of ownership, and how ownership can be transferred through payment and transactions.

Many of the social interactions described in previous chapters, such as grooming between primates, and the transfer of social information, are about transactions; individuals exchanging information or objects with each other. Transactions and sharing are a critical part of interaction, both from the point of view of resource allocation and the building of trust. But here we will argue that several aspects of ownership and transaction are being fundamentally altered in the digital public space, or have new implications which must be considered.

The notion of transaction is intimately related to that of ownership and reciprocity. If you have ownership of something, then you also have authority to transfer ownership to someone else, and you might expect to get something in return (whether now or at a later date). The history of ownership, in many different forms, is complex and unclear, however it seems likely that agricultural societies with fixed dwellings must have needed to delineate which items could be used by whom. Certainly, property is a principal component of many modern societies; and with property comes the notion of property exchange.

Before the widespread use of currency, exchange likely consisted of reciprocal systems where goods or services were exchanged for equivalent services that could be provided by others – not always a direct exchange, but through the building up of credit which could be redeemed when suitable (Graeber, 2011). The development of currency is hypothesised to have provided formalisation of this credit and debt into an abstract system of worth which could be assigned to anything, and allowed movement beyond the simple 'score keeping' system that could exist in smaller communities. Through this process, any valued service or object could be exchanged for any other item of similar value through a chain of transactions involving money. However, the lines of ownership and value are not always straightforward, especially when information is involved. Many digital goods distributed and shared in digital spaces are 'non-rival' goods, that is, ownership by one person does not negate it being used by others. Unlike a piece of cake, which can only be consumed by one person, if I consume a piece of music via an MP3 it does not reduce its availability for you. Digital public space provides a medium by which valuable content can be created and distributed, however given the ease by which digital 'objects' can be copied and shared with no loss of quality, the value of the content may be obscured. There are risks in the attribution of value according to tangible attributes, in that things which are intangible and more difficult to quantify are undervalued – priceless is not the same as worthless. This is an issue seen with regard to intangible tacit knowledge, as described by Krogh et al discussing sharing knowledge between collaborators: 'The whole notion of "transacting" something that is hard to articulate means that anything that cannot be specified has little perceived value' (Von Krogh et al, 2000). What is the value of an individual digital copy that can be effortlessly reproduced, as opposed to that inherent in the design and creativity that produced the original? Economic theory calls these costless information goods 'public' but traditionally suggests that if they are distributed freely, the market will not allow their production because there is no incentive to spend time on their creation. For this reason, copyright, licensing and patent law exists to provide recompense to creators, and incentives for production. This gets even more complex if the creative production does not take place on the expectation of direct financial reimbursement, or is based on prior existing media that is owned by others, as we shall see below.

Ownership of digital spaces

In Chapter 3, the issue of ownership arose in relation to the digital spaces which exist in virtual worlds. Linden specifically gave ownership of digital items created in *Second Life* to the players who made them. But ownership refers not just to property but also to authority over spaces that are not public. In Chapter 1, there was discussion of the definition of public space versus private space, and the fact that in the physical domain, lots of space that is perceived as being 'in public' such as shopping centres and restaurants, are privately owned spaces that allow the public to use them as long as they conform to rules set out by the owners. In fact even 'public' space which is owned by the people through the edifice of government has usage rules set out by public convention and enforced by said government.

Blizzard, the company who created *World of Warcraft*, retains their 'ownership' of the space they have created, and have sued players for violation of the rules laid out in the space. In this sense the boundary lines are clear. But some large organisations do not act in such a proprietorial way and appear to be attempting to create archives and digital spaces that are for the wider public benefit, and theoretically exist in perpetuity as a shared resource for all. What are the implications of this?

The digital public space was, as discussed in Chapter 1, a concept initially developed by the BBC. The BBC are an organisation funded (via the licence fee) by the British public, and their remit has long included the creation of publicly shared assets. By aiming to create shared archives of content, they could be said to be creating digital public space; however, they are still a British institution. The traditional broadcasting media used by the BBC were radio and television, and these formerly had a specific geographic radius. It was difficult to access these resources if you were not a member of the British public, that is, not in Great Britain. Even the BBC World Service, providing radio broadcasts accessible worldwide, was initially set up to cater to citizens of the Commonwealth. In the words of Director General Sir John Reith in its inaugural broadcast, it provided 'a connecting and co-ordinating link between the scattered parts of the British Empire' (BBC, 2007). However, the internet has no geographic boundaries other than those arbitrarily enforced by region locking. Unless specifically prevented from doing so, anyone can access content from anywhere in the world, and even region locking can be overcome (illegally) by means such as proxy servers which provide a false IP address from an allowed region. This 'public' space is ineffectively restricted.

Then there are commercial organisations who have taken the mantle upon themselves to provide digital services which, on the surface, appear to benefit by their existence the public good. A good example of this is, again, Google. Their services are so ubiquitous and 'free' to use that you would be forgiven for forgetting that they are a highly successful commercial company. They do indeed provide services which are not a part of their commercial activities and have the remit of benefit for all. Launched in 2011, the Google Cultural Institute is a good example of this, a project to make exhibition and archival content available online in 'an effort to make important cultural material available and accessible to everyone and to digitally preserve it to educate and inspire future generations' (Google, 2015). They digitally archive material from many of the most prominent international museums including the British Museum, Museo Galileo in Florence and Yad Vashem in Israel. This appears to have a similar ethos to the BBC's digital public space project, but although it is run as a not-for-profit branch of the organisation, Google is ultimately a privately owned business rather than a public organisation. Nicholas Carr, in his book *The Shallows* reminds us that Google's business is selling advertising, and thus their business model ultimately rests on encouraging people to spend more time on the internet so that they see more advertisements (Carr, 2011).

If Google is hosting and curating the content, who 'owns' it? Who has made the decisions about which content to include, and what are the processes by which this decision was made? An example of the controversies that can arise with regards to these ownership issues is demonstrated, again with Google at the heart of it, with another service: Google Print (later called Google Books). The mission of this project was to scan all books and make the text available for searching, described thusly by Eric Schmidt of Google: 'Imagine the cultural impact of putting tens of millions of previously inaccessible volumes into one vast index, every word of which is searchable by anyone, rich and poor, urban and rural, First World and Third, en toute langue—and all, of course, entirely for free' (Schmidt, 2005). When it was first proposed it was backed by partners including prestigious libraries (Oxford's Bodleian, and the New York Public Library) and many of the major trade and academic presses (including Houghton Mifflin, McGraw-Hill, and the university presses of Oxford, Cambridge, and Princeton) who agreed to grant Google access to scan books either in the public domain or included in the 'publisher programme'. In the first year, Google scanned an estimated one hundred thousand books (Carr, 2011).

However, some of these books were still in copyright, and lawsuits were brought against Google in 2005 by the Authors Guild and the Association of American Publishers. This was because of Google's decision not to track down the copyright holders before scanning the books and adding them to their database, and to only exclude scanned books from the database if they received a formal written request from the copyright holder. The result of this action, three years later, was compensation of $125 million to the copyright holders whose work had already been scanned, and a cut of advertising revenues from the book search. But this led to larger issues of ownership as Carr describes:

> In return for the concessions, the authors and publishers gave Google their okay to proceed with its plan to digitize all the world's books. The company would also be 'authorized to, in the United States, sell subscriptions to [an] Institutional Subscription

> Database, sell individual Books, place advertisements on Online Book Pages, and make other commercial uses of Books.'
>
> The proposed settlement set off another, even fiercer controversy. The terms appeared to give Google a monopoly over the digital versions of millions of so-called orphan books—those whose copyright owners are unknown or can't be found.
>
> (Carr, 2011)

By being the sole reliable source for, in effect, the world's digital libraries, Google has a huge amount of power and control over access. While they might appear to benevolently allow access for all, there is a level of trust involved in this. This also interacts with issues of bias, whereby it is entirely in Google's hands whether any particular books are scanned or searchable. There may be inherent biases which are not immediately visible (for example, are books in the English language given priority?). Possible bias effects such as this are explored in Chapter 7.

We can also look at the management of 'public' digital access across physical spaces. There is perhaps a perception that publicly provided services are 'free' (at point of use). However, while there may be no direct transactional aspect at the point of use, services which are provided by government are paid for by the public, usually via the medium of taxation in some form. If something is provided ubiquitously and made available to everyone, it may become 'invisible' as the link between payment and reciprocation is broken. This is the case, for example, with public services such as parks, and public toilets, and perhaps to infrastructure such as road networks and sewerage systems. The expectation will be set that such services should be provided to an adequate level by the government, and on a day-to-day basis their use fades into the background for most people. But this extends back to the analogy we were making earlier with digital services – many of these 'free' 'public' systems are in fact privately maintained. In the UK at least, clean drinking water is perceived to be 'free'. Most restaurants and food retail outlets will provide tap water free of charge. This is despite the fact that very few households are not connected to the mains water supply. The supply is paid for, and may be metered or unmetered but is not paid for at the point of service (that is, you do not have to put money in your tap for water to come out). Public toilets are provided in most areas, some of which charge a nominal fee for use (the origin of the euphemism 'spend a penny') but there is a public perception that toilet access should be cheap or free. Consider how it is expected that retail outlets (particularly those serving food) will provide toilet facilities, and that these are accessible. Often the public use these facilities without using the retail outlet – it is discouraged, but socially understood as an option. In these cases, the cost of provision of such services is built into the running costs of the establishment, on the understanding that people who use the bathrooms or drink water will make other purchases or accompany those who do.

Recently this perception has begun to extend to internet access – free wifi is a big draw for those on the move, and such facilities are becoming so endemic that establishments may be perceived as lacking if they do not provide it, unless they make a feature of its lack as a marked difference; somewhere that is cut off from digital life as a haven of tranquillity. But for the most part, wireless internet access has joined tap water and toilet facilities as 'expected' free services which customers feel entitled to. This weight of expectation has led to free wifi provision in increasingly wide coverage, including many international airport terminals and the slopes of Mount Fuji, provided by the Japanese government (BBC, 2015). These facilities are not really free, but the costs are hidden; and this is also the case with 'free' digital services such as informational websites and social networking platforms.

As ubiquitous computing becomes more pervasive, and digital aspects are added to public spaces, it may be that these too are considered necessary. It might be that a public park without, say, digital augmentation is seen as unfit for purpose.[1]

Copyright and protecting value

As we have established, digital public space exists not only as archives and collections of public objects but also a space in which interaction can take place. Much of this interaction involves transactions between people sharing information and digital content, the latter of which is often subject to copyright which may affect its rights to be shared.

Copyright and intellectual property laws were developed as a way of protecting value assigned to ideas and concepts so they cannot be used without recompense to their originators, but it is a relatively modern way of looking at things: intellectual property was not codified in law until the eighteenth century. The Statute of Anne, also known as the Copyright Act 1709, was the first time in British law that copyright was legally held by authors rather than publishers. In America, the Copyright Act of 1790 was the first to grant intellectual property rights to authors, and referred specifically to books, maps and charts (Lessig, 2008). Intellectual property rights are particularly important in asserting ownership of digital content, since it is often by nature intangible, and the products are ideas-based rather than physical objects. If the rights holders wish to be rewarded for what they have created, they must control how it is distributed. But as Lawrence Lessig points out, digital content creates new challenges for copyright law, particularly because of the way that the law was changed in 1909 so that the word 'copy' would refer generally to the rights of any copyright holder, rather than the specific commercial production activities of 'printing, reprinting, publishing and vending'.

> The law regulates "reproductions" or "copies." But every time you use a creative work in a digital context, the technology is making a copy. When you "read" an electronic book, the machine is copying the text of the book from your hard drive, or from a hard drive on a network, to the memory in your computer. That "copy" triggers copyright law. When you play a CD on your computer, the recording gets copied into memory on its way to your headphones or speakers. No matter what you do, your actions trigger the law of copyright. Every action must then be justified as either licensed or "fair use."
>
> (Lessig, 2008)

Several measures have been taken to implement and protect digital copyright. Digital Rights Management (DRM) is a term which describes certain technological solutions to prevent copying and sharing except where expressly allowed. This can be challenging to implement effectively when the sale of a digital object relies on it being transferable between individuals, but further subsequent transfer is a violation of the law. It may also be made more complex by the non-geographical nature of the digital public space which can cross national boundaries and thus run into legal differences affecting copyright and ownership. For example, UK law does not have a concept of 'fair use', and the equivalent laws regarding 'fair dealing' are much more specific and deal with particular exemptions including non-commercial research, criticism and review, parody and pastiche, or reporting current events (Collins, 2015). There is also a difference in copyright between the UK and the US in regard to how long copyright is retained. In the UK this is 70 years (either from the death of the author or publication date) while in the US it is 95 years. This means that some properties may be public domain in one region and not in another. Enforcement in digital public space is complicated by these differences.

Some of the measures to restrict illegal sharing have implications for ownership by legal purchasers. Take for example e-books. When you buy a digital book, you purchase the right to download said book to your digital device. But theoretically the book could be removed at any time by revoking ownership rights. This happened in 2009 when Amazon deleted books from users' devices, because the books had mistakenly been sold illegally. This caused much outcry, and Amazon subsequently issued a public apology (Bezos, 2009). If archiving and cataloguing of your content is handed over to those you bought it from, ownership might even be revoked accidently, as appears to be possible with Apple's iTunes; 'duplicate' songs might not be correctly stored if they are in fact different recordings of the same song (Statt, 2016). Therefore, when you purchase a digital copy of a book or an album are you really 'buying' the object, or simply renting it? When you buy a physical copy of a book you are not laying claim to the ideas contained within it, but the object is yours. As long as it is not reproduced, it can be used as you like, lent to others and even resold. This difference between physical and digital purchasing has implications for the nature of ownership and transactions when it comes to digital goods. It may be that we need to find a new paradigm for the distribution of digital copies of media content as isolated from their physical storage media such as books, vinyl records or DVDs.

Value in digital objects

Part of the reason that these physical artifacts command value as objects is that they are not readily reproducible by individuals without a loss of tangible attributes that give them their desirability. The pages or words of a physical book can be copied, but it takes significant effort, usually more than the cost of a commercially produced copy, to bind these together in a way that would have the tactile qualities of the original book. With older technology, audio-visual content retained some of this reproduction hurdle by being difficult to copy with high fidelity. With the development of media systems that could be copied by users, such as cassette and video tapes, there were worries that the media industry would be damaged by people illegally copying and distributing content. However, despite the fact that illegal film piracy was significant, the film industry offset this with revenue from selling and renting high quality recordings of content; these were preferable to pirated copies because of the loss of quality that came from repeated copying.[2] Even when the copies were made from original high quality versions, each transfer reduced the quality of the recording incrementally, so there was a limited number of times a copy could be made before it became unwatchable; the cost and effort of making a copy of the same quality as the original was higher than that of purchasing a new copy, and this discouraged copying except for those to whom the value of the content itself was higher than that of the desire for quality; for example distribution before the official release of a film. But with purely digital files there is no loss of quality however many times the object is replicated or passed between individuals, therefore the negative aspects of obtaining a non-legal copy are purely ethical and legal. One of the major features of the digital public space is the fact that digital objects can be shared, copied and distributed freely, in a 'public' openly accessible arena. In particular with images and shorter video clips, which are becoming more and more a part of online discourse, perception seems to be that once a digital file is 'released' into the digital public space, it becomes disconnected from its creator and there is less onus to provide payment to the originator (who may not be the person from whom the file is directly received). Thus the value of objects in the digital public space is separated from the cost to create them and financial compensation for their provision.

With the rise of digital media, new controversies and discussions have therefore arisen around piracy and illegal sharing of media content that has not been purchased from the owners. While large corporations which make their money out of in-copyright properties understandably wish to protect their interests, there are some, such as legal academic and political activist Lawrence Lessig, who suggests that waging a 'war' on copyright infringement and piracy is counter-productive and harmful. Lessig does not suggest that some form of ownership protection is not necessary in order to fund creation, but instead recommends scepticism on how far this should reach and how it is enforced (Lessig, 2008).

New processes are being developed in which the distributable nature of such media is an advantage rather than a hindrance: note for example the rise of streaming media services where it is access to the content rather than the physical objects that are being purchased. Many of these services such as Netflix and Spotify work on a subscription service, so that rather than charging payment for individual files (which would in many cases be small) content is paid for at the provider level and made available on demand as part of a subscriber's catalogue. In this way, access to content is paid for as a service rather than transaction of an object, allowing people access to media immediately that they might otherwise be tempted to obtain illegally. These people may still purchase physical media copies later, if they desire the object tangibility.

This process of media leasing may still be subverted, for example by people sharing access to accounts, or stealing digital services (such as using someone else's internet connection without asking), but the link to physical provision of media as objects has been broken. Media subscription services, although they are not 'public' in that you must pay to be given preferential access to them, provide an impression of an open library, a digital space where once you have a key you can explore at your leisure. This overcomes the question that was asked earlier about whether you 'own' a digital copy that you purchase, by highlighting the fact that you do not: here the fact that you are purchasing access via subscription is clear.

It is important to note that a common reason for disconnection between accessing a file in the digital public space and repayment to its originator is that it is easier to obtain access from a third party; this is especially the case where the legal owner has restricted sharing and access in the name of copyright protection. For example, much illegal distribution takes place for films and television programmes which are released at intervals in different regions. Those in territories which have to wait to see it legally, obtain it illegally. But anecdotally, it seems that many who access illegal file sharing sites would be quite willing and might prefer to use legal means to access the content if it existed, and only turn to illegal methods if they have no other option. Therefore, if companies provide access in a timely manner in the digital public space, they can be the preferred content providers, retain hold of their copyright and be recompensed accordingly. An argument towards this is the popularity of the subscription services mentioned above such as Spotify, Amazon Prime and Netflix which provide instant access to a wide variety of popular culture. The musician Amanda Palmer, who is known for exploring various new forms of gathering support for her work, puts this succinctly, describing how in her experience many would be happy to pay artists for content they have obtained 'free' in the digital public space, if only given the opportunity: 'Maybe we should stop asking how do we get people to pay for music, and start asking how do we let them pay for music?' (Palmer, 2013).

There might also be more radical solutions to the ephemerality of digital objects and thus their perceived lack of value by the consumer. Rather than be the sole providers of timely content (which might be difficult with instant distribution via third parties) the originators may instead invest additional value in versions which are obtained directly from them. Since purely digital items can be potentially duplicated as soon as they enter digital public space, this might involve attaching digital content directly to physical objects and locations, as described in Chapter 3 when discussing digital physical hybrid spaces. It might be the case that digital features are sold alongside tangible objects that hold the same information. For example, when buying a physical CD of music from Amazon, their 'AutoRip' service now often provides a 'free' digital download of the music that you have purchased, allowing instant access. Alternatively, value might be added in the quality of the physical objects themselves. It is possible that resurgence will be or is already being seen in the creation of objects with high aesthetic values. This might include hardback and leather-bound books (in reaction to the increased popularity of eBooks) and vinyl with elaborately illustrated sleeves. Another option is that more novel physicality might be given to the digital objects so that they can interact in both the digital and physical worlds, as is the case of the Physical Playlist project being developed by the BBC and the Creative Exchange (see Case study 5.1).

An even more extreme solution might be to enforce degradation and thus replicate some of the features of physical objects in the digital. This might initially sound like something that nobody would be interested in – why would you wish to add the loss of quality as a feature? However, there might be benefits in terms of how we use our 'digital memories' as described in Chapter 4. Gulotta et al (2013) designed three interactive systems to explore the ideas of patina and decay in digital objects. Two of these, BitLogic and DataFade, involved degradation of digital images over time, either as a function purely of time passing, or through the influences of weather and 'handling' – that is, additional quality changes through exposure. The analogy used in these instances was to physical artifacts such as photographs. But if we are using digital storage as an extension of our mental storage, then the correct analogy should be that of memories. Memories do not get damaged by access, but are instead reinforced. It is memories that we do not access which fade over time. Perhaps we should instead look at systems by which associations, context and memories can be added to media objects depending on how we access them (like the bent spine of a paperback). This might also involve the fading of files which are less frequently accessed, and are thus likely to be considered 'less important'.

Case study 5.1: The Physical Playlist

For two decades, between 1970 and 2000, the dominant medium for music was the compact cassette, or cassette tape. And this medium was not only the preferred way to purchase and experience music, but contributed to specific social practices: most noticeably, the mix-tape; a selection of songs transferred onto a single tape. The creation of these tapes could be done collaboratively as a shared experience, or a completed tape could be given as a gift to a friend or loved one. In both cases the tape acted as an enabler for communication, and sharing and creation of memories.

Creating a mix-tape required care, attention and time. The songs had to be selected, and then put on the tape one by one – usually taking at least the same amount of time to transfer as to play. Receiving the physical object of a tape was to receive an investment in someone's time, but also required attention from the receiver. It was necessary to listen to the songs in the order they had been placed on the tape, since although it was possible to move forward and backwards through the content, it was not usually easy to find the start and end of any particular song.

Although the practice of sharing songs and music has not declined, and has in many ways been made easier by digital music formats, it is difficult to replicate the experience of a carefully selected song list concretised in a physical form. The songs included on a playlist created on Spotify or 8track might have been selected with similar care, but with a fraction of the effort expended in the compilation. Creating a tape often involved recording songs played on the radio, and thus waiting hours or days for the desired song to be aired. A digital playlist also has neither the same tangibility, nor requires the same pre-determined experience, since you can skip back and forth through the songs at your leisure. It also does not have the same weight of impact as a tape that has been made as a unique object that, while possible to copy as content, would not be identical.

The Physical Playlist project attempted to address some of these questions by considering what the mix-tape might look like in the digital age, recreating some of the restrictions and affordances of the cassette tape using new technologies. Working with Ian Forrester of BBC R&D, academics Paul Coulton, Dan Burnett, Adrian Gradinar and Joel Porter of Lancaster University developed a prototype which included playlist 'charm bracelets' and a specialised player with which to use them.

The bracelets consist of a series of interlocking 'charms' which each contain an NFC tag: enabling a piece of digital media content (a song or video) to be encoded within. To create a physical playlist, a user must choose the media content and use a mobile app to register one item on each charm. The resulting bracelets can then be given as a gift. Like the mix-tapes, they can be rewritten, but doing so adjusts the original intentions of the person that created it.

The content of the bracelets can be read in two ways. The first is by using a mobile app which reads the chip as the user scans over it. But the bespoke, 3D printed player maps more closely to the 'mix tape' experience. Using the player, the bracelet is suspended next to a reader which travels along each charm in turn, moving to the next one when the content is complete, with no 'skipping' possible. In this way, the playlist is experienced in the exact order and timeframe that the person who created it intended.

By creating a tangible, unique object which has a set way to be experienced, the physical playlist creates something which can be owned and gifted in a way that is distinct from the individual digital items it contains, and even from a purely digital expression of the curated list. The association with the physical object provides a different experience for the creator and giftee, one that has aesthetic factors and a physical form. This object can evoke emotion and memory by its physical presence in a way not usually possible for purely digital objects. Although the content is not unique, the collection in its new medium of the charm bracelet is. Using the player requires one to experience the playlist in the way that was intended by its curator, and requires to set aside a particular period of time to do this.

However, it must also be considered that the creation of mix-tapes was for a long time in a legal grey area, and 'home taping' and copying of music between media was considered by the music industry to be a major threat. The prototype physical playlist provides content via existing online services such as Spotify, and does not create additional (legal or illegal) copies: what is owned is the physical object and not the content. This does mean that the persistence of the functioning object depends on the songs still being available, and raises interesting questions of what constitutes an owned physical/digital hybrid. If a version of the Physical Playlist were to hold actual copies of the songs and media content, these would have to be either licenced or copied illegally. This might influence how people perceive and use such an object… or it might not.

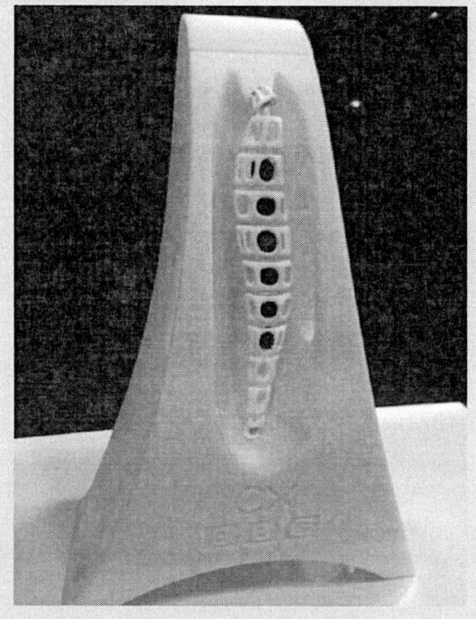

Figure 5.1 The Physical Playlist
Source: The Creative Exchange

Sharing and co-creation

The discussion here so far has primarily been about digital content such as music, books, films and television that has been produced commercially and provided for sale to consumers through relatively traditional means. But it is also important to consider the effects that the digitally enabled shared experience and prosumer culture have on conceptions of ownership, and the economics of the creation of online material. As an example, we will first concentrate on the collaborative creation of online cultural multimedia content.

There is a growing culture, especially among young people, of what Henry Jenkins calls 'grassroots production practices' (Jenkins, in James, 2014); creating and sharing multimedia content as an intrinsic part of digital life, through the creation of images, videos and other creative output that spreads, mutates and evolves. Internet 'memes' such as 'lolcats' or 'doge' are not created by any one individual but emerge from cultural participation and sharing. Images, text and videos are freely shared and distributed, having been accessed on sites such as YouTube, Google image search or Tumblr. Some creators, particularly those participating in fandom and remixing culture, may not own the original content on which they are basing their work. These secondary creative outputs are known as transformative works, and the acceptance of them by copyright holders is extremely varied. In these cases, creators of new work such as fanfiction or mashup videos might find it difficult (if such was the desired outcome) to take financial recompense for what they have made because the copyright to the original property lies elsewhere. The legal case for 'fair use' often rests on a lack of commercial utilisation. However, many of these creative expressions are in fact produced anyway, not for financial benefit but to participate in community expression and achieve peer recognition. Henry Jenkins, who famously explored the phenomenon of fandom and appropriation in his book *Textual Poachers* describes it thusly in his introduction to an updated edition in the internet age:

> So many in the media industries are arguing that acts of "piracy," of "stealing content," are undermining their motives to create, and it seems important to hold onto the idea that people create for many different motives, only some of which are pecuniary, many of which are social, and some of which are enhanced through sharing rather than selling content. As legal scholars such as Lawrence Lessig (2008) and Yochai Benkler (2006) have suggested, fandom offers us a powerful model for understanding how widespread grassroots creativity may persist despite (or perhaps even because of) limited opportunities to directly profit from one's own labor. Not all kinds of cultural exchange should be commodified. It isn't just a question of who gets to profit from their labor – some relationships are damaged when they are reconstructed through commercial logics.
>
> (Jenkins, 2012)

It is not the case that there are never any ways in which financial gain can be made from transformative material, but new methods must be found for this as discussed below, since it is usually frowned upon to sell appropriated content via traditional avenues. This does however introduce ethical dilemmas of ownership; should the original creator be reimbursed if new value is built on the basis of their original work? Is this the case even if significant effort has also been put in by the creator of the transformative work? But equally, is there value beyond the financial that should be attributed to the author of the transformative work, especially if this is then being consumed by a further audience? Many have suggested that legal definitions of fair use must be re-examined in this new creative context where ideas can be created and circulated so easily (Jenkins, 2012, Fiesler & Bruckman, 2014). Lessig (2008) suggests that hybrid economies combining 'sharing' (gift) economies and traditional commercial economies are the future of internet transactions, recognising the difference between value and money and attending to both.

These questions are further complicated by the fact that there is often lack of clarity in the copyright status of material found online. Quoting Carrie James:

> The Web offers a compelling platter of content, often without lucid information about creators' and owners' rights and intentions regarding access and use. Moreover, because online content often becomes disconnected from the site to which it was originally uploaded, the identities of its creators and owners may be lost or require significant effort to track down.
>
> (James, 2014)

Projects such as the Copyright Hub[3] are attempting to create infrastructure which makes it easier to find out this information and, where applicable, source an appropriate use licence from the creator. However, beyond content which is specifically watermarked, tracking sources is still currently extremely difficult in many cases.

There are ways in which creators can attempt to exert more control over the use of their work without restricting use; Creative Commons[4] for example has designed a range of licences which allow for re-use and redistribution. Should they wish to do so (with an awareness of the collaborative ecosystem in which they sit) creators may openly allow use of their work with specific caveats. Sites such as Flickr incorporate these licence options for their users. Photographs on Flickr licensed under these terms may be distributed and used for commercial or non-commercial purposes, within limits defined by the type of licence chosen. In almost all cases, whether use is restricted to non-commercial purposes or not, attribution is required. However, they also offer an option which allows copyright holders to waive all rights and place their works in the public domain free of restrictions, creating content for a true digital public space. Use of this process could enable the creation of open repositories. These could be publicly managed as envisaged by the BBC Digital Public Space project, or collaboratively produced as in the case of Wikimedia Commons. This latter, as part of Wikipedia, collates media which is in the public domain or available under freely-licensed terms such as Creative Commons licences, and currently holds over 34 million files.

The concept of creating a vast public repository of creative works has huge potential benefits, as Tim Cowlishaw suggests in his response to Tony Ageh's proposals:

> By ensuring that intellectual and creative works available through the Digital Public Space are freely licensed for transformative re-use by default and by providing the education and access to the technological infrastructure required to enable such re-use, such projects can ensure that we move beyond a general right to access the network, to a fully-fledged, transformative, Lefevbrian "Right to the Network", enabling humanity to collectively to (*sic*) shape the whole of digital space for the common good.
>
> (Cowlishaw, 2015)

It may seem idealistic to propose that creative works are made freely available, but as mentioned above this already happens in many cases, where creative endeavour is undertaken with no expectation of payment and reciprocation. This may be in part because of a rise in

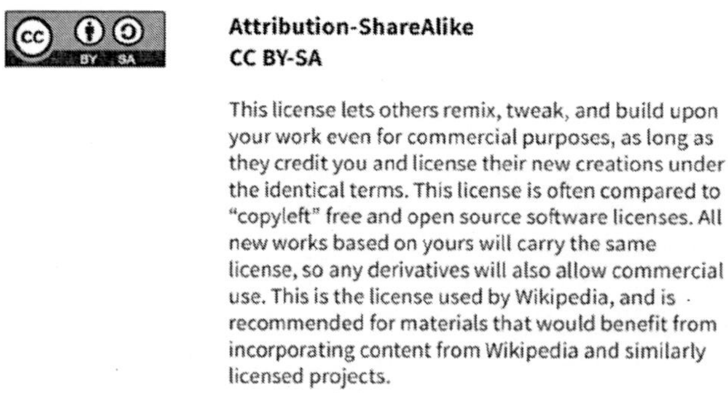

Figure 5.2 Example of a Creative Commons licence

Source: http://creativecommons.org, used with attribution under a Creative Commons Attribution 4.0 International license

sharing culture and non-financial benefits (such as community building and personal expression) as hypothesised by Graham Murdock:

> This extension of the philosophy of public goods has been accompanied by an upsurge of intellectual and creative production on the internet based on horizontal networks of peer-to-peer exchange regulated by an ethic of reciprocity. I post something that I think might interest or benefit you. I do not ask for any payment but I do expect that you, in turn, will post material that might be useful to me. It is a variant of the moral economy of the gift adapted for virtual transactions.
>
> (Murdock, 2004)

Murdock cites Wikipedia as an example of how individual effort is put into building something for the benefit of all in the digital commons, with sometimes large amounts of time and effort contributed to adding information, for no payment. Clay Shirky describes this same phenomenon, putting it into the context of a 'social contract' between people who use 'cognitive surplus' to create shared content (be it of civic value like the Ushahidi maps mentioned in Chapter 3, or lolcat meme images) on the expectation that if they put effort into producing something, others will do the same: 'Communal value on the networks we have is everywhere – every time you see a large aggregate of shared, publicly available data, whether it's photos on Flickr or videos on Youtube or whatever' (Shirky, 2010).

This practice of communities of users contributing to something which is shared freely is not new, and can form part of an economically productive market. Examples of these complex economies highlighted by Yochai Benkler include Free Software practices, and academic publishing.[5] In the former case, computer software is shared under licences which allow it to be used freely, adapted and modified as long as any copy or derivative product is distributed under the same licence terms. There are however many companies whose business is based on providing support for users of this free software, or is facilitated by its existence (for example much of Microsoft's and Amazon's cloud services are based on open source Linux platforms). Benkler (2006) suggests that three characteristics of modern information technology are particularly suited to produce an information production system which can function effectively as a market that is not based on proprietary claims (i.e. where the products are donated to the 'commons' and shared without financial transaction). These are: an almost universally distributed means of production and communication (i.e. low cost personal computers connected to the internet); raw materials for information production that are public goods, (i.e. existing information, culture and knowledge); and the ability to structure solutions to problems in a modular way that can be divided up easily. The key feature of these is minimal capital outlay for production; work can be contributed for a variety of reasons in 'spare time' for purposes that do not directly maximise financial returns but may create wider structures and benefits. It is important to recognise that if resources are being 'gifted' as a matter of course, then imposing financial aspects can damage relationships – in the same way that a friend would potentially be offended if, rather than bring a bottle of wine to a dinner party, you gave them the cash value of the gift. Similarly, this relationship can be disrupted if over-zealous copyright enforcement limits creativity by restricting access to prior knowledge, as Benkler (2006) states:

> Preserving the capacity of industrial cultural producers to maintain a hermetic seal on the use of materials to which they own copyright can be bought only at the cost of disabling the newly emerging modes of cultural production from quoting and directly building upon much of the culture of the last century.

Many large organisations working in the area of digital public space archives are encouraging principles of open sharing, collaboration and re-use. The BBC encourage sharing and utilisation of their content, which can be seen by such initiatives on their R&D 'BBC Tester' site as 'Scrubbables' which turns selected content into touch interactive, six-second videos. These can be annotated with Post-it-style text pop-ups and shared across social media platforms.

This implicit expectation that objects and information made available in the digital public space will be appropriated and re-used, raises interesting questions about the purpose of such repositories, and how analogous they are to traditional public archives such as museums and galleries. These, unlike digital archives, are used by the public to view content, but not directly take it with them and use it to create things that are new. There is scope to investigate whether there are differences in the usage rates and expectations people bring to digital archives, between those that are for 'viewing' (such as those provided by museums and galleries, which do not encourage re-use of the often copyrighted material) and those which encourage re-use.

New forms of distribution

The benefits of the digital public space to creative production are not just limited to the ability to share content in a way that allows it to be adapted, but also the very distribution itself and the ways that digital sharing allows real time collaboration. This can range from a small group collaboratively editing a document in real time on cloud based services such as Google drive (see Case study 5.2), to the construction of large constantly evolving wikis and community projects. Digitally connected spaces allow real time communication and transactional activity to take place unlimited by geographic distance or boundaries, and mean that collaboration can take place much faster and more easily.

Digital technology also allows for easier connections for sharing and resource management. Many recently emerging services such as Uber and Airbnb work on the principle that individuals are able to connect easily to those who wish to pay for resources (taxi rides or spare rooms) that they have available. For these companies, it is the technology to allow these connections that allows them to be so successful. 'These companies are indescribably thin layers that sit on top of vast supply systems (where the costs are) and interface with a huge number of people (where the money is)' (Goodwin, 2015).

It is also the case that digitally enabled tools and processes can break down tasks into smaller component parts and spread them across a wide range of individuals. In this way, roles traditionally needing to be completed by one person can be accomplished by a team. This can provide huge amounts of computational power, either in terms of digital computation (like the SETI@home project which uses idle computer time to run distributed software to analyse radio telescope data) or human effort (such as *Foldit*, a game via which players are solving protein folding problems for biotechnology innovation). 'Entire categories of work have been dissolved into algorithmic testing and hidden user labour, individual worker roles have been broken down into complex systems of activity and use, bound together with digital resources and computation' (Dalton & Fass, 2014). These contributions may be made altruistically (as in the examples given above) unknowingly (via obscure terms of service or malicious botnets) or in return for payment (such as in the case of Amazon's Mechanical Turk). Mechanical Turk allows people to sign up to complete small tasks (such as translating a word) and receive micro-payments in return. This type of distributed activity and fragmentation has the potential to revolutionise the nature of work and the workplace.

Case study 5.2: Collaborative creation

In 2014, Creative Exchange doctoral researchers Susannah Haslam and Tess Denman-Cleaver were working as a team on a project critiquing knowledge exchange and collaboration, called THESAURUS and Preface. While working together in a physical co-working space, they also collaborated digitally, using Google Docs. A recording was made of the ten hours of collaborative work that was carried out in this digital space, and this formed the basis of a new artwork by Susannah Haslam which was presented at the Designing Digital Now exhibition held at FACT in 2016.

> If Google Docs is a space of intimate proximity, is a support, an avenue, a virtual studio, office, desk, knee, a space of production and a means of communication, essentially, we are then able to observe quite intimately this new sense of intimacy on an un–geographical scale between the local and the psychic.
>
> Such intimacy marks out and exposes the gradual emergence of a hybrid, operative territory. Composed as such across an ambivalent private self, extremely public self and digital, coded, physical versions of self. An exploration of this territory as a channel of communication, a shared space of productivity and a space of communion – rather like a dance – prompts us to question an ethics of co–authorship. It begins to skew traditions of thinking behind the conventions of work, productivity, togetherness, participations, relations, relationships and notions of the embodied and disembodied self and other selves so chaotically post–rationalised today.
>
> *Source:* Haslam, S., 2016. Towards an ethics of intimacy – as though we were together. *Designing Digital Now.* FACT.

Digital collaborative technologies, which allow people to work together more effectively, have become a vital part of modern working life. Digital sharing of documents allows people based remotely to work together in real-time, and those separated by time zones to know exactly what the other contributors have added or removed. Tools such as email and transferrable digital documents have significantly increased the speed at which collaborative writing can take place; compare this description of how a remote collaboration (to write a novel) was conducted in the late 1980s by authors Terry Pratchett and Neil Gaiman: 'In those pre-email days, the two writers would phone each other and post a floppy disk back and forth, until right at the end when they both owned modems, though they were working at a speed so slow that Gaiman and Pratchett could have dictated the story down the phone faster than it uploaded.' (Campbell, 2014). Networked computer file systems and cloud-based storage such as Dropbox now allow near-instant access to the same documents without creating multiple conflicting copies (as can happen if a document is emailed to more than one person), with 'track changes' showing you other people's updates.

But above and beyond document sharing, real-time collaboration tools also allow people sharing the same physical space to work together on one 'piece of paper', everyone contributing and having joint ownership at the same time without causing confusion. The most established and broadly used example is Google's suite of cloud-based document tools, called Google Docs (or more recently, Google Drive).

Haslam says: 'I am quite fascinated by the capacity of Google Docs. On the one hand it's insanely striking technology… we're oblivious to it because it's a form of infrastructure. Infrastructure is a hidden framework, second nature and we become blind to it' (S. Haslam, 2016, personal communication, May 26). Platforms that mediate collaboration may be doing something more fundamental than swapping drafts: providing us with a jointly owned space that can be used for new types of working together, and potentially new types of friendship and intimacy. Haslam continues: 'Thinking about intimacy in the context of a Google Doc; so often there is a sociable element to it, a social side to it which sits in parallel to being a site of productivity. We are generating a social space that transcends our understanding of proximity.'

By creating a space for shared construction, shared ownership is enabled: when new things are created together in this way, they are truly shared because it is impossible to separate contributions. The whole truly is now greater than the sum of the parts.

The ease of distribution in the digital public space, while a problem for owners of copyrighted information who do not wish it to be distributed freely, is a boon for amateur as well as professional creators who want to share material they have produced; allowing them to do so at little or no cost beyond that entailed in production of the initial content. While hosting multimedia material for sharing can entail bandwidth costs for multiple downloads, sites like YouTube provide hosting that is 'free' for the producer and the consumer, with costs paid for by advertising. In this case, the act of sharing requires no direct transaction with consumers, who can find content available free by browsing the internet. Why might producers wish to share their creations with others for no financial recompense, by uploading them to the digital public space for wider sharing?

Some artists have found that the very act of making content free increases sales in the paid-for versions. This might be because physical sales (such as printed books, CDs or vinyl) have extra materiality as described above that makes them desirable objects, so that people who are exposed to the free digital content are drawn to purchase the 'enhanced' version in a different medium. Other, paid-for versions, might also have additional content such as artwork or bonus tracks that are not available in the free downloadable version. Or it might simply be that access to materials acts to draw people to the author or creator and leads them on to other work. In 2008, bestselling author Neil Gaiman provided full access to the e-book of his back-catalogue novel *American Gods* for one month. Publishers HarperCollins analysed the results and found that:

> In the Bookscan data reported for Independent [bookshops] we see a marked increase in weekly sales across all of Neil's books, not just American Gods during the time of the contest and promotion. Following the promotion, sales returned to pre-promotion levels.
>
> (Gaiman, 2008).

Overall, sales of the author's books in these independent bookshops increased 40% during the promotion. Gaiman also notes that 'The reason that independent booksellers were the only places they could see it having an effect was that some of the chain stores were doing a promotion that my books were also in, which fogged the results for them'.

However, there are risks with this approach. With the increasing availability of 'free' content, there is a danger that the link between creation, ownership and payment is broken. Consumers may come to expect free content and resent paying for things that they see as low cost to create, or simply only seek out services which are free (Ritzer & Jurgenson, 2010). If benefits exist, they may be longer term while immediate negative effects can be highly damaging if current systems do not adapt. Recently, there has been public expression of concern from several authors who have found that piracy of ebooks has a significant impact on sales, particularly at launch, which puts at risk future publishing contracts. (Flood, 2017).

So far we have spoken mainly about distribution of purely digital objects. But digital public space, because of how it enables transmission of information, is (alongside new production technologies) also enabling transformation of manufacturing. 3D printers now allow individual bespoke printing of physical objects, which can be developed from digital designs. This is important for allowing rapid prototyping. Designs can be tested quickly and cheaply without requiring access to high level manufacturing, which might have previously required a very large initial run to test a design. Since each print is also unique, designs can also be easily customised to create bespoke products. An aspect which has had much public interest is local production; you can design and create a physical object (or download the files necessary to recreate one that someone else has designed) anywhere, as long as you have access to a 3D printer. This has led to claims that 3D printers will grace every home, and there will no longer be any need to transport goods because people will be able to print their own. It should however be noted that design skills are necessary to create quality products, and therefore just because anyone could easily access the software and hardware to design and make 3D printed objects, this does not mean that anyone can create anything: sewing machines and fabric can easily be purchased but clothing design and fabrication skills are required in order to do anything beyond following patterns that have been created by others. The key innovation is with regard to the digitisation of distribution, rather than creation. It is therefore perhaps more likely that bespoke printing services like Shapeways will dominate the market.

More significantly, because designs can be shared digitally and manufacture does not require significant infrastructure, the link is broken between distribution and production. Not only can the end product be manufactured remotely, rather than where it has been designed (which already occurs with, for example, cheaper manufacture in China for UK designs), but it can be manufactured at its destination, negating the need for transportation. The potential power of this is highlighted by the fact that a tool has been 3D printed on the International Space Station. Space-based manufacture could extend capabilities of space exploration significantly, and be critical on longer missions, including to Mars (NASA, 2014). The capabilities of 3D printing even stretch to being able to print the parts for a printer of your own, as in the case of RepRap,[6] a low-cost community developed open source 3D printer.

New forms of payment

Returning to digital content, much online 'free' content is funded by advertisements; this creates a tension for consumers who do not wish to be served intrusive advertising, and yet expect the benefits of the content that could not exist without it. Many people use 'ad-blocker' services, software that removes advertising from view, but this negatively affects the content providers by decreasing their revenue. As reported in *The Guardian* newspaper, a recent YouGov survey found that less than half of UK adults were aware that most of the free content they consumed online was funded by advertising, with more than a fifth of UK adults using ad-blocking software (Jackson, 2015). This is a concern because without this revenue, many sites would be unable to operate. Google appears to be looking at potential ways to combat this problem with its recently released (currently in the US only) 'Contributor' service,[7] which allows individuals to remove ads but still provides compensation to website owners. It does this by taking a monthly subscription from users which is split between the site owners of all replaced adverts. Again, this comes back to Amanda Palmer's maxim of 'letting' people pay for a service (Palmer, 2013).

There are also ways in which implicitly 'free' content can lead back to a transactional relationship with the consumer, and provide financial recompense for the creators outside of the advertising model. This works particularly well when a meaningful relationship can be built between the creator and the consumer. An example of this can be found in the thriving arena of webcomics. These are regularly updated comics which are available online. In most cases, these can be accessed for free (with advertising revenue) by anyone who wishes, and can have millions of regular readers. Liz Dowthwaite has undertaken work surveying webcomic readers and creators about their activities funding those who provide 'free' content. She notes that the basic business model of webcomics is to subsidise the content through use of adverts, but also to trade on audience loyalty to sell merchandise. She found that most of the respondents claimed to have bought merchandise of some kind. However, surprisingly, despite this she found that merchandise (such as books and artwork) was only sold by around half the creators surveyed. Alternate models of remuneration included inviting readers to donate money, either through one-off payments or via a subscription model. A relatively high proportion of reader respondents said that they had donated in this fashion which 'suggests that readers are quite happy to 'reward' an artist that they feel deserves it, and to pay for things they already receive for free' (Dowthwaite, 2014).

There are currently many systems which are being developed and deployed to enable new forms of transaction and payment between creators, and communities that have developed to support them. These range from micropayments to crowdfunding, and other new mechanisms. In most of these cases however, it is necessary for creators of content who wish the public to provide them with financial recompense to establish a relationship and trust with the people who will provide this support. It is not enough to simply open a payment system and expect it to be used, without this social network building and maintenance.

For some projects with large start-up costs, particularly for those involving design and manufacture of physical products but also those with up-front production costs such as a studio recorded album, the issue is not just to be able to sell goods directly to consumers, but to have enough initial investment to be able to produce the items to sell. Crowdfunding, popularised by platforms such as Kickstarter, allows potential customers to 'pledge' an amount upfront with the promise of goods being delivered once enough initial capital has been raised to start the development and production process. Kickstarter has a notable feature whereby funding is only collected from backers once a project has achieved a specific target. Because of this, if not enough people think that it is worth investing to reach the initial capital necessary for production to begin, no money is taken at all and those who wished to register their support do not lose out.

Rather than offering equity, as in traditional investment, crowdfunding platforms simply offer the opportunity to place pre-orders for items that do not yet exist, thus allowing their production in the first place. This is not a particularly new phenomenon – for example Simon Garfield relates an early instance of crowdfunding in 1668 by the mapmaker John Ogilby, who wished to create a multi-volume survey of England and Wales, called *Brittania*. Ogilby started by announcing a lottery for a stake in a mystery project. Once the project was announced, 'New investors, whom Ogilby called 'Adventurers' were called upon to repair to

Garaway's Coffee House near the Royal Exchange, where they 'may put in their Money upon the Author' – and if they paid enough their name would appear on a cartouche of one of the maps' (Garfield, 2012). However, the connections and wider transmission of such ideas through digital public space means that such campaigns can travel further, faster, and that the financial transactions can be handled through third parties. This has significantly increased the popularity of such crowdfunding in recent years, through platforms such as Kickstarter and Indiegogo which have been set up specifically to manage such campaigns.

In some cases, this issue is not one of pre-ordering for production, but that the perceived monetary 'value' to consumers of the creative outputs is low. If the audience for the output is large, this does not necessarily mean that the creator cannot be fairly recompensed for the effort they have put into their work, however this has not traditionally been a feasible economic model because the administrative costs are high for taking small payments from a large number of people. In the 1990s, development was attempted on new technology which enabled 'micropayments' of extremely small monetary values, but it was not until at least 2010 that services of this type became commonly deployed, with the allowance of micropayments by electronic payment services such as Paypal, and dedicated services such as Flattr.

Flattr[8] incorporates its payment model into currently existing platforms which allow feedback and praise, a common component of content sharing platforms. It functions as a micro-donations system, allowing people to choose an amount monthly that they wish to donate to online content creators. Creators uploading content to sites such as YouTube and Instagram can add the facility for consumers to 'flattr' their work, meaning that a fraction of the monthly donation is gifted to the creator of that piece of media. This allows consumers to provide monetary payment for things that they consume online while keeping to a regular expenditure and not overspending. It is also a similar process to the non-monetary process of giving praise on such sites, such as 'upvoting' or 'liking' a piece of content.

But these systems are still reliant on the organic access of content by a perhaps inconsistent audience, and as such can be unreliable income streams, especially for smaller content creators. Patreon[9] is a system which attempts to address this by emulating the traditional patronage system for the online age.

Patreon functions by allowing content consumers to subscribe to an ongoing relationship with the content provider, or artist. This can either be a regularly scheduled donation (for example per month) or based on rates per output (for example per song by an artist). The consumer can set the level of donation that they feel appropriate, and the producer can also set 'rewards' at different levels, so for example, those who donate $1 per month may have access to information, $5 per month access to content downloads, and $50 per month personal interaction with the artist. However, in this system there can also be an agreement that content produced will not be made available exclusively for those who pay, but also to all: the patrons fund the artist to continue to create work but others may benefit from the outputs. The benefit of this model is that it can ensure a more reliable income stream, but as with many of these strategies it depends on the artist having an existing stock of social capital (Davidson & Poor, 2015), building and maintaining a relationship with patrons, and retaining their interest so that they feel that they are getting a fair recompense for their donation, since they can withdraw support at any time.

These new economies of distribution and payment are disrupting traditional models. Because creators are more and more easily able to offer their art to consumers, these consumers expect to be able to access content easily; for example on demand television programmes, music on streaming video sites and eBooks downloaded directly to a reader. But as seen above this does not mean that they are not willing to pay the creators for the content, just that this payment system must be as straightforward and effortless as any other part of the process.

Not only new payment processes but also new forms of currency are being explored in digital spaces. In Chapter 3, we discussed Linden Dollars, the currency of the virtual world *Second Life*. This virtual currency allows purchases in-game, but importantly has an exchange rate with traditional currencies allowing work within the game to be translated into real financial recompense for those who earn lots of virtual money. This type of process of exchange between real and virtual currency means that, for example, there is a growing industry of people, particularly in low income nations like China, making a living by so-called 'gold farming' activities – undertaking repetitive in-game work to earn commodities which can be sold for real money to other players. This can be extremely lucrative (Alexander, 2013).

Some digital currencies however, are not part of games (which are at risk of being hacked or shut down by the owners, or can be affected by actions of the central game authority), but are instead being developed specifically as alternative forms of currency and payment services. The most well-known of these is bitcoin, which is 'a purely peer-to-peer version of electronic cash [which] would allow online payments to be sent directly from one party to another without going through a financial institution' (Nakamoto, 2008). The key innovation which allows this to work is that all transactions are publicly recorded and viewable, but accounts are not tied to individuals so they are private and cannot be traced. This is useful for those who wish to carry out anonymous transactions, but also allows secure micropayments and has other advantages: an example of this might be the fact that since all transactions are publicly recorded, it may provide a new way to verify real estate purchases. This would do away with many of the fees that are currently required to provide deeds, titles and record the sale (Gallippi, 2013). Independent currencies that are not controlled by governmental central banks may return some of the security that cash, as an anonymous transaction, offered. Jem Bendell (2015) suggests that a truly cashless society, while convenient, is potentially dangerous because of the power it gives to the central controllers of such, citing examples such as the fact that Visa and MasterCard have over 90% of the market. Political power can also be wielded by cutting off electronic currency transactions. Examples of this include the financial blockade placed on WikiLeaks in 2011 after pressure from the US government (Gillmor, 2011), or the suspension of secure SWIFT transactions to Iranian banks by the EU in 2012 (Matonis, 2012). By providing means for independent transactions to operate in digital space which are not subject to such restrictions, power is placed back in the hands of the public, for good or ill, lending a stronger argument to this being a digital public space.

Ownership of personal information, productivity and assets

Much of the discussion above relates to ownership of and payment for existing assets and IP; designs, books, product designs or cultural artefacts that are stored in digital format on a particular platform; that may be compiled for public or private benefit. But digital assets do not just consist of such tangible 'objects' or even ideas.

Another form of asset is the productivity of those who contribute work. Consider for example, the fact that Facebook is built upon the content provided and uploaded by its users; vast amounts of individual time and labour creating content without which the service would be useless. The art piece 'Wages for Facebook' (Ptak, 2014) which highlights that this 'unwaged work' increases Facebook's profit, explores the implications of this by raising questions about how we feel about the benefit that others gain through our contributions. Contributing content to Facebook provides value to the company as 'human computation' but this is hidden from its users, who are in effect acting as workers to produce creativity, filtering and documentation. In return, users of such services are provided with benefits such as storage and infrastructure for social interaction, but the transfers of ownership are hidden 'deep within the clickthrough small print of the online terms and conditions' (Dalton & Fass, 2014). Some online services are specifically designed to serve the dual purpose of providing a function and eliciting productivity, for example 'ReCaptcha' log in tests, which both allow users to verify that they are humans rather than algorithmic bots attempting to gain access, and also provide services that are difficult for computers such as image identification and correct digitisation of scanned text.

There is also significant value inherent in the data that is produced through use of the digital public space. 'Big data' technologies mean that by collecting data generated by the movement of every individual through the digital information space as discussed in Chapter 4, connections can be made and meta level information can be obtained. Each data point may be recorded by individuals, or companies, or governments, but it has value when collected together. This data generation is how many digital operations make their money while offering 'free' services: rather than being paid for by taxation as in the case of public services, these are paid for by a data tax; the trail of information that you contribute by using them.

This data goes beyond the content that we are aware that we provide. The very modes of usage of such sites can provide information that we may not even be able to articulate ourselves, such as where we are at any time, what we like and do not like, who our friends are and what our political motivations are.

This information may simply be used for advertising purposes, for example, in order to sell advertising that is specifically targeted at particular groups of users. But there are many other uses to which a detailed data set could be put and therefore this is high-value information that can be sold for lots of money. Who owns this data; the individual that produces it, or the company that collects it? Take for example quantified-self devices such as the Apple Watch or Fitbit. If they are able to track where you are and what you are doing (whether you are walking or sleeping, what you eat each day, or where you access wifi) are they justified in being able to sell this data, and do you sign all rights to it away when you agree to use their services? A commonly repeated mantra is, if the service is free, then you are not the customer, you are the product.

How are these companies able to utilise data that we would think might be owned by ourselves and that we have full rights over? In many cases, the rights to use the information uploaded to digital services are signed away in licence agreements and privacy policies that the user is required to agree to before being allowed access to the product. An example of this can be seen in the controversy that arose when photo-hosting site Flickr began to sell prints of photographs uploaded to the site. Many of these were uploaded under creative commons licences which included commercial usages, therefore legally entitling Yahoo! (the parent company) to sell prints, retaining all profits. This is in contrast to photographs in which the decision was made to retain the rights to commercial usage, where instead the licensed images were used with a 51% share of the sales going to the photographer. Many users of the site were surprised by the fact that this was allowed under the licences in question, or debated whether this was an appropriate use as opposed to the supposed commercial use by small companies and third parties (Dredge, 2014).

Sometimes it is not just inattention, or misunderstanding of the content of licences, but whether they are read routinely by consumers at all. These policies can be extremely long, written in complex legal language, and the majority of users may only skim the top level of such agreements, or not even attempt to read them. McDonald and Cranor (2008) found that to read fully all the privacy policies (regarding use of data) that are signed by the average person would take around 250 working hours every year, or around 30 full working days. This is clearly not practical, implying that nobody is truly able to read all such agreements.

This suggests that while consent may be given for use of data in all sorts of contexts and for commercial exploitation, it is not informed consent.

A more illustrative example of the realities of how people approach such privacy agreements is the art project Chattr, which took place in two contexts; at the FutureEverything festival in Manchester, and TodaysArt in the Netherlands. Before being allowed access to a preferential area with benefits over the other lounge areas at the events, participants were asked to sign a data use policy. This confirmed that they agreed to carry a recording device with them while using the space, and for all conversations to be recorded, transcribed and archived in a publicly accessible database to remain permanently in an online public space. The participants did not appear to be unduly concerned with the content or consequences of what they signed: 'Despite Chattr efforts to convey the contents of the DUP, participants would often only take a superficial look before signing and join Chattr without having a clear sense of how Chattr would operate' (Salinas et al, 2016).

If we were not to simply give away our data ownership, might we instead consider charging companies for it? If others are going to make money out of the data we produce, some suggest that individuals should also have direct control over where their data is shared, and take some measure of the profits. CitizenMe, a company originating in market research has developed an app to let people see what data they are sharing, and pays users small fees in return for sharing (anonymised) data with brands and researchers (Hodson, 2016).

This tendency to give up rights to personal data and information does not just have implications for ownership, but also for privacy, as will be seen in Chapter 6. In a purely digital situation, where many such agreements are completed regularly, it appears that people are not as protective of their online rights and ownership of their personal information as they might be – and this is before taking into account identity theft and the re-use of material made available online for fraud or other malicious purposes. Identity theft is often considered in terms of single financial fraudulent acts such as credit card details being used to make payments, but it can be subtler and more personal. Ruth Palmer, for example, found that a false profile had been set up on various dating and social network sites over a long period, using photographs and other material of her and her friends, without her knowledge (Kleinman, 2015). It is difficult to fully control access to material which is placed online, even if it is in what appears to be a restricted context. It is also important to consider that services which offer to store data 'securely' may not actually be private, if the owner of the service can access the content. People routinely discuss 'the cloud' as an ephemeral, detached storage of information but 'the cloud' is really just a remotely linked server owned by someone else. This has led to what feel like invasions of privacy when unexpected access occurs; for example, the outcry that took place when the band U2 in association with Apple provided their album free to all iTunes users. The music was uploaded to all relevant accounts, highlighting the fact that control of these music libraries was not exclusive to the user. Similar intrusiveness is felt when, for example, covers can be changed unexpectedly on digital books on the Amazon Kindle. This is a fundamental reminder of the question of how much digital content you 'own' if it can be manipulated by others, but brings us to Chapter 6, which will discuss notions of privacy in digital public space.

Key points

- Transactions are related to ownership and exchange. Property is a key aspect of most modern societies and their function.
- Digital goods are often 'non-rivalrous' and are not consumed by their use. They can often be copied and distributed with no loss of quality. This creates issues with attributing value and compensation.

- Ownership of digital spaces involves authority about their use. Public spaces are constructed for the common good, but may still be restricted.
- Archives of content in digital public space raise questions over ownership, copyright and curation.
- Services which may appear to be 'free' usually have associated costs. If the direct link is broken this may be through taxation or indirect benefits to the provider. If a service becomes ubiquitous and 'invisible', costs may become hidden.
- Copyright enforcement protects creators, but can be complex, especially across national boundaries, and can restrict legitimate use and creativity.
- When objects are difficult to create or copy, they have inherent value. This is not always the case with digital objects, so a different model must be sought (for example, subscription for access) or value added through attachment to physical objects.
- Customers may be willing to pay for access if this is as easy as or easier than illegally obtaining the content in a timely manner.
- Transformative works may exist in a grey area of copyright and ownership, and introduce ethical issues of ownership due to the effort put into their creation, and the (not necessarily financial) value inherent therein.
- Ownership of online content is not always clear, or possible to source.
- Creative Commons, the Copyright Hub and digital public space are potential solutions to solving some of these issues and providing digital content for the 'public good'.
- Digital technology also allows new forms of collaboration and sharing for collaborative working, joint ownership and resource management.
- It may also encourage new forms of payment, marketing and distribution, benefiting creators, who may even see sales increase because of, not despite 'free' content being available.
- Digital transmission of information may also allow new forms of manufacture and distribution of physical objects, such as via 3D printing.
- New forms of payment may exist in the digital information space, in terms of transactional relationships, micropayments, crowdsourced payment and patronage, and new digital currencies.
- Ownership of assets created by users of digital spaces and services must be considered; commercial companies may be taking advantage of unwaged labour by their users, or the data they give up knowingly or unknowingly.

Notes

1 See Chapter 9.
2 This is less the case in developing countries where the costs of legitimate goods are prohibitive (Yar, 2005).
3 http://www.copyrighthub.co.uk/
4 http://creativecommons.org
5 Many debates are currently underway as to whether the current academic publishing model, where academic journal publishers are paid but authors and peer-reviewers are not, is sustainable, particularly given the increasing demands of public funding bodies for research results to be openly accessible without payment.
6 http://reprap.org/
7 https://contributor.google.com/
8 https://flattr.com/
9 https://www.patreon.com/

References

Alexander, W., 2013. How to make thousands of pounds a month playing computer games. [online] *Vice.* Available at: www.vice.com/en_uk/read/i-make-thousands-of-dollars-a-month-from-playing-computer-games [Accessed 17 October 2016].

BBC, 2007. *75 years BBC World Service: A history. The 1930s.* [online] BBC World Service. Available at: www.bbc.co.uk/worldservice/history/story/2007/02/070123_html_1930s.shtml [Accessed 14 October 2016].

BBC, 2015. *Japan: Free wi-fi for Mount Fuji climbers.* [online] BBC News From Elsewhere. Available at: www.bbc.co.uk/news/blogs-news-from-elsewhere-33426338 [Accessed 14 October 2016].

Bendell, J., 2015. What price a cashless life? *New Scientist*, 226(3024), pp.24–25.

Benkler, Y., 2006. *The wealth of networks: how social production transforms markets and freedom.* Yale University Press, p.9.

Bezos, J. 2009. Announcement: an apology from Amazon. [online] *Amazon.com*. Available at: www.amazon.com/tag/kindle/forum/ref=cm_cd_ef_tft_tp?_encoding=UTF8&cdForum=Fx1D7SY3BVSESG&cdThread=Tx1FXQPSF67X1IU&displayType=tagsDetail [Accessed 14 October 2016].

Campbell, H., 2014. *The Art of Neil Gaiman*. Harper Design, p.65.

Carr, N., 2011. *The shallows: what the internet is doing to our brains*. WW Norton & Company.

Collins, 2015. Fair use: copyright differences in the UK and US. [online] *The Design and Artists Copyright Society.* Available at: www.dacs.org.uk/latest-news/us-fair-use-uk-fair-dealing-differences-law?category=For+Artists&title=N [Accessed 14 October 2016].

Cowlishaw, T., 2015. The right to the network: radical urbanism of digital public space. [online] *Contributoria.* Available at: www.contributoria.com/issue/2015-05/551000321045c8eb71000132/ [Accessed 14 October 2016].

Dalton, B., and Fass, J., 2014. Work and wellbeing in digital public space. *In*: Myerson, J. and Gee, E., eds. *Time and motion: redefining working life*. Liverpool University Press, p.148.

Davidson, R. and Poor, N., 2015. The barriers facing artists' use of crowdfunding platforms: personality, emotional labor, and going to the well one too many times. *New Media & Society*, *17*(2), pp.289–307.

Dowthwaite, L., 2014. Getting paid for giving away art for free: the case of webcomics. [online] *CREATe.* Available at: www.create.ac.uk/blog/2014/02/25/webcomics-dowthwaite/ [Accessed 17 October 2016].

Dredge, S., 2014. Flickr takes flak for selling Creative Commons photos as wall-art prints. [online] *The Guardian.* Available at: https://www.theguardian.com/technology/2014/dec/02/flickr-creative-commons-photos-wall-art [Accessed 17 October 2016].

Fiesler, C. and Bruckman, A.S., 2014, February. Remixers' understandings of fair use online. *Proceedings of the 17th ACM Conference on Computer Supported Cooperative Work & Social Computing.* ACM, pp.1023–1032.

Flood, A., 2017 'We're told to be grateful we even have readers': pirated ebooks threaten the future of book series. [online] *The Guardian.* Available at: https://www.theguardian.com/books/2017/nov/06/pirated-ebooks-threaten-future-of-serial-novels-warn-authors-maggie-stiefvater [Accessed 9 November 2017].

Gaiman, N., 2008. The results of free. [online] Available at: http://journal.neilgaiman.com/2008/07/results-of-free.html [Accessed 17 October 2016].

Gallippi, A., 2013. The present and future impact of virtual currency. *Statement before the Subcommittee on National Security and International Trade and Finance and Subcommittee on Economic Policy of the United States Senate Committee on Banking, Housing and Urban Affairs.* Available at: www.banking.senate.gov/public/index.cfm/2013/11/the-present-and-future-impact-of-virtual-currency [Accessed 17 October 2016].

Garfield, S., 2012. *On the map: why the world looks the way it does.* Profile Books.

Gillmor, D., 2011. WikiLeaks payments blockade sets dangerous precedent. [online] *The Guardian.* Available at: www.theguardian.com/commentisfree/cifamerica/2011/oct/27/wikileaks-payments-blockade-dangerous-precedent [Accessed 17 October 2016].

Goodwin, T., 2015. The battle is for the customer interface. [online] *TechCrunch.* Available at: https://techcrunch.com/2015/03/03/in-the-age-of-disintermediation-the-battle-is-all-for-the-customer-interface/ [Accessed 14 October 2016].

Google, 2015. Google Cultural Institute: Frequently asked questions. [online] Available at: www.google.com/culturalinstitute/about/ [Accessed 1 February 2015].

Graeber, D., 2011. *Debt: the first 5000 years*. Melville House Publishing.

Grinberg, R., 2012. Bitcoin: an innovative alternative digital currency. *Hastings Science & Technology Law Journal*, *4*, p.159.

Gulotta, R., Odom, W., Forlizzi, J. and Faste, H., 2013, April. Digital artifacts as legacy: exploring the lifespan and value of digital data. *Proceedings of the SIGCHI Conference on Human Factors in Computing Systems*. ACM, pp.1813–1822.

Jackson, J., 2015. Less than half of UK adults are aware ads fund free content online. [online] *The Guardian.* Available at: www.theguardian.com/media/2015/jul/01/less-than-half-of-uk-adults-are-aware-ads-fund-free-content-online?CMP=share_btn_tw [Accessed 17 October 2016].

James, C., 2014. *Disconnected: youth, new media, and the ethics gap*. (Foreword by Jenkins, H). MIT Press, p.53.

Jenkins, H., 2012. *Textual poachers: television fans and participatory culture*. Routledge, p.xxxi.

Kleinman, Z., 2015. Who's that girl? The curious case of Leah Palmer. [online] *BBC News.* Available at: www.bbc.co.uk/news/technology-31710738 [Accessed 17 October 2016].

Lessig, L., 2008. *Remix: making art and commerce thrive in the hybrid economy*. Penguin.

Matonis, J., 2012. The payments network as economic weapon [online] *Forbes.* Available at: www.forbes.com/sites/jonmatonis/2012/03/27/the-payments-network-as-economic-weapon/#2b6cc33b2182 [Accessed 17 October 2016].

McDonald, A.M. and Cranor, L.F., 2008. The cost of reading privacy policies. *Journal of Law and Policy for the Information Society*, *4*, p.543.

Murdock, G., 2004. Building the digital commons: public broadcasting in the age of the internet. *The 2004 Spry Memorial Lecture*, pp.3–7.

Nakamoto, S., 2008. Bitcoin: A peer-to-peer electronic cash system. [online] Available at: https://bitcoin.org/bitcoin.pdf

NASA, 2014. Space Station 3-D printer builds ratchet wrench to complete first phase of operations. [online] Available at: www.nasa.gov/mission_pages/station/research/news/3Dratchet_wrench [Accessed 17 October 2016].

Palmer, A., 2013. The art of asking. [online] *TED Talks 2013*. Available at: www.ted.com/talks/amanda_palmer_the_art_of_asking [Accessed 14 October 2016].

Ptak, L., 2014. Wages for Facebook. *Presented at University Art Gallery, University of California* and available at: http://wagesforfacebook.com/

Ritzer, G. and Jurgenson, N., 2010. Production, consumption, presumption: the nature of capitalism in the age of the digital 'prosumer'. *Journal of Consumer Culture*, *10*(1), pp.13–36.

Salinas, L., Coulton, P. and Dunn, N., 2016. Using game design as a frame for evaluating experiences in hybrid digital/physical spaces. *Architecture and Culture*, *4*(1), pp.115–135.

Schmidt, E., 2005. Books of revelation. *Wall Street Journal*, *18*, p.A18.

Shirky, C., 2010. How cognitive surplus will change the world. [online] *TED Talks 2010*. Available at: www.ted.com/talks/clay_shirky_how_cognitive_surplus_will_change_the_world [Accessed 10 October 2016].

Statt, N., 2016. Apple says it doesn't know why iTunes users are losing their music files. [online] *The Verge.* Available at: www.theverge.com/2016/5/13/11674388/apple-music-itunes-file-deletion-bug-update [Accessed 14 October 2016].

Von Krogh, G., Ichijo, K. and Nonaka, I., 2000. *Enabling knowledge creation: how to unlock the mystery of tacit knowledge and release the power of innovation*. Oxford University Press on Demand.

Yar, M., 2005. The global 'epidemic' of movie 'piracy': crime-wave or social construction? *Media, Culture & Society*, *27*(5), pp.677–696.

6 Challenges of the digital public space

Privacy and Security

Chapter 5 focused on ownership and transactions as a fundamental aspect of human interaction and culture, which is being affected by digital public spaces and related technologies. The discussion mainly focused on potential benefits conveyed by new contexts, although we touched on potential negative effects. This chapter and Chapter 7, in contrast, will examine two areas of digital life which are raising particular concern amongst both researchers and the general public. While the tone of this book so far has been generally optimistic with regard to digital public space, we should take care to examine as cautionary examples areas which might have significant negative impact on our interactions unless carefully managed. The specific set of challenges which will be addressed in this chapter are those connected to privacy and security.

Privacy in digital space

> Some thoughts are so private that you only share them with a therapist or 17,000 people on the internet.
>
> (lordoftheinternet, Tumblr, 2013)

At the start of this book, we spent some time discussing what is covered by public space, and public interaction. Publicness is usually held up in contrast to privacy, and the nature of what we are able to keep private is also something heavily affected by digital technology. We have touched on this several times already; for example the enforcement of rules within privately owned digital spaces that are occupied by the public, or ability to keep your identity private while interacting online, and the effect this can have on responsibility and behaviour.

Privacy has been defined as 'the right of an individual to be alone' (Acquisti et al, 2007). It can also cover rights over management of access, including the access individuals, groups or institutions give others to information about themselves (Westin, 1967) or 'the selective control of access to the self' (Altman, 1975). Most people consider that they have a right to some form of privacy, and are sceptical of technology that appears to compromise these rights. However, privacy is not a single concept, and has many different aspects which must be considered. Susen (2011) notes three different facets or spectra of privacy; 'society versus individual ("collective" versus "personal"), visibility versus concealment ("transparent" versus "opaque"), and openness versus closure ("accessible" versus "sealed")'. Each of these is separate, meaning that something can be private without being closed, as well as public without being visible.

This complex dynamic of individuality, concealment and accessibility is reflected in the nature of online privacy in digital public spaces. Public and private are not always antonyms, and it can be difficult to draw lines between them. Since they are not binary on/off positions, and it is also possible for something to be both, confusion can occur if there is a difference in expectations. While public space is accessible to all and may be used by anyone, there may still be an expectation that activity within this space is personal and private, in that it is not shared with everyone, in the same manner as a private conversation in a public park. In digital public spaces, these boundaries between public and private are often much less visible, and this can lead to situations where misunderstandings arise or advantage is taken of the shareable nature of content that was perceived as private. It is even the case that the nature of what is 'public' has changed, as digital publics create long lasting persistent records that can be shared long after the initial interaction.

> Under existing notions, privacy is often thought of in a binary way – something is either private or public. According to the general rule, if something occurs in a public place, it is not private. But a more nuanced view of privacy suggests that [a particular instance of public shaming on the internet] involved taking an event that occurred in one context and significantly altering its nature – by making it permanent and widespread.
>
> (Solove, 2007)

Digital information space is constructed through sharing of information, much of which is personal. While we have already discussed that there are many benefits to connectedness, there are risks to allowing others unrestricted access to information about yourself. There must therefore be consideration of who any particular digital content is made available to, what can be done with it, and whether it might be 'overheard' by people other than the intended recipient. In a comparison with physical public space, some activities which people carry out in public in large conurbations are only acceptable because of a sense of anonymity; that as part of the crowd, you are unidentified and (potentially) untraceable. If this anonymity is lost, the activities may become riskier or less desirable. In Chapter 4, profiles were discussed, and how any interaction in the digital public space requires creating an online identity that might be tied to personal information and could be used to identify individuals with their 'real world' identities. This, along with the persistence of digital information, means that seemingly fleeting interactions may become concrete and traceable, at the detriment of privacy. Intrinsic to the digital public space is the ease of sharing of information, but this brings up many questions about what should be shared, and what should be kept private.

Security, as well as privacy, is a critical issue for consideration: how information that is deemed to be private is maintained as such. The interconnectedness of digital systems makes it easier to cause large effects across wide groups of people, which opens up opportunities for criminals. Marc Goodman has spoken about how the connected world has created a 'crime singularity', citing for example the Sony Playstation Hack in 2011 that compromised the banking details of 100 million people (Goodman, 2012). For the first time in human history, it is possible for one person to perpetrate a crime of theft against millions of other people in one go. If the security of a database fails, it compromises the privacy of many people. Equally, damage can be caused when private information enters the public space and becomes spread and shared: the practice of 'doxxing' which is carried out by groups wishing to attack individuals, involves sharing personal details such as names, addresses and family information which can lead to those targeted fearing for their personal safety. Perpetrators of such information distribution often appear to hold a perception that if such information can be uncovered by a determined individual, then there is no moral barrier to spreading this to the public space, with little consideration for the potential effects.

Particularly in terms of digital communities and social networks, there is a lot of grey area between things that are posted 'publicly' (with the expectation that anybody can see it) and things that are posted 'privately' (for a small privileged number of receivers). Information on social networks is often conveyed 'publicly' within a specific group but there is an expectation that it will not be viewed beyond these boundaries. This is maintained either explicitly by 'privacy settings' that require logins and passwords to access the content, or by what danah boyd calls 'security through obscurity' mentioned in Chapter 1: that nobody outside of the intended audience will be interested in the content and therefore will not seek it out (boyd, 2007). There is also a significant social factor of trust involved: that secrets shared with trusted friends are not distributed beyond that circle. There are many moral and ethical dilemmas regarding what we share online about ourselves and about others.

When people think about private digital information, they often consider things like personal details, credit card and account information such as in the above PlayStation example, or emails and documents that are intended to be sent to a specific, small audience only. Security of this information is very important, and is the objective of many tools and technologies to ensure that only the intended recipients can read, say, an email that you send to your family. But because there is the facility to record every process and interaction in digital public space, it means that privacy can be compromised through gathering and distribution of information that appears innocuous, or by use of information which is created by the very act of using the digital public space. The time at which you accessed an individual website may not be considered private, but your privacy may be compromised if this data over time reveals your movements each day. These considerations mean that a different awareness of privacy is necessary when existing in digital public space. When designing and working with digital spaces is it important to consider how privacy might be upheld, what risks to privacy might exist; and whether giving up privacy is an acceptable loss for the benefits entailed by systems that rely on mass data collection and thus cannot function in an entirely private manner.

There is some evidence that this awareness is already beginning to be adopted by those who have grown up living with digital public space. Carrie James, in conducting research with young people, describes how they consider what privacy means to them and how this changes their approach to the internet: 'In their accounts, privacy is about controlling content about yourself and the audiences for that content. Madeline, age 21, said, "Privacy, to me, means that people I don't know can't find out where I am. I don't want people that I don't know knowing where I am or what I'm doing."' (James, 2014). She found that 'A little more than half of the tweens (52 percent) and teens and young adults (55 percent) that we interviewed asserted that privacy is diminished online' and quotes a twelve-year-old who explains that 'basically anything in the cyberspace is, when you put it on something electronic, it's gone. Your privacy is pretty much gone'.

Because the digital public space offers new ways to structure and manage our social interactions, as discussed in Chapter 4, we can often fall back onto actions and behaviours that would have been appropriate in non-digital space but now carry greater risks because of the changed nature of digital interaction; something written in a group email or posted on a forum is not the same as a conversation in a pub because it persists and can be shared with others. This means that the repercussions of such behaviour can be far more widespread and persistent: 'Invaded privacy, stolen words, racist speech – offenses such as these have existed in human life for eons. Yet when they are committed in networked publics in a globally interconnected world, the stakes are arguably higher, the harm arguably deeper or at least more lasting' (James, 2014).

Some of these risks may not be immediately apparent to people using the digital public space as an arena for interaction:

> In daily off line life, these boundaries [for controlling privacy and disclosure] tend to be obvious. We are aware of who we are talking to, through vocalization or bodily posture and gestures; who we write to; what we have heard and from whom; who can see us walk down the street; who can see us use the toilet; if cameras are pointed at us (although some closed-circuit television requires actively looking for it); who or what has touched us; and who and what we have touched (whether friendly or unfriendly).
>
> (Houghton & Joinson, 2010)

This attentiveness to where our social boundaries are and how information is distributed is an intrinsic property of our behaviour as described in Chapter 2, and thus the fact that the boundaries can be much less clear in digital space may cause disconnects between our perceptions and the reality.

Digital trails: data created unknowingly

> The scattered bits of data in the electronic universe can seem to be 'nothing more than the odds and ends of our lives – data lint that only the perverse would bother collecting.' What makes current attacks on privacy so insidious is the fact that few of us have any idea how those bits of lint are being gathered into a lint ball of truly remarkable dimensions.
>
> (Sykes, 1999)

A surprising amount of supposedly 'private' knowledge can be gained from information that the majority of people do not even think about giving up, or are unaware of the power of. By carrying out our lives in the digital public space we may be revealing more about ourselves than we expect, and this can lead to violations of privacy.

Most digital interactions leave a trace, and these can be used in remarkable ways to overcome seeming anonymity within the digital public space. The following examples are simply that, examples; and it would be almost impossible to list all the ways in which information about us is recorded living in a world overlaid with digital public space. Traffic analysis, for example, uses your movements in digital space to recover information about you. Traffic data (collected regarding online usage) does not contain information on the content of messages that you send, but simply the fact that you sent them, and to where. When sending information over the public internet, whether it be emails, what you enter into web fields or simply connections to a website, this traffic data may be visible publicly and can be collected, and analysed. This traffic analysis can be useful for users, for example to inform search engines by examining how people navigate between pages and which links are most popular. It may help identify patterns to prevent criminal activity such as credit card fraud. But it can also be used to infer sensitive information which may be used negatively, for example allowing a company to know whether customers have been looking at the sites of competitors, and offer a lower price only if that were the case. To avoid such traffic monitoring, it is possible to use services such as Tor which conceal traffic and location information, but while some argue that such encryption services should be built as standard into networks (Danezis & Clayton, 2007), others express concern that the use of them may conceal illegal activity. Indeed, concerns over crime, and in particular terrorist activity, have led to several governments introducing rules (such as the EU Data Retention Directive 2006, or the Communications Data Bill proposed by UK Home Secretary Theresa May, nicknamed the 'Snooper's Charter') whereby internet service providers must keep customer data for a significant amount of time and turn this over to the police in the case of an enquiry. Some organisations such as the Open Rights Group[1] have argued that this sort of surveillance may contravene privacy rights.

In Chapter 5, we discussed the common practice of giving up ownership of personal data collected by 'free' services. Many websites use cookies: data files which store individual information in web browsers. This allows the pages to 'remember' you, to present you with appropriate content, to prevent you having to log in repeatedly, or to enable you to build a 'shopping cart' that retains your selections while you browse other parts of the site. In 2009, an EU directive was introduced which as implemented in the UK meant that all websites using cookies must make visitors aware of this fact and give them an opportunity to opt out should they wish. But because cookies work to make the browsing experience smoother and more useful, there is not a great deal of visible incentive to opt out, despite potentially compromised privacy. Many sites found it difficult to comply with this directive and retain their functionality. Enforcement was limited, and the scope of the law was clarified in 2013 to include exceptions.[2] Cookies do not record large amounts of personal information, however if shared between servers they could contribute to a large body of data from which personal information could be inferred that could compromise privacy.

This kind of information, about digital activity rather than the content of the activity itself, is encompassed by what is known as metadata. This can be remarkably powerful, even at an individual level: for a simple demonstration of this, consider what might be inferred about someone who visits a football team's website, then a ticket sales page, then makes a telephone call to a friend, then visits their workplace's sick leave policy page, before sending an email to their boss. None of the content of their communication is known, but a pattern of information is revealed. Metadata also means it is extremely difficult to fully anonymise digital information, because these kinds of links function in such a way that unique patterns of behaviour can be identified. This tendency will only increase, as connectivity becomes more ubiquitous and pervasive, and everyone carries with them or is surrounded by technology which generates metadata about their activity. This reduces the possibility that any such data can be made truly anonymous, because of the detail captured and links that go from it to other information.[3]

As mentioned in Chapter 5, many people sign user agreements giving up ownership and thus privacy rights in order to use free services, often with no realisation that this is what they are doing. This may be either because they do not fully understand the ramifications of the agreements that they sign, or do not read them in the first place. It is also important to remember the distinction between security and privacy. Many people use secure cloud services to store personal files, which allow them to access these files from any computer. But although the 'cloud' sounds like an impartial, ephemeral space that you can purchase a part of, what cloud services really are is storage on networked servers. As put by the Free Software Foundation Europe: 'There is no cloud, just other people's computers'.[4] Although users instil trust in services such as Dropbox and Gmail which store their information in a way that cannot be accessed by those who are unauthorised, the owners of these services are still able to access the content, should they wish (though this would perhaps be breaking the trust of the users).

But while people may not be aware of occasions when the privacy of their documents is potentially compromised, at least it is easy to understand what a violation of privacy means in this context; someone unauthorised accessing your files. On the other hand, they may not even realise how much personal data they are producing just by moving in the digital public space and carrying out activity there, or what it might be used for and what implications this might have for their privacy. This data may not appear to have much weight or relevance on its own, but can be extremely powerful in aggregate. Some of this may arise from the sheer weight of information, and the fact that it can be connected to draw conclusions: the dossier effect. When all information about a person gets cross referenced, it can reveal significant amounts, and there are many uses of this both lawful and unlawful.

What other kinds of apparently inconsequential metadata might you be generating in the digital public space? One category of information is location data. There are several ways this might be collected, for example the IP address which locates your PC, or GPS data collected by mobile phones. This GPS information might be attached to files that you create (giving information on where and when, for example, a photograph was taken). Alasdair Allan and Pete Warden found that Apple products including iPhones track and keep location data, even migrating it across devices (Allan & Warden, 2011). This data is not encrypted, meaning that it can be accessed by anyone, and potentially used without your knowledge to track your movements. Garfield (2012) describes how 'Allan and Warden had no problem translating the recorded coordinates into maps, and one particularly striking screengrab from their presentation showed a train trip from Washington DC to New York City, with Allan's whereabouts being registered every few seconds'. It is not clear why Apple keeps this data, but it does appear to be a conscious decision, and their Licence Agreement implies that they may use this 'to provide and improve location-based products and services' (Garfield, 2012). But extremely detailed pictures can be built in this way not only of the lives and activities of individuals but also patterns of behaviour across a wider population.

Another, older example of this kind of data harvesting is supermarket loyalty cards. By offering rewards for their use, these are able to keep track of every purchase that is made at the store, and can be used to identify trends both on an individual and more wide-scale level. This may allow the store to target particular items to individuals based on what they have bought previously, often in highly analytical ways. For example, algorithms can understand that people who buy pregnancy tests, and then maternity clothes, and then infant formula and nappies, might be prime targets for pureed baby food and teething rings. It is not necessarily that a greater amount of information about individuals exists, but that it can now be collected on a large scale, connected with other information and analysed in great detail through the mechanisms of data mining. This large scale interconnectedness of vast data sets and the computing power to analyse them is part of the previously mentioned 'big data' capabilities.

An example of the power of this big data analysis has been shown by Kosinski et al (2013) who were able to accurately predict a range of personal information about Facebook users, including sexual orientation, ethnicity, religious and political views, simply from items that they 'liked' using the social media service. The fact that these individual simple actions cumulate to form a powerful profile of an individual is probably not something that most people are aware of, and they may be quite horrified to realise that this is information that they are providing to Facebook as a company, even if not more widely.

As the digital public space extends into more aspects of our lives, there is the potential that even more of our private information could inadvertently be made available. In Chapter 3, the internet of things was discussed, with technology currently in development which will connect objects and items that we purchase and use, so that information can be shared at various aspects of their life cycle. This however means that metadata will be available from these objects, and this could have privacy implications. If objects have embedded RFID tags built into them in order to simplify supply chain monitoring, this identification will not necessarily stop once they are purchased. Tracking of merchandise via embedded RFID, as is already used by organisations such as Wal-Mart and even the US Department of Defence (Hayles, 2009) leads potentially to the tracking of the owners of said merchandise. There are implications for privacy if this information is shared and utilised by organisations with interests counter to our own. 'Increasingly, we face a world where the things on our person, near to the body or means of transport, will be communicating with the network of embedded chips in the environment, allegedly for our benefit' (Featherstone, 2009). This connectedness of our possessions is already coming into effect with identification documents such as passports, and even NFC[5] enabled payment cards. Because these can be passively read, and readers (and the tags themselves) may not always be obvious, they carry an inherent risk that they can be 'overheard' to reveal information that we may prefer to stay private. This fear is already leading to measures being taken by some who fear this surveillance, such as wallets lined with aluminium foil to defeat unauthorised reading of RFID encoded cards. As our objects become Bruce Sterling's 'spimes', (see Chapter 3) contributing to an information mesh of the digital public space, we also by our ownership of them become part of this information space.

Personal information might also be revealed about us through logjects we own temporarily, which record their activity and transactions. Although objects which record their history might be extremely useful for supply chain, and encourage responsible use and recycling, it may be the case that you do not wish to have particular products associated with you. If their entire history can be read, including location data, it might mean that privacy for owners is compromised if, for example, your alcohol consumption can be tracked by the life cycle of the bottles that you purchase, and how quickly they are emptied and disposed of. If objects are tagged as belonging to you, should you reserve the right to

Figure 6.1 RFID embedded object (Oyster card)

have them 'forget' if you do not want to be listed as part of their history? It may be that these logjects move us away from built in obsolesce as we become more responsible for how we treat things we own, knowing that they will retain our history of use. But if this information can be read by external organisations, it increases the chance that anonymity will no longer be possible.

This networking of objects becomes even more critical when they are objects whose function has serious, continuous integration and impact on our lives, the disruption of which could be catastrophic. The most immediate example of this is health related devices, which can collect extremely personal data intrinsically linked to wellbeing. Compromised privacy in the data on these devices could mean information being released to those who we do not wish to have it, for example employers. While medical devices such as pacemakers are not new, there are now many more that routinely collect data and transmit it digitally to healthcare providers or make it accessible to patients in order to monitor their own health. Examples include insulin pumps for diabetics which provide constant monitoring and display blood glucose levels. Connecting these devices in order to allow them to transmit data to a central store may offer benefits to healthcare providers, but adding connectivity does potentially allow for abuse, especially if there is a two-way connection allowing the device to be accessed remotely, as is sometimes already the case. If this connection is accessed by someone unauthorised, it could lead to attacks whereby an individual was harmed by someone stopping their pacemaker or giving them too much insulin.

This assumes a criminal intent; someone hacking maliciously into what is private data. However, even well intentioned action can be harmful. In November 2014, the support organisation Samaritans released an app which tracked tweets and analysed them for potential evidence of those who might be struggling to cope and in need of support. There was widespread criticism of this app (*New Scientist*, 2014), with concerns both for the collection and sharing of mental health information without the consent of those concerned, and also the potential exposure of vulnerable people to trolls and others who wished to react negatively.

This type of data sharing also applies to wearable devices which may be tracking basic physical data such as heartrate and activity levels. The potential effects of this may be less severe or immediate, but of a greater current impact across a wider range of people. Devices such as the Fitbit are designed to turn individual data into shareable social information, in order to promote competition in achieving health goals. But this information could potentially also be shared without consent, added to the total digital profile of an individual, and could potentially be used for purposes which may compromise privacy. For example; someone with a wearable device which records sleep patterns may not wish to have that information accessed by their employers if it shows that they are only getting four hours sleep per night, leading to questions over whether they are fully able to perform their role adequately. Information on movement, sleep, even what you eat could be used for discriminatory purposes if accessed by insurers or employers.

It may even be the case that private data, which relates to your own health and wellbeing such as blood test results, is available to others but not to you. Although it can be accessed by clinicians, and potentially on a larger central database which may not be secure, medical test data is not always available to individuals who want to access it without lots of asking. And even if the data is accessible, it may not be in a format that promotes understanding by the patient. This is question of data ownership as well as privacy.

Case study 6.1: Blood data visualisation and use of patient data

There are many challenges and stresses to patients who, for whatever reason, have experienced kidney failure and need to be given regular dialysis treatment. But communicating with your doctor should not be one of them. And challenges can also exist for physicians, trying to help patients adapt their behaviour to improve their health. They may have huge amounts of data at their fingertips, but to translate this into things that will make sense can be a struggle without support.

In an attempt to support these doctors and help patients, the Creative Exchange 'Kendal Blood Data Visualisation' project created a digital app which takes the large volumes of (mainly numerical) data and displays it in easy to understand ways. These might show patients how, for example, their improved diet and eating habits over the course of a month have affected the amount of potassium phosphate in their blood results; too high or too low a level of which can lead to heart and bone disease. Patients can increasingly take the role of 'partners' in treatment, responsible in part for their own wellbeing. This can be empowering for some patients.

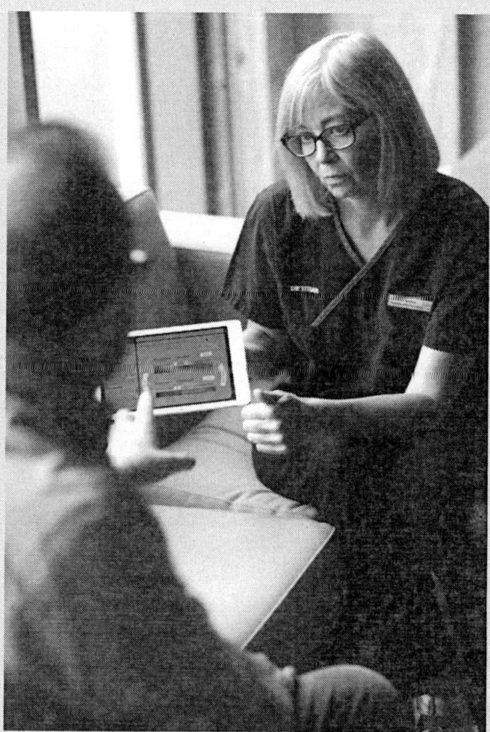

Figure 6.2 Blood data visualisation project

What is interesting here is that there is no new data being generated by the app; it is simply a tool for easier communication between doctors and patients. The visualisation techniques tease out the most important bits of information for the patient, and help the clinician explain them in a way that is straightforward to grasp for someone who might not be able to quickly interpret a large table of seemingly meaningless numbers. However it is important to remember that the original data might be interpreted in many ways by many different people.

In this project, development of the prototype was possible only because one of the researchers (Jeremy Davenport) is himself a dialysis patient working in partnership with clinicians, and was able to use his own data, and give consent for it to be used in the prototype. Jeremy commented that he was happy for his data to be used in this way given its value in building a credible prototype, although he did reflect on privacy issues; the prototype will be disseminated with anonymised data (J. Davenport, 2016, personal communication, 29 May).

Ethics of data usage are taken very seriously by the National Health Service (NHS), and to be able to gain access to medical blood test data from patients for research or development of such tools and processes requires significant justification. A detailed ethical approval application process must be carried out to ensure that patient privacy is protected. In theory, all of this data is anonymised, but with increasing computing power it becomes more difficult to ensure that patients cannot be identified: especially if, for example, it is a small subset of people with a particularly rare condition. While ethics of privacy are taken very seriously, it is conceivable that new innovations in the digital public space might mean that data can be used for purposes which were not considered when releasing it was approved.

In May 2016, *New Scientist* magazine obtained access to documents detailing an agreement between the Royal Free NHS Trust (which runs three London hospitals) and Google-owned artificial intelligence company DeepMind (Hodson, 2016a). This followed an announcement in February the same year that DeepMind was working with the NHS to build an app to help hospital staff monitor patients with kidney disease. But the document revealed that not only did DeepMind get access to current records of kidney health, but also to five years' back records of full medical data revealing many aspects of health information including HIV status and details of abortions and drug overdoses. The aim of such projects is that the information from an individual patient could be compared to millions of other cases, and potentially identify if they are in the early stages of a disease with no symptoms if their course maps to that of many others. Tests could be run to confirm the preliminary diagnosis. An alternative use for these large data sets could be predicting outbreaks of infectious disease. However concerns have been raised (Hodson, 2016b) that Google and the Royal Free did not obtain regulatory approval for the use of this data, which could lead to privacy violations since identifiable patient information is included. Additional to this, there are concerns over the fact that Google is taking ownership of this data, from which they could potentially profit, as a closed private organisation with no obligations back to the public whose data they are using. Such privacy concerns must be carefully considered, especially when the data concerned has major implications for the wellbeing of large numbers of people.

Networked information from your devices and objects may also include personal details even if they are not on your person – such as the time you set your heating to come on being related to the time you arrive home from work – and may not be securely stored. When we add connected things to our homes or networks, we assume that the only people who will be able to manage how these objects will act will be ourselves. But breaches in this information security could enable crime to be committed based on information about your personal habits. It is even possible to imagine scenarios where your smart home could be 'hacked' and take actions that you do not want – such as the one mentioned in Chapter 3 where the doors are locked and you cannot leave.

Permanence of information leading to a lack of privacy is also, ironically, potentially an issue with Bitcoin, despite privacy being among its defining aspects because of the cash-like anonymous nature. As described in Chapter 5, Bitcoins can be exchanged securely, and because all transactions are recorded within the code there is no risk of 'double spending', meaning that they can be treated like cash. Consequently, they are extremely useful for transactions that need to be anonymous online, while also popular with whistle-blowers and for the trafficking of secrets. However since all transactions are embedded within the block chain, there is a permanent record of how each coin has been used.

As this selection of examples shows, there are numerous ways in which our actions create information about ourselves that can make its way into the hands of others by legal or illegal means. Craig Mundie lays out another critical issue, that of consent to use:

> Today, there is simply so much data being collected, in so many ways, that it is practically impossible to give people a meaningful way to keep track of all the information about them that exists out there, much less to consent to its collection in the first place.
>
> (Mundie, 2014)

Even if were possible to give an individual the opportunity to consent or not to every piece of personal information requested or collected in the digital space, this would take vast amounts of time and energy. In addition, most people are not easily willing to opt out, since they see no immediate impact and want to be able to access services that rely on this consent; often free in exchange for targeting advertising. Sensitive information may also be collected from completely public sources, and can be used in conjunction with dossiers collected from digital spaces to build up a detailed personal profile of an individual. This passive collection does not necessarily provide an opportunity for consent. And if the passive collection becomes a ubiquitous and intrinsic feature of modern life, there will potentially be large societal and individual benefits that come with being included in it; choosing to opt out could have significant costs, increasing the likelihood that consent will be given with all the associated negative impacts. A possible solution to this could be to change the point at which consent is necessary: it is not the collection of the data itself that matters, but the use of it. Mundie suggests that one option might be a 'wrapper' of metadata, encasing any personal information and requiring individual authorisation for each request to open it and make use of it for specific purposes. This wrapper would describe the origins of the data and the rules governing its usage, allowing blanket authority to be given for beneficial purposes while minimising potential misuse. This would however require that all data collected complied with this protocol, which seems optimistic given the myriad ways in which our personal data haze surrounds us and bleeds into every interaction we make in digital public space. This data generation is also not limited to passive collection of metadata but might be linked to our online profiles, personal identity or identities that we create for ourselves as the public-facing aspect of our existence in particular parts of digital public space, as discussed in Chapter 4.

Digital social spaces: speaking privately in digital public space

We have spent some time talking about the fact that the digital connected world consists of a variety of digital public spaces, with different properties and qualities, rather than one amorphous digital public space. While these are all to some extent 'public' spaces, in many cases they have boundaries which mean that while anyone *could* still access these areas, in practice they are restricted either through the shibboleth of cultural understanding or by arbitrary limits on participation. These are public spaces where the notion of 'public' is restricted to those who fulfil certain criteria. With the right properties, these can become 'safe spaces' where members of a particular community can expect to maintain a certain level of privacy, and information shared there will not escape into other online spaces or the 'real world'.

Jessa Lingel writes about a specific example of this type of behaviour, in members of online community spaces and support groups for those interested in extreme body modifications (EBM). Because this behaviour is stigmatised, there is a tension between providing information in online forums to others to whom it would be beneficial (and who might find this information difficult to acquire elsewhere), and keeping the information exclusively within the community and sharing secrets only with trusted 'insiders'.

> This is the clearest expression of the politics of information: Information is not a stable, passive artifact in accounts from our participants; it is interactive, collective, and performative. Information is thus political in that it serves as the means of deciding who can be trusted and who cannot, who is a member and who is not.
>
> (Lingel, 2013)

However, there is a limit to how private and 'safe' such spaces can be, by the very nature of the fact that they need to be accessible to any member of the group, and that there is no foolproof way of confirming eligibility. There is also a measure of trust given to participants that the information will not be taken outside the confines of the specific space. Because this is fallible, and total trust may not be given, the area may not be considered an entirely safe space; behaviour and speech may still be regulated. The level to which this occurs may be dependent on the platform and type of space. For young people, who have grown up using online spaces to communicate, there now seems to be some acceptance of the idea that even if an attempt is made to restrict information sharing to supposedly exclusive digital public spaces, like a restricted Facebook group, there is a risk that by doing so the information may make its way further than intended and to other spaces. But it is easy to become complacent in a seemingly safe space, especially if others are sharing private information freely.

Some of the wariness of young people might be because the ownership of the space within which information is posted is held by corporations; so content may be shared not only with peers but with the curators of the information. Jared, age 24, said to Carrie James:

> Online you almost give up your privacy. When you send an email, that email is actually the property of Yahoo or Gmail or whoever, or Harvard or the museum, whoever's site you're using. I mean, typically they're not going to look at these things, but, you know, they have that right.
>
> (James, 2014)

It is increasingly the case that digital spaces which were considered safe, anonymous or private are bleeding into other contexts and this can cause significant issues. One example of this is the phenomenon of employers monitoring or utilising the social media profiles or other online presence of current or potential employees. Several media reports have covered individuals who have been fired because of things that have been posted on their personal social media profiles, and it can now be common practice to search these profiles before interviewing candidates or offering them a position. What might previously have been a closed off personal life, undertaking activities that are not immoral or illegal but simply inappropriate for a work context, might now bleed over and affect other spheres of life: because they are no longer entirely private and can easily be found and connected to the individual. It may be that this propensity for searching online material means that the next generation learn to restrict their behaviour or curate their online presence more, or alternatively that there will come a time when acceptance of youthful indiscretions posted online means that such material is no longer considered as a barrier to employment. But it is not yet clear which of these will take place.

The reason this material can be so damaging may not only be because the content being examined is inherently inappropriate, but because context of what was considered a private space can be lost. There may be social cues, communicative norms and understanding associated with a particular group (for whom a message was intended) that get lost when the message is taken outside that group. This can particularly be the case with jokes; issues can arise when statements intended to be joking are removed from an ongoing context and thus unspoken cues associated with humour. In some cases, this does not affect the outcome: if the content reveals attitudes which are incompatible with the role then it could be argued that the context is irrelevant, as in the case of police officers disciplined for racist speech on Facebook (BBC News, 2011). However a different view may be given to cases such of that of Paul Chambers who was charged with 'sending a public electronic message that was grossly offensive or of an indecent, obscene or menacing character contrary to the Communications Act 2003' for the following tweet:

> Crap! Robin Hood airport is closed. You've got a week and a bit to get your shit together, otherwise I'm blowing the airport sky-high!!
>
> (Chambers, 2010)

The tweet, which was posted after cold weather forced closure of several UK airports, was found during an unrelated search by an off-duty airport manager (Wainright, 2010), and was reported to the police as a threat against the airport. Paul Chambers was fined £385, ordered to pay £600 costs and lost his job, though the conviction was later quashed after three appeals. The judgement in the final, successful appeal concluded that, in regard to the provisions of the Communications Act, 'a message which does not create fear or apprehension in those to whom it is communicated, or who may reasonably be expected to see it, falls outside this provision, for the very simple reason that the message lacks menace' (Collingwood & Broadbent, 2015). Two key points to note here are the expectations of audience, and equally the 'creation of fear and apprehension'. The judge highlighted the importance of context and means of conveying a message in ascertaining its intent. Large public support for Paul Chambers, including many celebrities, tried to make the key point that almost everyone reading the tweet would from context conclude that it was a joke rather than a threat; indeed the case became widely known as the 'Twitter Joke Trial', referencing the humorous nature of the content. But it is often very hard to pinpoint humour and intent on a contextless medium such as Twitter, when ambient intimacy (as described in Chapter 4) has not been established and a single point of contact is examined. Another example of this was the case of PR chief Justine Sacco, who was the centre of a Twitter 'storm' over a tweet she made on 30 December 2013 before boarding a plane:

> Going to Africa. Hope I don't get AIDS. Just kidding. I'm white!
>
> (Sacco, quoted in Ronson, 2015)

While she was on the 11-hour flight, the tweet spread virally and became a top trending topic, with thousands of calls for her resignation. She was indeed fired, and the results of her very public shaming had a major impact on her subsequent career and life. Having met with her, Jon Ronson reflects on the meaning behind the tweet:

> Read literally, she said that white people don't get AIDS, but it seems doubtful many interpreted it that way. More likely it was her apparently gleeful flaunting of her privilege that angered people. But after thinking about her tweet for a few seconds more, I began to suspect that it wasn't racist but a reflexive critique of white privilege – on our tendency to naïvely imagine ourselves immune from life's horrors. Sacco… had been yanked violently out of the context of her small social circle.
>
> (Ronson, 2015)

Should there be an argument that these tweets were 'private' because there was no expectation that anyone outside the immediate social circle of the tweeters' 'followers' might see them? Not really, since they were broadcast in the 'public' realm of an unlocked Twitter account. Content that is uploaded to social media platforms, particularly those of the open variety such as Twitter or Tumblr that are designed for sharing and spreading, should not be considered private and therefore their content should be judged accordingly before posting. This is certainly the view of the US Navy, who claimed in 1998 that Master Chief Petty Officer Timothy R. McVeigh violated their 'don't ask, don't tell' policy by having an online profile which identified him as gay. 'The Navy insisted that by posting his biography on-line, McVeigh had waived his own privacy. In effect, they said, he had "told" the world of his sexual orientation. But what the Navy insisted was public was considered private by critics' (Sykes, 1999). Such issues, particularly in the case of the 'jokes', arise both from the ambiguous public or private nature of online space, and (because of the novel nature of these spaces) a lack of societally accepted standards for what is acceptable behaviour in such 'public' spaces.

There is another key to this difference between digital social interaction venues, and previously existing physical public spaces where people met and discussed sensitive topics, or made jokes that could be taken as offensive or risqué: record of activity. A group meeting in a pub to discuss their participation in semi-legal activities will not (leaving aside covert surveillance, which can occur in either context) leave evidence beyond the say-so of the other members of the group, who have been trusted enough to participate in the first place. However in digital public space, all information transfer is recorded; potentially including metadata such as exact time and location. This could be shared outside the group in full, verbatim and leaving a permanent record. This reduces the safety of conversation, because of the knowledge that it can always be retrieved by someone.

This permanence also affects our public information across time. It might be the case that we no longer consider things we said in the past to represent the people we are now, and thus we would prefer them not be public or publicly linked to our current selves. Ayalon and Toch (2013) looked at how people's attitudes to publication of information on social networks changed over time. They found that there was a significant negative correlation between how long ago status updates were posted on Facebook, and how happy people were for them to be shared. There was also a significant effect of major life changes – it seems that the participants did not consider the older updates relevant or representative, and no longer wanted them to be shared, with many opting that they would choose to remove or change them. The permanence of digital content means that it is harder to 'forget' things from our past or put them behind us. Retrospective privacy is important, and the new properties of digital public space mean that we must consider the implications of what we put online not only now, but in our future: and this may be something we are not yet equipped to do.

A solution to this may be to maintain secrecy rather than privacy. Increasingly, digital personae present a maintained identity that contains only selected aspects of the self. These are then, if not necessarily designed specifically to be public, more amenable to being exposed with less damage to reputation than a fully representative version of oneself. Individuals might even prepare several different versions of these for different audiences, and keep them separate. However these may be difficult to maintain in practice because it can be difficult to isolate social groups and remember exactly which contacts are in each group. This is especially the case once these groups grow past a certain size – it may be the case for example that you forget that a particular individual has been given access, and post something via that persona that you would rather they did not see. This is a particular hazard of the online space, as opposed to face to face interactions where the people present can be immediately seen and recognised.

Another problematic aspect can be if groups contain a mixture of different types of acquaintances. This can blur the boundaries of what is acceptable for that group, making it complex to create a suitable, consistent persona. There is also a limit to how many separate accounts can be maintained, leading to inevitable crossover. 'Even with private accounts that only certain people can read, participants must contend with groups of people they do not normally bring together, such as acquaintances, friends, co-workers, and family' (Marwick & boyd, 2010). There is also the risk, as outlined above, that things posted through a particular faceted profile may not stay within it and can quickly spread to unintended audiences, which might include those for whom the profile was deemed inappropriate.

In some cases, extremely personal information is willingly shared entirely anonymously in online spaces, disconnected from persistent identities, perhaps to gain support or simply to express sensitive or controversial opinions. As mentioned in Chapter 4, the art project PostSecret consists of a website where contributors submit (physical) postcards which are posted online anonymously. Each postcard reveals an intimate confession, which would generally be considered extremely private if traceable back to their originator. But despite being publicly posted, privacy is maintained by anonymity, and a large audience can still be reached. The ability to hold an anonymous profile and access digital space without linkage to a legal or public identity may be compromised by the increasing insistence on 'real names' by the companies who own the infrastructure of many digital spaces, who wish to link all aspects of online life. While this creates a fuller profile which is useful for marketing and personalisation, it means that it is more difficult for fragmented profiles to be established and therefore privacy between them to be maintained.

Why does personal information posted in these spaces become spread and distributed, when we might be uncomfortable with our own information being communicated in this manner? For this we must return to the basis of social media in our social communicative history; in gossip. We desire privacy for ourselves, but the propensity for gossip, to share privileged information, is strong. We have a high level of interest in the lives of others, particularly when the information is considered to be secret or illicit, therefore vicariously sharing other people's private experiences is something that should be an expected component of any medium that enables this information to be transferred.

Some people give up their right to privacy for notoriety, for the opportunity to be known, for fame. Sykes (1999) talks about the 'exhibitionist society', and uses examples such as live webcams, comparing this to television equivalents: 'tell-all' reality TV shows such as Jerry Springer. The ease with which content can now be published means that anyone can make themselves into a celebrity, but with this comes the risks of loss of privacy that puts celebrity lives under public scrutiny. This kind of semi-consensual loss of privacy is not restricted to digital public space, however the rise of what Jenkins et al (2013) call 'spreadable media' means that it is much easier for those who put themselves in public view to find they are giving up more of their personal life than they may have intended. It may be that a constructed front was put forward as the image to be spread and the 'personality' to be in the public eye, but if this cuts too close to the personal, private information then privacy may be sacrificed without knowing it. In Stuart Evers' speculative fiction story 'Everyone Says', technology is described by which the public can 'link' to the lives of others to experience their senses, thoughts and emotions. The appeal seems to be both to vicariously enjoy 'celebrities' who lead risky and hedonistic lives, and the simple novelty of experiencing someone else's life, even if it is apparently unexciting. But the pressure of this scrutiny when too many people become interested in one's life is palpable and leads to tragedy (Evers, 2015). While such technology is not likely to be realised any time soon, there are some aspects which ring all too true. By opening up our lives to others we leave ourselves open to scrutiny and the possibility of large public attention, which for people like Paul Chambers and Justine Sacco can have negative consequences.

Part of the reason for this possible creep between the information that we deem public and that which we would prefer to keep private, is that it is not just social content that we ourselves upload that can have impact on our privacy. A consequence of the social aspect of much of the digital public space is that others may, by virtue of digitising their social experience, share on our behalf without fully considering the consequences. 'None of us are exempt from this social fact – people who elect not to join online social networks are often unconsenting participants on Facebook, YouTube, and the like, since both well-intentioned and mal-intentioned users share photos, videos, and comments featuring these nonusers' (James, 2014). In Chapter 5, we looked at how there may be many conflicting claims to contested digital objects and their copies. This applies not just to information put into the digital space by individuals about themselves, but by others based on non-digital interactions. Privacy may be violated if, for example, photographs placed online contain images of individuals who may or may not have given consent for this usage. People using social media might include personal information about their own activities which can be used to draw inferences about others; for example describing being at a party with a friend who does not have a social media presence. When this third party content is uploaded about an individual it can be linked to their personally constructed online dossier and profile, and this can warp the intended digital presence; containing information that the individual might prefer to remain private, but which is then irrevocably linked with their online persona.

This extends beyond information recorded or passed on by those who we may know outside of the digital space – it can also take place between strangers if information that was collected in a physical public space is uploaded. Recording devices are becoming more and more ubiquitous; camera phones are carried by the majority of people, video recorders may be mounted on cars to record potential traffic accidents and determine fault, and there is an increased push towards cameras for officials such as law enforcement. This phenomenon has come to be known as 'sousveillance' (Thompson, 2013). The word, coined by Steve Mann, plays on the notion of surveillance and means the constant monitoring of all by all. Many of these digital recording devices also have the built-in capability to share the captured information instantly in digital space, with videos and photographs uploaded and shared on smartphones in almost real time in some circumstances. This has its own implications for privacy, since there is often not time to fully consider the repercussions that uploading information might have on individuals who might not even be known to the one doing the sharing.

An example of this can be seen in one of the earliest 'memes' and viral videos which spread across the internet. Artist and filmmaker Matthias Fritsch shot a video at the 2000 'Fuck Parade' festival in Berlin which he hosted on his website for several years, and uploaded to YouTube in 2006. The video shows a tall, dynamic, Scandinavian-looking man confronting a drunken groper and then apparently leading a troupe of techno dancers. It was picked up by online communities, gained millions of viewers, was shared widely across the internet, and spawned a huge amount of remixed and reimagined material including figures, t-shirts and other merchandise. In 2009 the man playing the starring role, who came to be known as 'Technoviking' sent a cease and desist notice, and later sued Fritsch over commercial use of his image rights without authorisation. The German courts ruled that Fritsch must remove the man's image from any material displayed in public, and pay him the €8000 earned from YouTube advertising on the video plus legal fees (Fritsch, 2015). But it is in many ways too late: the meme and its associated imagery has spread far and wide and cannot be contained. This situation is another example of controversies and limitations of copyright law for media in which work is shared and remixed and distributed. But it also highlights privacy concerns, as the video can be seen as an invasion of privacy for 'Technoviking', who would have at the time had no knowledge that his actions would be shared with millions or become a worldwide phenomenon. The case has raised many questions over the nature of image rights in a world where such material can be distributed so widely, so easily. While this is in part because of the timing, and the fact that it occurred many years before such viral distribution became a well appreciated act, similar viral videos with unknowing stars still regularly occur, suggesting that public behaviour is not limited by the knowledge of sousveillance (though it may have changed for many).

Sometimes such sharing is done not out of lack of thought for the rights of the individual, but in active pursuit of behaviour change either from them or to set an example to others. This digital 'public shaming' (similar to the backlash against the 'joke' tweet of Justine Sacco) is becoming more common, and can apply not only to antisocial behaviour committed online, but also in physical space; translated and transmitted online through images or video. Solove (2007) has written at length about 'dog poop girl' whose actions not cleaning up the mess left by her pet led to her image being shared widely online. The general opinion of many commentators seemed to be that since she was in public, she should not expect privacy, and there was no ethical dilemma in the sharing of her image. This is despite the fact that she gained notoriety to the point of strangers recognising her in the street, for an incident that prior to widespread digital sharing would not have spread beyond those who immediately witnessed the incident. There are ethical dilemmas associated with the use of digital public space not just in regard to your own privacy, but how you treat the privacy of others by sharing information about them, which you might be transferring from a physical public space to the digital public space.

Privacy implications of digital/physical world: surveillance and data

The examples above of 'real life' activity crossing over into the digital space and being distributed there are a critical reminder that when talking about digital public space we are not just talking about the digital information space, but also how, as discussed in Chapter 4, digital aspects are overlaid and intertwined with the physical world. This includes such technologies as the internet of things, augmented reality, and data collection and delivery captured within and through physical public spaces.

During the trial period of Google Glass, many people were worried about the privacy implications of, in effect, a camera that could record everything that someone sees as they go about their daily life and travel in public spaces. Public perception was that privacy might be negatively impacted by being recorded by someone wearing Glass, despite the fact that in reality constant recording was not practical due to the battery life of the device. These concerns may in part have led to the shelving of the project, though future versions may be in development. Whether these will address the privacy concerns is yet to be seen.

These fears belie the fact that much of our lives is already recorded. This is due not just to the sousveillance referenced above, which might be transient and associated with people passing through the space, but also to digital infrastructure increasingly built into (especially urban) environments. In particular it is important to note how this data collection can impact on privacy, through surveillance and technological innovation that can collect much more information than people might be aware of.

The first thing to consider is that physical proximity and access can have a surprisingly significant effect on digital privacy. A good example of this is the controversy that arose in 2010 when it emerged that Google Maps cars, travelling the world since 2007 to record photographic imagery to add to their maps, were also gathering electronic data and personal information as they went.

> If you were on the Internet as one of Google's Subarus rolled by, Google logged the precise nature of your communications, be it emails, search activity or banking transactions. As well as taking photographs, the cars had been consciously equipped with a piece of code designed to reap information about local wireless services, purportedly to improve its local search provisions. But it went beyond this, as another program swept up what it called personal 'payload data' and led the Federal Communications Commission in the US and other bodies in Europe to investigate allegations of wiretapping.
>
> (Garfield, 2012)

There is no evidence that Google used the information, and they claim it was collected accidently because of legacy code embedded in their street view car equipment. They did however admit that 'it was a mistake for us to include code in our software that collected payload data' (quoted in Garfield, 2012). But this situation demonstrates how easy it is to collect private information simply by driving along a street close to wifi routers, and picking up signals that are openly accessible.

Wider sweeps of less detailed data from larger ranges of physical space can also lead to invasions of privacy. By collecting data from large areas and using algorithmic computing power, it is possible to undertake complex analysis that would be impossible without ubiquitous digital connectedness. This can reveal information about both individual movements and activities, and co-ordinated groups. The connectedness of physical public spaces might include the surveillance afforded by CCTV cameras and other recording devices (both publicly owned, and owned by individuals who upload the content to the digital public space) which can give wide coverage of activity, and provide a potentially permanent audio-visual record of what happens. The UK is particularly well known for significant amounts of CCTV coverage: Cuff (2003) suggests that by 2001 the average British citizen was captured on camera 300 times each day. The stated aim of this coverage is generally crime prevention, a task at which it appears to succeed, with reports of a 20% to 40% decrease in the crime rate following installation of cameras (Bowyer, 2004). But is this worth the associated reduction in privacy?[6]

The availability of camera footage in all spaces can lead to circumstances like those of the 'dog poop girl' mentioned above, where individual acts become spread and shared and have consequences for the people involved. But sophisticated analysis can also be used on surveillance footage to gather demographic and other data. For example face recognition technology has been the focus of much work, especially with heightened alerts in the wake of terrorist activity in many countries. These technologies might track particular groups, or conceivably allow any individual to be identified and tracked across multiple public spaces, and their activity monitored by, for example, governments. Certainly it is already possible to identify individuals from such footage, and not always for purposes of law enforcement: note the case of the comedian Michael McIntyre whose image was tweeted by the National Police Air Support Unit. While the publication of the photograph was potentially a breach of the Police guidelines and a legal invasion of privacy, it appears that taking the photograph in itself was not. A statement from a Metropolitan Police spokesperson said that 'this tweet does not, as far as we know, constitute a breach of data protection legislation' (BBC News, 2016). Controversy similarly arose around a Russian face recognition service called 'Findface' which allowed people to match photographs, perhaps taken on the street, to social media profiles. Concerns were raised that it compromised privacy and could be used for unscrupulous purposes such as by debt collectors. These fears seemed to prove justified when it was used by members of an online community to track down women who had appeared in pornographic films and spam their friends and family (Rothrock, 2016).

In Cory Doctorow's 2008 speculative fiction novel *Little Brother*, he describes a near-future scenario which includes, as well as facial recognition, gait recognition. The young characters, in order to play truant and skip school, place stones in their shoes in order to disguise their gait and not be identified by the monitoring systems throughout the campus. This is not too farfetched, given that technology to identify individuals based on their gait already exists (Wang et al, 2003). While attempts have been made to address the issues inherent in this kind of mass surveillance capability, a code of conduct has not yet successfully been developed: in June 2015 talks to address this broke down after privacy advocates left in protest at the lack of engagement. They were disappointed in the conduct of industry representatives who are, for example, implementing systems designed to be able to identify when high value customers enter a shop (Lynch, 2015; Hodson, 2015).

There have been several responses to this constant thread of identification in public places including art pieces, research and recommendations (see Case study 6.2). An example of this type of work is the CV Dazzle project, which has been developed by artist Adam Harvey and uses hairstyle and makeup specifically designed to confuse face recognition algorithms and prevent identification. The name derives from dazzle camouflage used during World War I by naval vessels, which, to quote the company's website,[7] 'used cubist-inspired designs to break apart the visual continuity of a battleship and conceal its orientation and size'.

Case study 6.2: Computer vision invisibility

For a long time, recognising faces was one of many tasks that were extremely difficult for computers but easy for humans. However these days facial identification software is more advanced and can often match people's identity; in some cases even from images which are blurry or indistinct. This means however that anywhere which is covered by camera surveillance should change our expectations of being anonymous and lost in the crowd.

Ben Dalton, as part of his work in the Creative Exchange programme, has experimented with artworks which explore these notions of being identifiable by computer systems. Two particular works contrast different aspects of making oneself 'invisible' to ubiquitous surveillance. The first, the 'Wildermann', is based upon outfits worn in traditional festivals:

> The wild man reoccurs as a motif in festivals throughout Europe, and is echoed in characters and costumes across the world. The wild man often takes on a role that muddies social order, mischief and the wilderness. Traditionally built from wild materials like branches, grass, animal bells and furs, the materials of modern wilderness are not moss and straw but mass-production and military-industrial detritus.
>
> (*Dalton, 2015*)

The Wildermann camouflages identity and identification by obscuring the features which are recognisable by computers such as body outline, walking pace or number of limbs; similar to the approaches explored by the CV Dazzle face makeup project. However, while the wearer might look invisible to human-tracking algorithms, they look highly distinctive and extraordinary to humans. There is a social awkwardness cost to wearing a personal invisibility outfit.

Contrasting with this is a second 'invisibility design experiment' project which generates images that trick computer surveillance systems not by being invisible, but by being visible when they should not be. Dalton has produced a series of t-shirts which show images such as a cat, a diagram, or a piece of architecture, which contain key features that algorithmically match those in the face of Elvis Presley. As far as the surveillance software is concerned, this is Elvis' face, but humans do not even notice. The more the t-shirts are worn in public, the more his face is visible, perhaps in many different places at once – sightings of Elvis on the increase! People buying the t-shirts are collaborating in confusing tracking systems, and are therefore shaping another form of invisibility. If this technique were to be used with markers for the face of an individual person travelling in the world, their actual location would be hidden amongst a sea of irrelevant data.

These experiments are extremes, but help to explore individual and group responses to the challenges of privacy in modern public spaces. Our traditional expectations of anonymity in the crowd are subverted by surveillance networks, and yet the algorithmic biases encoded by the makers of these systems suggest new ways of retaining control over our visibility, both in physical and online spaces.

If we want to avoid detection we might increasingly have to make use of such techniques to subvert what is becoming a standard part of life in a digital world – that our privacy is no longer ensured just because no human is watching us at that particular time.

Despite these fears surrounding surveillance and lack of privacy in a digitally connected world, it must be tempered with the fact that current technology is limited in its capabilities. It is generally only possible to match individuals to a 'gallery' of target faces, and any system will most likely generate errors, which may be false positives or false negatives depending on the sensitivity of the system (Bowyer, 2004). This then, is technology to identify specifically targeted individuals rather than track every person who travels in a public space (at least at current levels of technology) and the privacy implications should be judged accordingly.

Privacy solutions

Clearly privacy in digital public space is a major concern for many people, and it seems evident that individuals are right to be concerned over how data can be gathered and used, because the consequences could impact on their right to privacy. But since there are also benefits to digital public space that are bound up with such data collection, the critical question might instead be: how can privacy be maintained while still accessing the benefits of digital public space? Central to this is the understanding that if the default is to be connected (by ambient computing or digital public space), then to preserve private space and time there must be the facility to disconnect, to reserve an isolated state where we can be alone. Additionally, it is desirable to limit the amount to which our digital footprint can be used against us to provide benefits to others (such as marketing) rather than contributing to our own information stream. In effect, if we are using the digital public space as cognitive augmentation, we need to be able to control what enters our 'minds' and maintain private areas which are fully controlled by us and not subject to the wishes of others. The other side of this is that we must be able to contain the flow of information outwards from our person, and take control of what about us becomes available to others.

Awareness is an important factor: by alerting people to the fact that their data may be stored, and potential ways in which it might be used, it may be that they are able to make more informed decisions about what they share and what agreements they sign. Some of the concerns above relate to information which is stored on servers belonging to commercial companies, and the trust which must be placed in those organisations to keep your information secure. There is a movement developing of people who resent that these large organisations have an almost complete monopoly on our information through the fact that they enable connected digital life. Ind.ie is a company which is pushing back against this trend. Aral Balkan, the company founder, explains in his blog: 'We've built a world where our everyday things track our every move, profile us, and exploit those profiles for monetary gain. A world with a wholly privatised public sphere. A world of malls, not parks. A corporatocracy, not a democracy' (Balkan, 2015). Ind.ie are developing new networks and systems which instead operate on a peer to peer basis so that rather than relying on, say, your email being stored by Google in order to send it to someone else, files and documents are sent directly to the recipient without passing through any third-party intermediaries.

Another potential solution to privacy concerns is to build privacy maintenance into the structures of the technology itself. A critical aspect is the anonymisation of data, so that it cannot be traced back to individuals. Vaidya and Atluri (2007) discuss privacy-preserving profiling, using algorithmic clustering to conceal details of any individual and draw out conclusions from general user population while maintaining encryption and privacy. In this way, marketers can see trends and create profiles which can be used to create better services, but without being able to access individual details of the data which was collected, or even the structure of the profile itself.

Allowing for removal of personal information from the digital public space also goes some way towards ensuring privacy, such as the 'right to be forgotten' which was implemented in 2014. The Court of Justice of the European Union ruled that companies including Google, Microsoft and Yahoo must implement this, giving individuals the right to ask these search engines to remove information about them from search results 'if the information is inaccurate, inadequate, irrelevant or excessive'. (European Commission, n.d.). But this kind of data 'cleaning' can itself have implications. It may by its existence reveal private information, in other words the fact that there is something to be 'forgotten'. Information accidently made available in Google's source code revealed in 2015 that over 95% of removal requests were from members of the public to remove 'private, personal' information (Tippmann & Powles, 2015).

This kind of manual curation of data collected about ourselves, making changes to the corpus of information that comprises the digital public space, may allow us to protect individual privacy by attaining control over our online selves. But by putting in the hands of individuals this power of choosing which 'truths' to display, are we damaging the integrity of the digital public space and causing bias? This will be explored in Chapter 7.

Generally speaking, privacy in the digital public space must be maintained both by design in the technologies which underpin it (to ensure that privacy is maintained wherever possible), and by social factors allowing the public to be more aware of their privacy rights, and what the implications might be of using connected technology.

Key points

- Publicness is often placed in opposition to privacy, but in actuality both exist as a non-binary range of concepts and can co-exist.
- Digital information space is constructed from shared information, which may be personal; it is therefore important to consider who it is shared with and who can 'overhear'.
- Activities carried out in digital public space may be less anonymous than those in physical public space due to the nature of digital content, its traceability and persistence. Actions and behaviours which may have been acceptable in non-digital space may carry greater risks and consequences.
- Security must be considered in conjunction with privacy: how your private information is protected and kept out of public space.
- Expectations of privacy in certain spaces may be based on false assumptions (such as the 'security of obscurity') and must be carefully considered. Boundaries between private and public, and who might be watching, may be less clear than in non-digital space.
- We may create large amounts of data by our actions online that we are not aware of, which may be used without our knowledge or consent. This can be seemingly trivial information which is powerful in aggregate.
- Sometimes relinquishing data is a prerequisite of using digital services which may be difficult to give up.
- Connected physical objects in the internet of things may also reveal private information about our behaviour, and even critical information about our health and wellbeing.
- It may not be possible to evaluate every piece of data collected about us and give consent for its use, an alternative would be blanket authority for specific purposes only.
- Public spaces with boundaries restricting entry may become 'safe' spaces with community rules and accepted behaviours, especially if they cater to marginalised groups.
- However these spaces may not be as private as they initially appear, and issues can occur when information leaks out.

- Spaces may be separated for different audiences, and problems can occur if information moves out of these to inappropriate audiences (such as employers) or without context. It is dangerous to treat content on social media as 'private'.
- Digital information may persist over time and become inappropriate and thus less publicly acceptable if we want to provide a representative version of our 'current' self.
- Maintaining secrecy by having separated digital personae may help with this, but is difficult to maintain, especially with the growing preference from social media networks for 'real' names.
- Privacy may be sacrificed for notoriety, but this can occur without consent if information about you is distributed by friends or strangers; more possible in the 'sousveillance' world of constant recording.
- Digital privacy aspects invade the 'real world' with increased digital surveillance and sousveillance, as well as privacy implications of data analysis such as face recognition.
- Solutions might include radical steps such as obscuring faces from computer vision, controlling more closely what information about us enters or is preserved in the digital space, or improving the infrastructure so that privacy maintenance is built in.

Notes

1 https://wiki.openrightsgroup.org
2 http://ec.europa.eu/ipg/basics/legal/cookies/
3 See Weise, Hardy et al, 2012; Conti et al, 2012.
4 Free Software Foundation Europe provides downloadable materials bearing this slogan at https://fsfe.org/contribute/spreadtheword.en.html#nocloud
5 Near-Field Communication, which allows devices to communicate when brought into close proximity and allows 'contactless' payments.
6 It is worth noting that the reduction in crime might not solely be attributable to the surveillance, but also to the perception of its existence. A sense of being watched, especially when images of eyes are used, has been shown to affect crime rates (Nettle et al, 2012).
7 http://cvdazzle.com/

References

Acquisti, A., Gritzalis, S., Lambrinoudakis, C. and di Vimercati, S., eds. 2007. *Digital privacy: theory, technologies, and practices*. CRC Press, p.348.

Allan, A. and Warden, P., 2011. Got an iPhone or 3G iPad? Apple is recording your moves. [online] *O'Reilly Radar*. Available at: http://radar.oreilly.com/2011/04/apple-location-tracking.html [Accessed 18 October 2016].

Altman, I., 1975. The environment and social behavior: privacy, personal space, territory, and crowding. Brooks/Cole Publishing Co., p.2.

Ayalon, O. and Toch, E., 2013, July. Retrospective privacy: managing longitudinal privacy in online social networks. *Proceedings of the Ninth Symposium on Usable Privacy and Security*. ACM, p.4.

Balkan, A., 2015. Ethical Design Manifesto. [online] Available at: https://ind.ie/blog/ethical-design-manifesto/ [Accessed 19 October 2016].

BBC News, 2011. 150 officers warned over Facebook posts. [online] Available at: www.bbc.co.uk/news/uk-16363158 [Accessed 18 October 2016].

BBC News, 2016. Did aerial photo of Michael McIntyre break privacy rules? [online] Available at: www.bbc.co.uk/news/magazine-33535578 [Accessed 19 October 2016].

Bowyer, K.W., 2004. Face recognition technology: security versus privacy. *IEEE Technology and Society Magazine*, *23*(1), pp.9–19.

boyd, d., 2007. Social network sites: public, private, or what. *Knowledge Tree*, *13*(1), pp.1–7.

Chambers, P., 2010. My tweet was silly, but the police reaction was absurd. [online] *The Guardian: Comment is Free.* Available at: www.theguardian.com/commentisfree/libertycentral/2010/may/11/tweet-joke-criminal-record-airport [Accessed 18 October 2016].

Collingwood, L. and Broadbent, G., 2015. Offending and being offended online: vile messages, jokes and the law. *Computer Law & Security Review*, *31*(6), pp.763–772.

Conti, M., Das, S.K., Bisdikian, C., Kumar, M., Ni, L.M., Passarella, A., Roussos, G., Tröster, G., Tsudik, G. and Zambonelli, F., 2012. Looking ahead in pervasive computing: challenges and opportunities in the era of cyber–physical convergence. *Pervasive and Mobile Computing*, *8*(1), pp.2–21.

Cuff, D., 2003. Immanent domain. *Journal of Architectural Education*, *57*(1), pp.43–49.

Dalton, B., 2015. The Barrow Woodwose. *Royal College of Art Work in Progress 2015.* Available at: http://soc2015.rca.ac.uk/ben-dalton/ [Accessed 19 October 2016].

Danezis, G. and Clayton, R., 2007. Introducing traffic analysis. *In*: Acquisti, A., Gritzalis, S., Lambrinoudakis, C. and di Vimercati, S. eds., 2007. *Digital privacy: theory, technologies, and practices.* CRC Press.

Doctorow, C., 2008. *Little Brother*. Tor Books.

European Commission, n.d. Factsheet on the 'Right to be Forgotten' ruling (c-131/12). Available at: http://ec.europa.eu/justice/data-protection/files/factsheets/factsheet_data_protection_en.pdf [Accessed 19 October 2016].

Evers, S., 2015. Everyone says. *In*: Page, R., Amos, M. and Rasmussen, S., eds., 2015. *Beta-Life: stories from an A-life future*. Comma Press.

Featherstone, M., 2009. Ubiquitous media: an introduction. *Theory, Culture & Society*, *26*(2-3), pp.1–22.

Fritsch, M., 2015. *The Story of Technoviking.* Available at: https://vimeo.com/140265561 [Accessed 18 October 2016].

Garfield, S. 2012. *On the map: why the world looks the way it does.* Profile Books.

Goodman, M., 2012. A vision of crimes in the future. [online] *TED Talks 2012*. Available at: www.ted.com/talks/marc_goodman_a_vision_of_crimes_in_the_future [Accessed 18 October 2016].

Hayles, N.K., 2009. RFID: human agency and meaning in information-intensive environments. *Theory, Culture & Society*, *26*(2-3), pp.47–72.

Hodson, H., 2015. Face recognition row over right to identify you in the street. [online] *New Scientist.* Available at: www.newscientist.com/article/dn27754-face-recognition-row-over-right-to-identify-you-in-the-street [Accessed 19 October 2016].

Hodson, H., 2016a. Google knows your ills. *New Scientist*, *230*(3072), pp.22–23.

Hodson, H., 2016b. Did Google's NHS patient data deal need ethical approval? [online] *New Scientist.* Available at: www.newscientist.com/article/2088056-did-googles-nhs-patient-data-deal-need-ethical-approval/ [Accessed 18 October 2016].

Houghton, D.J. and Joinson, A.N., 2010. Privacy, social network sites, and social relations. *Journal of Technology in Human Services*, *28*(1-2), pp.74–94.

James, C., 2014. *Disconnected: youth, new media, and the ethics gap*. MIT Press, pp.2, 38 & 27.

Jenkins, H., Ford, S. and Green, J., 2013. *Spreadable media: creating value and meaning in a networked culture*. NYU Press.

Kosinski, M., Stillwell, D. and Graepel, T., 2013. Private traits and attributes are predictable from digital records of human behavior. *Proceedings of the National Academy of Sciences*, *110*(15), pp.5802-5805.

Lingel, J., 2013. 'Keep it secret, keep it safe': information poverty, information norms, and stigma. *Journal of the American Society for Information Science and Technology*, *64*(5), pp.981–991.

lordoftheinternet, 2013. Some thoughts are so private that you only share them with a therapist or 17,000 people on the internet. [Tumblr post] Available at: http://lordoftheinternet.tumblr.com/post/44788412914/some-thoughts-are-so-private-that-you-only-share [Accessed 18 October 2016].

Lynch, J., 2015. EFF and eight other privacy organizations back out of NTIA face recognition multi-stakeholder process. [online] *Electronic Frontier Foundation.* Available at: www.eff.org/

deeplinks/2015/06/eff-and-eight-other-privacy-organizations-back-out-ntia-face-recognition-multi. [Accessed 19 October 2016].

Marwick, A.E. and boyd, d., 2011. I tweet honestly, I tweet passionately: Twitter users, context collapse, and the imagined audience. *New Media & Society*, *13*(1), pp.114–133.

Mundie, C., 2014. Privacy pragmatism; focus on data use, not data collection. *Foreign Affairs*, *93*, p.28.

Nettle, D., Nott, K. and Bateson, M., 2012. 'Cycle thieves, we are watching you': impact of a simple signage intervention against bicycle theft. *PLOS one*, *7*(12), p.e51738.

New Scientist, 2014. Twitter health. *New Scientist*, 2994.

Ronson, J., 2015. How one stupid tweet blew up Justine Sacco's life. *New York Times*. Available at: www.nytimes.com/2015/02/15/magazine/how-one-stupid-tweet-ruined-justine-saccos-life.html [Accessed 18 October 2016].

Rothrock, K., 2016. Facial recognition service becomes a weapon against Russian porn actresses. [online] *Global Voices*. Available at: https://globalvoices.org/2016/04/22/facial-recognition-service-becomes-a-weapon-against-russian-porn-actresses/?platform=hootsuite [Accessed 19 October 2016].

Solove, D.J., 2007. *The future of reputation: gossip, rumor, and privacy on the internet*. Yale University Press, p.7.

Susen, S., 2011. Critical notes on Habermas's theory of the public sphere. *Sociological Analysis*, *5*(1), pp.37–62.

Sykes, C.J., 1999. *The end of privacy: the attack on personal rights at home, at work, on-line, and in court*. Farrar, Straus, and Giroux, p.28.

Tippmann, S. and Powles, J. 2015. Google accidentally reveals data on 'right to be forgotten' requests. [online] *The Guardian*. Available at: www.theguardian.com/technology/2015/jul/14/google-accidentally-reveals-right-to-be-forgotten-requests [Accessed 19 October 2016].

Thompson, C., 2013. *Smarter than you think: how technology is changing our minds for the better*. Penguin.

Wainwright, M., 2010. Wrong kind of tweet leaves air traveller £1,000 out of pocket. [online] *The Guardian*. Available at: www.theguardian.com/uk/2010/may/10/tweeter-fined-spoof-message [Accessed 18 October 2016].

Wang, L., Tan, T., Ning, H. and Hu, W., 2003. Silhouette analysis-based gait recognition for human identification. *IEEE Transactions on Pattern Analysis and Machine Intelligence*, *25*(12), pp.1505–1518.

Weise, S., Hardy, J., Agarwal, P., Coulton, P., Friday, A. and Chiasson, M., 2012, September. Democratizing ubiquitous computing: a right for locality. *Proceedings of the 2012 ACM Conference on Ubiquitous Computing*. ACM, pp.521–530.

Vaidya, J. and Atluri. V., 2007. Privacy enhancing technologies. *In*: Acquisti, A., Gritzalis, S., Lambrinoudakis, C. and di Vimercati, S., eds. *Digital privacy: theory, technologies, and practices*. CRC Press.

Westin, A., 1967. *Privacy and freedom*. Atheneum Press.

7 Challenges of the digital public space

Bias

In Chapter 4, we describe the 'sea of information' that surrounds us. But while it is tempting to think that this information sea is made up of equal, impartial, objective facts, this is not the case. It is also not a randomly aggregated collection: the digital space, private or public, is curated. It needs to be, because there is simply so much of it – otherwise it would be useless because nobody would be able to find the information that they needed. The fact that we can now sort and reference large amounts of information is one of the major advances that has led to the advent of the information age (as coined by Castells, 2011) and is what gives digital computing systems their power. However, the way in which information is catalogued influences how we use digital public space, and what we can do with it. When choosing to display or interpret information, choices must be made. And these choices are not neutral, even with the best of intentions. For example the first port of call for most people to find something on the internet is to use Google's search engine – in fact 'to google' has entered the lexicon as the terminology for an online search. But Google's search engine, like any algorithmic delivery of information, may be subject to inherent bias.

In Chapter 2, we touched upon how perception can affect the interpretation of the world, and how what is perceived is not necessarily always the same as what is seen, or what is 'really' there in an objective version of the world. Like the gorilla on the baseball pitch, if we are not expecting to see something, we may not perceive it even if it is 'right in front of our eyes'. And while this is beneficial because it allows us to focus on things that we really need to be paying attention to, it can blind us to things that might be important. This applies not just to visual information but also abstract knowledge. Because of our expectations and worldview, we can have unconscious biases inherent in how we interpret information we are given. This can affect our problem solving and reasoning, and lead us to make assumptions based on our prior outlook. When designing for digital public space, it is important to consider how people interpret information and ways in which their assumptions might lead to bias, but also to look at how biases of the designers themselves can create unfortunate implications in the products and services that emerge. These biases might mean they are not suitable for use by everyone, or not fit for purpose.

To understand how choices made in delivering information can cause biases, and how these biases may have larger effects, we can first look at an example from an earlier technology – that of cartography.

Bias in maps

> Google Maps as 21st C. Mercator Projection: a map so engrained all other versions simply seem a bit strange.
>
> (Kelly, 2013)

In the 1750s, many maps were being created in Britain for many different purposes, each with different systems and standards. As Simon Garfield explains:

> A great many people were out at all hours… creating maps for commercial or land interests, or assessing tax liabilities ('cadastral' maps). Prominent (and accurate) county surveys were also conducted by expert plotters such as Carrington Bowles, Robert Sayer and John Cary. But each of these maps was particular to the demands of its patrons. Many of the maps were symbols of influence, coverage was patchy, and there was no agreement over what was included or ignored.
>
> (Garfield, 2012)

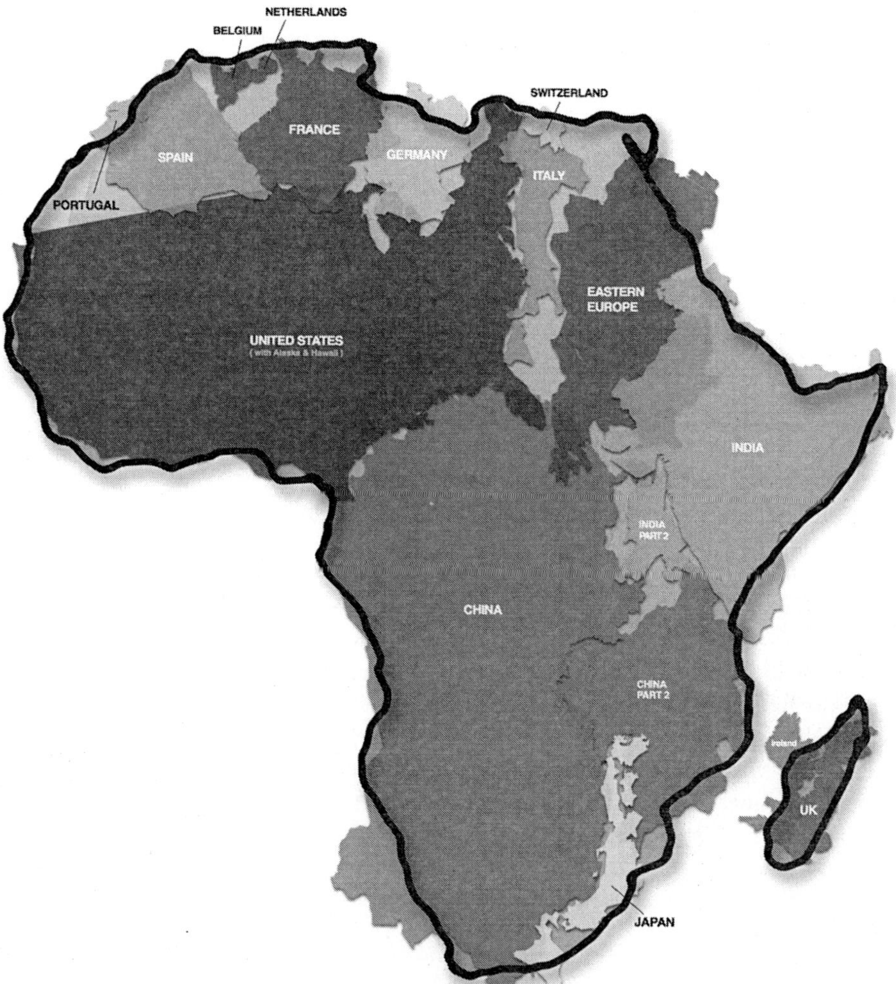

Figure 7.1 The 'Size of Africa' graphic designed by Kai Krause highlights the deceptive nature of the Mercator projection.

Source: Kai Krause

This changed with the widespread adoption of the Ordnance Survey map, which created a whole new 'language' of symbology, which influenced how people from then on thought about mapping space and the culture of place. This even had implications for how legal decisions were made: Garfield paraphrases the 1963 'Red Book' of surveyor's regulations: '*Legal Value of OS Mapping:* Indisputable. Two court judgements in 1939 and 1957 ruled that anything appearing on an OS map is prima facie evidence of its existence on the ground; if it's on the map, it's in the world' (Garfield, 2012). By putting something on the map, its existence was confirmed. However sometimes decisions are taken when compiling OS maps which, for the sake of legibility or consistency, might slightly move features such as roads and mean that the maps are no longer a strictly 'true' representation of the geographical features.

Similar decisions must be taken when creating world maps. Since the earth is three dimensional and (roughly) spherical, to represent the land masses and geographic areas on a two dimensional paper map in a way that is readable requires some work, and is never going to be a truly accurate representation of the reality. There are several ways to go about this translation into two dimensions, known as projections. Many different projections have been used over the years, and each makes compromises on the accuracy of what it displays, in terms of area, distance, shape or direction. The world map that most people are familiar with is known as the Mercator projection, and we think of it as giving a good representation of 'reality'. However in actual fact, the true relative sizes of the continents are vastly different than those displayed on the Mercator projection. For example, Africa appears to be of similar size, or smaller, than North America; despite the fact that it is in actuality much larger.[1]

Other projections may represent area more accurately, but compromise by distorting the shapes of land masses. An example of this is the Gall-Peters projection.

Figure 7.2 The Gall-Peters projection

An episode of the television series *The West Wing* ('Somebody's Going to Emergency, Somebody's Going to Jail', 2001) portrays a fictitious organisation who are campaigning to have the Peters projection taught in schools as the primary map. They also argue that the emphasis on land masses in the Northern Hemisphere in the dominant Mercator projection has affected how western society views the rest of the world, has influenced social attitudes and biased interactions. They suggest flipping the map so that North is at the bottom. While this may be an extreme caricature, it is nevertheless the case that the focus and positioning of maps might well impact how we perceive our place in the world.

Similar biases may affect our view of the world presented digitally. In the same vein as the Mercator projection (on which they are based), the maps produced by Google Maps are also giving a particular view of the world – this time one centred on the location of the user rather than any particular overview. By the very fact that so many people use Google Maps, one might assume they represent the 'true' nature of the world. But such a thing cannot really exist because every presentation will include some form of decision making, and thus bias.

Consequences of bias

If the presentation of the digital public space is also a matter of making choices, then biases may also enter this and affect the outcomes. Wikipedia is as powerful as it is because it allows updates by anyone, but certain articles on controversial subjects can be the subject of 'edit wars' where two factions each try to present their view of the subject to the exclusion of those who disagree. In such cases there may be no objective 'truth'. One argument suggests that perhaps, rather than relying on decisions made by human curators with biases, we should rely in the digital public space on the algorithmic power which facilitates its very existence. There is far too much complexity for any individual to manage: we are benefiting from 'big data' analysis but cannot undertake it without large computational power. However, although algorithmic decisions may seem on the surface to be an answer to bias because they are logical and impartial, this is a misconception. They are also susceptible to bias; because of the biases of the people who have created them.

When using Google's search engine, for example, very few people look beyond the first few pages of results displayed. The order in which pages are ranked is determined by Google's algorithm and is influenced by many factors. Which factors are given priority may affect the order of the search results, and thus the view of the world experienced by users. Some of these biases might be intentional on the part of the developers (for example returning results which are more closely related to the locality of the person searching) and some may be inherent in the assumptions and the understanding of those who created the programme. Computers, and the algorithms which run on them, have no intelligence of their own and only do what they are told. The way that people set these up is influenced by their personal politics and experience of the world. For example, a sensor network might be deployed to monitor air quality, and the person creating the network might hold the belief that air quality does not differ depending on height from the ground. If this base assumption is wrong and air quality 10 centimetres from the ground is different from 5 metres from the ground, their placement will affect the result and incomplete or inaccurate data may be gathered. It is not that individual sensors return false data, but that the overall dataset will be biased by the assumptions in the way that it was gathered.

As with the Mercator projection, sometimes biases (which the developers may be unaware that they even have) may influence the design of systems that handle and display information, and therefore the experience of users, in significant ways. This kind of bias was demonstrated in 2015 in a gym in Cambridge, which used a digital system to assign numeric pass codes to access the changing rooms (Challis, 2015). Dr Louise Selby was assigned a number but found that it did not work in the women's changing room. When she queried this, she was told that because her title was 'Dr', the system categorised her automatically as male and gave her a pass code that worked only in the men's changing room. The system did not provide a method by which this could be overridden and she could be assigned to the correct room. It can be surmised from this that the programmers of the system were biased and did not consider that someone other than a man could be a doctor, so did not build the capabilities accordingly – leading to significant issues.

Sometimes issues of bias in digital public space can have significance on a much wider scale. In Chapter 3, the way that Google Maps has changed people's experience of navigation and physical space was discussed. Inaccuracies in navigation information have caused accidents when people have trustingly followed directions that turned out to lead them into rivers or across dangerous areas. Maps may contain intentional 'mistakes' in order to improve the experience of using them, or even to protect copyright and prevent plagiarism. They can also retain errors from earlier versions, even significant ones such as 'Sandy Island' which was shown on Google Maps until 2012 when Australian researchers visited the area and found there was nothing there (Seton et al, 2013). The island was 'discovered' in 1876 and added to charts in 1908. Although it was not found by subsequent expeditions and removed from maps in the 1970s, it was erroneously included in transitions from physical to digital maps and therefore included when Google compiled theirs.

Disputes over the content of maps have also caused political problems. When Microsoft released world maps with their Windows 95 operating system, they made a decision on how to colour the Kashmir Province, using a slightly different shade of green for the 8 pixels representing the region, out of 800,000 for India. This region is disputed territory, and the difference in colour was taken to show Kashmir as non-Indian, leading to the banning of the product in India. Microsoft was forced to recall all 200,000 copies of Windows 95 at a cost of millions of dollars (Best, 2004).

This was a decision with political ramifications which, again, the developers were likely unaware of. But sometimes there are political or commercial motivations of such decisions, and bias is introduced intentionally. A controversy arose in 2014 between Amazon and the publishers Hachette, who clashed over the pricing of eBooks. While the discussion was ongoing, Amazon introduced what might seem minor restrictions to Hachette books: increasing the time taken to fulfil the orders (Streitfeld, 2014) and removing the option to place pre-orders for some yet to be published Hachette books (Abbruzzese, 2104). Amazon, according to *USA Today*, accounted in 2014 for an estimated 40% of all new books sold and 65% of e-books (Weise, 2014). It is the first point of call for many customers. If customers are browsing books, the lack of discounts or availability for certain titles can significantly harm sales, because customers may simply choose another title rather than wait longer or pay more. They may not have been aware that the delays were biases introduced by Amazon rather than a restriction on the availability of the title from all suppliers, and therefore did not look for an alternative vendor. This negatively impacted the authors affected and caused uproar in the literary community. Pre-orders can account for as much as 25% of the total sales for a title so this restriction was a strong negotiating tactic for the dispute that was eventually settled in November 2014.

This example illustrates the influential power that biases can hold when using digital public space, via inherent preferences based upon ease of use and behavioural factors. These can then in turn have real-world implications that affect people's lives. Because large monolithic organisations are beginning to have control over large parts of the online world, some people are worried about the implications of private spaces online that appear to be public and are treated as such by users, but which have underlying biases created by their owners. In many cases the users may be unaware of the fact these spaces are not impartial.

In Chapter 3, the potential of the internet of things was discussed, and the fact that the inclusion of simple artificial intelligence algorithms (like changing the temperature in your house) could allow objects and environments to make choices without instruction and thus unload decision making. This agency of objects is beneficial, but users must be very careful that they fully understand the rules that govern the choices made, and that they do not include undesired biases of the designer. Otherwise these objects may start to make choices for us that we are not even aware of. An example of this might be a music player that delivers a tailored playlist based on songs that you have previously highly rated. But if, as in the Hachette example above, a particular label or artist was excluded from the algorithm, you might never discover their music even if it was precisely the kind of thing you like; they will not be included in the search results. In this way, the music player has decided for you in a manner that is biased and negatively affects you – and you would not even be aware that it was happening. It might even be an unconscious bias, for example if the player was not programmed to search for music outside of a particular country or countries.

One other aspect of search algorithms which deliver specific results is that they do not allow for the serendipitous discovery of content which might come from general browsing. Many people will recognise the experience of wandering through a book shop and having one's eyes drawn to a particularly interesting cover, or simply walking along the shelves reading random back cover blurbs until finding one that sounds interesting. While the selection of books on offer is as equally biased and non-random as online offerings (because the stock has been curated, chosen and placed by the bookseller), the physical space means that a wider and broader selection can be visible with opportunity for exploration and unexpected discovery. Online shops can attempt to replicate this by offering products on the main page or otherwise outside of the search parameters, but bias may mean that results returned are not a 'random' selection of things available. Often, this will be by necessity a much smaller range of options than the total available, and highly curated to maximise the chances of being appealing; thus losing opportunities for things that might branch out into a previously unforeseen area of interest, making connections that could not have been predicted. This lack of serendipitous discovery in the online space applies to areas other than retail, because of the way that digital information must be delivered directly rather than encountered along journeys to a destination.

Some designers are exploring ways that serendipity can be designed 'back in' to digital public space. This is a difficult challenge, for the reasons outlined above, but is a worthy goal which would provide a more organic experience of the digital public space as a place to explore rather than a predefined serving up of information. This is particularly important given that said information may be biased and lacking the depth that comes from unexpected juxtapositions of content, and that the discovery of 'unrelated' information can turn out to be apt.

Case study 7.1: Past paths

Serendipitous discovery is part of the joy of exploring a museum or art gallery: beautiful, fascinating, awe-inspiring or provocative objects or images, presented as a journey which you can travel through in many different ways. But while giving your museum presence in the digital public space can have huge benefits and increase the number of 'visitors', it can risk making this journey more like a conveyor belt: or alternatively a drive in a blacked-out car where you only get to see a tiny portion of the view.

Tyne and Wear Archives and Museums (TWAM) manage a collection of nine museums and galleries across Tyneside, as well as the Archives for Tyne and Wear. As a major regional museum, art gallery and archives service, they hold and manage large collections with a variety of artifacts in many media; in archives, art, science and technology, archaeology, military and social history, fashion and natural sciences. They have been exploring how to address this question in association with the Creative Exchange project and Microsoft Research. In order to reach a wider audience and connect with the public, their collection is being gradually digitised and being given some form of online representation; digital photographs of objects, digital copies of archive material, and information about the items. In the case of large holdings such as these, it is often challenging for museums and galleries to present these materials online in a way that allows 'visitors' to do the equivalent of walking around a gallery or museum and exploring the exhibits. An intrinsic part of visiting a good museum is, as described above, coming across surprising items that you did not know existed; so standard search tools are not useful for this purpose. And were the items to be presented in a linear fashion, it would take huge amounts of time to go through and find items that piqued particular interest.

Andy Garbett, Melina Boelmann and Peter Wright from Open Lab at Newcastle University worked with Microsoft Research to develop a new interface to TWAM's archive, moving away from traditional models of archive search to deliver a 'visitor' experience centred on the metaphor of a digital journey through the collection. This involves serendipitous encounters with families of objects that would not normally be discovered by traditional search-based interfaces.

The 'discovery engine' website[2] invites users to dive into the collection and enables the experience of serendipity while browsing through artifacts. If the user scrolls slowly, the interface assumes they are interested in what they are seeing, and shows related items, using sophisticated intelligence systems to find things that match those which have caught your attention and suit your particular interest. However if a user scrolls past much more quickly, it begins to present random new topics, in the hope of chancing across something more to your taste. In this way, the broadness of the archive can be explored but areas of interest can be investigated more deeply. In addition, the system offers a map view which presents all items that are stored and their relations to each other. This feature lets users find extra related content to things they are particularly interested in, and can act as a starting point for a new search.

Survey results collected from among the 13,000 visitors since the system has been online suggest that almost 70% of users discovered something they were surprised by and would not have expected during their visit. Visitors also spent almost four times longer (3.8 minutes on average) exploring items compared to the traditional search interface.

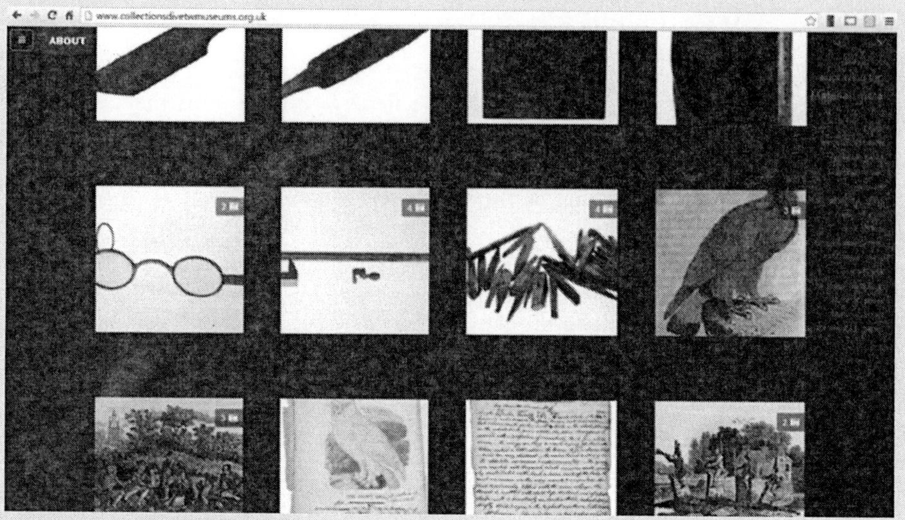

Figure 7.3 The Discovery Engine website at TWAM presents objects in the collection based on apparent interest. www.collectionsdivetwmuseums.org.uk/
Source: The Creative Exchange

This focus on the journey rather than the individual objects, and providing means by which random chance can lead to new encounters, is a first step back towards the experience one might get exploring a collection exhibited in a physical space. However, the fact that it is in a digital format extends the possibilities for browsing, and means that you might have the opportunity to see more things than could be viewed in a curated collection with more limited space.

Is truth possible in digital public space?

> Falsehood flies, and the Truth comes limping after it.
>
> (Swift, 1710)

Sometimes bias can exist in the corpus of information available rather than in the way it is sorted. This is especially the case with regard to reference information, which may include 'facts' which are not actually true. If false information gets spread disproportionately to the truth, because it is available first or is simply more interesting, it can be very difficult to correct later. This is in the nature of rumour and gossip, but as with many other aspects of human communication, digital technology is increasing the speed and spread of such things. In his book *Virtual Unreality*, Charles Seife (2014) talks of many instances of the ways in which the internet can facilitate the spread of untrue information with major ramifications. For example, large stock losses were seen when a false 'leaked internal memo' from Apple indicated the delayed release of the new iPhone, leading to a major financial impact on people who profited or lost from the market fluctuations caused. The website Snopes.com, founded in 1995 by David Mikkelson to provide information on hoaxes and urban legends, describes itself not inaccurately as 'as one of the World Wide Web's essential resources'.[3] The site undertakes extensive fact checking and research into online rumour and disinformation, and that it is so popular indicates how much content spread online is not immediately verifiable.

Even with such resources it is not always easy to find the 'truth', especially when what is true might in itself be subjective and shifting. In Chapter 4 we quoted Stu Card who marvelled at his new ability to check whether 'food chain' or 'foodchain' was correct, by looking up online how many people on earth used each of them. But such a search finds millions of uses of each. Is one usage 'correct' just because it is more frequently used, or might it just be an exceedingly common misuse? If both are used, can you tell which one is right, or in fact if either of them are?

Information associated with content can also be lost or distorted in being distributed. For example on Twitter, it is often the case that popular or particularly amusing tweets are 'stolen' by being copied and posted by another user rather than 'retweeted' with credit (which can be done either using the inbuilt system or quoting the name of the originator). In this way, followers of the person stealing the tweet are unaware that it is not original content and may retweet it to their followers. These people may in turn distribute with incorrect attribution to the thief rather than the person who originated it. Once this happens to a certain degree it can be almost impossible to track the originator; such tweets can have thousands of retweets. This type of loss of attribution also occurs on sites such as Tumblr where pictures and text are shared, often as images with no link to the original. Sometimes this content can be used in a context where it was never intended, with satirical jokes being spread widely as facts. An example of this is a well-known set of images of the Houses of Parliament, the first with mostly empty benches labelled as a vote on the living wage, the second packed with members and labelled as a vote on MP wages.

Figure 7.4 MPs debate what?

Unattributed image spread widely on social media.

The problem with this striking comparison, shared thousands of times, is that it is false. As deftly deconstructed by Isabel Hardman (2014), the second image is not related to said vote at all. This is a potential issue not only because of people basing action or opinion on things that are false, but also because the overall trustworthiness of online information declines. As Farhad Manjoo explains, it's not the proliferation of fake photos that's the issue; it's that 'true photos will be ignored as phonies' (Manjoo, 2008). If you pollute the currency too much it ceases to mean anything. This is the true danger in 'fake news' and its proliferation as a concept, which has grown in prominence in 2017 such that the phrase was assigned 'word of the year' by Collins Dictionary.

With that in mind, are there ways in which information can be evaluated and verified? Persistence of information in digital public space means that incorrect information can still be delivered even when it has been discredited. The 'right to be forgotten' ruling, as discussed in Chapter 6, means that individuals have the right to ask search engines to remove information about them from search results 'if the information is inaccurate, inadequate, irrelevant or excessive'. Does this count as creating bias in the system, or simply correcting for bias that already exists? If something 'wrong' is put online, and it spreads, it is very difficult to correct that information. Of course this is also true in the non-digital sphere, as anyone knows who has tried to quash a rumour spreading through a community of people. But it can be particularly frustrating if one is using the digital public space to try and find the objective truth.

Google is among several organisations using their own methods to try and support the delivery of accurate facts and information. Its 'Knowledge Vault' uses an algorithm to automatically pull in information from across the web, and create a knowledge base, which in 2014 included over 1.6 billion facts. Of these, 271 million were 'rated as "confident facts", to which Google's model ascribes a more than 90 per cent chance of being true' (Hodson, 2014). This fact checking is done by cross-referencing new facts with what it already knows. But these kinds of fact collecting systems are not infallible. A case in point might be the controversial answer returned by Google's systems for a short while to the question 'What happened to the dinosaurs?', which in its Google search card (short statements designed to give simple answers to questions without needing to follow links), gave a creationist response (Titcomb, 2015).

Google are also attempting to implement systems by which more 'truthful' pages are higher ranked in searches. The system uses the Knowledge Vault as a reference and similarly evaluates truth based on the unanimity of information on the web (Hodson, 2015). However there is a circular logic to this process. 'Facts' that are considered to be true will spread, and be reinforced in other places. There are plenty of things that 'everyone knows' which are actually false when looked into a little more closely. Sites like Snopes are useful in addressing these but still may be swamped by those perpetuating the falsehood.

Even if information is currently widely held to be factual and true at a certain point in time, this might not remain the case. If Google's measure of truth is built and based upon consensus, what happens if the consensus changes? 'Facts' which are considered to be true today may be superseded by new knowledge or scientific findings. Even if the web almost unanimously agrees that something is the case, it may be that the very small discordant information is correct, especially if it is new. Many facts widely held to be truths might be re-evaluated when new theories and evidence came to light, such as fixed continents (superseded by the theory of continental drift), the Divine Right of Kings (that monarchs obtain their rule directly from God and are not subject to any earthly law), or that the sun travels around the earth.

Biases of missing information

In Chapter 4, we explored the notion of digital storage acting as an augmented memory, whereby we can store records and knowledge and retrieve them at will, with a much larger capacity and higher fidelity than our brains can accomplish. But to use digital archives effectively for this purpose requires retrieval algorithms which can sort through our digital 'memories' and find what is appropriate. If these algorithms have inherent biases, we may find that our digital memories are not as trustworthy as we imagine (as is also the case with our biological memories).

Another possibility is that we only 'remember' things which have been digitised, and lower priority is given to things which are not retrievable in this digital context. The nature of 'facts' which we 'know' (if the digital public space is acting as a shared cognitive augmentation and knowledge base) may be affected not only in terms of information being correct or incorrect, but in the very sampling of what is uploaded in the first place. As discussed in Chapter 1, there are some things that cannot necessarily be adequately captured and retained, such as the 'liveness' experience of watching a theatrical performance in the company of an audience. Until we have ubiquitous information capture, which as examined in Chapter 6 has its own negative implications due to privacy issues, there will always be a subsection of life that is not included in the digital public space.

Even in areas where it appears that there is comprehensive coverage, there still may be many things that are not represented in digital public space. It is tempting to think of resources such as Wikipedia as being comprehensive and containing information on 'everything'; at least in terms of acknowledging items or areas of knowledge which exist, and noting where available information is limited. However this is not always an accurate representation, and presuming that it is can lead to skewed assumptions and biases. Areas of particular niche interest might have much more detail than others which do not have contributors with time, access or will to complete them to the same standards, and it may not always be obvious either to readers or contributors where gaps are being left unfilled. Wikipedia, because of the nature of how it is constructed, is particularly vulnerable to systemic bias. For example, in 2011 less than 15% of English Wikipedia's contributors were women, and this may affect the types or viewpoints of information included (Cohen, 2011). Cultural bias is also a pervasive feature:

> A lack of articles on particular topics is the most common cultural bias. Separately, both China and India have populations greater than all native English speakers combined, or greater than all of Europe combined; by this measure, information on Chinese and Indian topics should, at least, equal Anglophone or European topics. However, Anglophone topics dominate the content of Wikipedia. While the conscious efforts of WikiProject participants have vastly expanded the available information on topics such as the Second Congo War, coverage of comparable Western wars remains much more detailed.
>
> (Wikipedia, 2016)

Another example of how bias can be self-reinforcing relates not to the existence of information online but to its ease of access. Many academics undertaking research (including the authors of this book) now use Google Scholar to find relevant articles and access electronic copies where available. If broad sweeping literature searches are being undertaken, with the initial value of the references unclear, it can be very tempting to give preference to those papers and books which are available in full online, via institutional accounts or open access. Unavailable articles which appear useful based on the title or abstract may be made note of for later access at a library, but the chance of them being read is reduced. And this is even more the case for items which do not appear in searches at all, or are placed low in search results. Is this therefore creating a self-selecting research pool, where popular articles are cited more and thus become even more popular? This might be especially damaging if these placements in searches are unrelated to the quality or relevance of the items in question, for example because they are published in a different format or excluded by the publishers. The value of being searchable and readable is perhaps related to the drive to make, wherever possible, research content open access; the uptake of this being sped up by the requirements of funding bodies such as the UK Research Councils for all outputs of grants they have funded to be made available free of charge.

The digitisation process can also be affected by cultural and social factors. If digital public space contains content that is uploaded in spaces such as Facebook for public sharing, treating this as a public 'memory' that forms a comprehensive resource risks bias because this is likely an unrepresentative record. A life recorded on Facebook appears to be comprised of many momentous moments, positive and negative, without necessarily showing the banality of day to day living that people also experience in their lives. This is in part due to the nature of the things that people wish to share – extremes of happy or sad, things which you think will be interesting to your friends or are representative of the person you would like to be perceived as. There may be gaps in the record either because you did not think something was worth uploading (or did not want it to be part of your public record) or because records have been removed afterwards, again harking back to the 'right to be forgotten' ruling. Although our biological memories are not complete, and might also be biased because of our own preferences and reactions, this kind of editing of a collective memory might have other implications.

But there are also factors at work because social networks such as Facebook are not a 'public' space in the true sense, but a commercial operation with choices made in curation. One of the things that distinguishes Facebook is that, while users sign up to receive 'news feeds' from their friends, the news feed does not contain every update but a selected stream which has been curated by Facebook's algorithms. This means that there is no way to ensure that a particularly momentous update has been seen by others, even if they theoretically subscribe to your content. Conversely, it is very difficult to tell if you have received all the relevant news from a friend or the wider world rather than what Facebook decides is 'relevant' for you. Interaction conducted across owned social media systems is mediated, and therefore users must place a burden of trust that the mediation is being done for their benefit. Eli Pariser (2011) coined the phrase 'filter bubble' to describe this mediated world that is presented uniquely to each individual person, which might be chosen for reasons that are not necessarily in our best interests.

We must therefore consider that our corpus of 'known' information may not be objective, and that we must be cautious if our 'memories' are to be placed in digital spaces that are not curated solely by us.

Social bias

The 'filter bubble' phenomenon mentioned above describes the situation whereby people are presented with opinions and feeds that are similar to those they already hold, which therefore reinforce existing attitudes rather than providing an unbiased cross section of news and opinions. Algorithmic sorting can be blamed in part for this, but bias towards similarity is also something that naturally occurs in what we choose to read and who we choose to associate with. This is known as homophily; the tendency to choose things that are similar to yourself.

It is this tendency towards homophily which probably contributes to the increasing popularity of such algorithmic models. However what people prefer to or choose to look at, and what would contribute to their general wellbeing and knowledge of the world, are not necessarily the same thing.

As a consequence of this, networks which allow you to create your own selection of content, such as Twitter, do not escape the issues of bias. In this case however biases are the consequence of self-curation rather than algorithmic control. This is often encountered in terms of personal politics, where we are, again, most comfortable around those who are similar to us. By choosing to be surrounded by people who express views similar to our own, beliefs can be reinforced. When all you are presented with reflects a single viewpoint, it can create an 'echo chamber' of your own beliefs. Some of this is indeed due to self-selection but can also be reinforced if it is easier to encounter similar views, or if dissenting content is actively suppressed, as appears to be the case with Facebook's algorithm (Bakshy et al, 2015). It might also be the case that we mistakenly believe that the views expressed by those around us are common and shared by the majority of people, when in fact they are a unique consequence of the filter bubble in which you find yourself. Lerman et al (2016) have written about the 'majority illusion' which arises out of the structure of online social networks. As a consequence of some nodes of the network having many links and thus being of greater visible influence (in other words, highly connected people), certain beliefs or behaviours might appear to be more common among your peers than they actually are. It can appear that something is popular universally when in fact it is a relatively uncommon occurrence, because all your sources are linked. They found that this effect was particularly large in political blogs.

If this filtered experience is extended out into our wider activity within digital space, or even physical space, there could be repercussions. For example, in 2013 Google launched personal Google Maps which highlight 'the things that matter most' (Seefeld & Chawathe, 2013). These may, as Google suggests, provide a smoother, more customised experience; but they run the risk of reducing the serendipity of coming across unknown places, and may be subject to some of the biases described above. By reinforcing existing preferences, or preferences of those around you, such services create a distorted picture of the world. This idea of maps that direct you to places popular with those like you, runs the risk of resembling the reinforcing trails of ants. When ants explore an area outside of their nests, and find a food source, they return leaving a pheromone trail for other ants to follow back to it. As other ants follow the path and also return, they reinforce the trail and make it stronger. The closer the source is, the quicker this reinforcement happens, and thus the pathways to the closest food become clearer. This is an excellent strategy for the ants, as it means that they utilise the surrounding food sources effectively. But we are not ants, and may wish to judge which content we receive based on other criteria than popularity and speed.

In this way we could see self-fulfilling popularity loops happen in digital space; with more popular items being more easily located, becoming highlighted to more people and thus becoming even more popular: a virtuous circle of digital reinforcement. But this reinforcement does not necessarily include filtering for quality; and could marginalise excellent or relevant content that does not enter the loop at the start. One of the most remarkable qualities of the digital public space is its democratising effect; that contributions from all sides can be included. 'The current diversification of communication channels is politically important because it expands the range of voices that can be heard: though some voices command greater prominence than others, no one voice speaks with unquestioned authority' (Jenkins, 2006). To take full advantage of this opportunity, we must ensure that this is not compromised by giving undue prominence to those in privileged positions either through action or inaction.

This is also worth considering in terms of who populates different parts of the digital public space. In Chapter 6, when discussing privacy, we discussed the fact that there can be 'public' spaces which still have restricted access because members are welcomed only from specific demographic, cultural, or countercultural groups. This could be active gate-keeping, access barriers, or simply lack of appeal of the space to those outside the group. While an unrepresentative membership is part of the point of a forum for a particular special interest group or demographic, it can also be the case that such biases in membership of spaces occur more subtly, and without express intent. An example of this can be drawn from the very beginnings of digital social spaces, and the notion of 'Eternal September'. In the early days of the internet (and, prior to that, the ARPANET), one of the main platforms for public discussion was Usenet, a collection of forums each on individual topics to which threaded messages could be posted. Participation was restricted only in that access was required to a connected computer; but initially, this was common only to those on university campuses or who worked in technical roles at cutting edge technology companies.

Within the population of Usenet, behavioural and cultural practices arose, with accepted social behaviour and 'politeness' in terms of content and structure of what was posted. These rules often varied between groups, but were generally understood by most users in each space, with enforcement maintained by practices such as social exclusion of those who breached agreed etiquette. These agreed behavioural mores led to the development of such ideas as 'netiquette' and FAQs which would later spread to the larger internet. But each September, at the start of the academic year, there would be an influx of new members who had arrived at university and found themselves with an internet connection for the first time. It would take some time for these 'newbies' to learn the rules and either leave once the novelty wore off, or become valued members of the community. A certain level of initial leeway was generally provided to this cohort during their first month or two.

In 1993, internet service provider America Online (AOL) added Usenet access as part of its service to subscribers. There was subsequently a large influx of people to these forums who had not previously had access, which was a continuous process rather than a short-lived wave that died down after a short while. September 1993 therefore became known as the 'eternal (or endless) September', to describe the constant arrival of new uninformed users who did not know the 'rules'. Many have suggested that this fundamentally changed the nature and utility of Usenet, leading to an irrevocable downgrading of the quality of content and how it felt as a community space. Judith Donlath writes of this situation:

> It became clear that the concept of Usenet as an open public community had worked only because in the beginning, 'public' had not been very public. Admittance had required having already passed the admissions barriers to a high-end university or elite technical job. It had meant being a named person responsible for and protective of one's own reputation. Though Usenet had appeared to be an open public space, it had in fact been a protected garden set behind the walls of network access in the 1980s.
>
> (Donlath, 2014)

Usenet still exists, but has been superseded by other forms of online discussion and public discourse and is generally agreed to have been in decline for many years, not least because of the arrival of spammers whose content often drowns out that posted by users. It is however worth considering Donlath's point that the initial make-up of contributors was biased, and opening up access to a broader population of users may have been a contributing factor. There may be similar biases in other platforms, for example perceptions of typical user profile leading to different demographic groups choosing different social networks: Snapchat being more popular for young people (at the time of writing) than Facebook which is seen as more for adults.

Bias as a tool

It is worth keeping in mind that bias of digital information can be a powerful tool if used purposefully. Items can be hidden or information can be presented in such a way that tells a particular story or gives a particular message. If this is done at the level of government, it might enable censorship or propaganda. Discrimination can be applied based on algorithmic assumptions, even without your knowledge. The example earlier in this chapter of Dr Louise Selby's experience with gym changing rooms does not appear to have been a purposeful exclusion. But choices in digital information presentation can be made with an agenda. Digitally connected systems can give powerful control over the lives of others, and algorithmic control of this can lead to unfair discrimination based on models which make assumptions based on, for example, ethnicity. People's opportunities and experiences can be affected without their knowledge in various situations, and perhaps without consent. With large data sets, information can be gathered about a group and then used to influence behaviour, as in the case of 'nudges' which might be used by governments to move people towards particular desired behaviours based on psychological prompting (Hodson, 2013). Deciding whether 'nudges' are appropriate or even ethical is an important question in each circumstance, and putting such systems in place could open doors to their use by the less scrupulous; oppressive governments, terrorist organisations, or hackers.

Dodge et al (2009) give an example of how information collected about you could result in a biased response: 'when you arrive at the airport to board the flight you have no idea what risk and threat calculations have been applied to you since you booked your ticket'. If 'randomised' stop and search algorithms are designed to pull people out of a crowd based on specific profiling, this may negatively affect innocent people who fulfil some of the criteria, which may even be based on flawed initial assumptions. It could be argued that these false positives are worthwhile in the name of security, but there are some situations where bias is used with purposeful aims to deceive, for example as a marketing tool to present specific information, as in the example above of differing prices being offered based on previous search history and visits to competitors. There may be also more general infrastructural trends which increase inequality and marginalisation of some groups; not necessarily with defined purpose but in the name of targeting more lucrative or privileged groups for attention. This might for example include allocating premium road space on software-sorted public roads to those who can afford to pay for it, or internet service providers prioritising 'premium users' with better service, by actively rerouting bandwidth from less valued users and providing them a poorer service without them even knowing it (Graham, 2002). Extensive debate has taken place worldwide over whether laws should be put in place to enforce 'net neutrality', the principle that internet service providers are forbidden from blocking or giving privileged access to particular online material, for example providing 'fast lane' rapid access for those who pay a higher subscription.

Case study 7.2: Open planning

The Open Planning project, which was carried out as part of the Creative Exchange, had the aim of increasing public engagement with the planning process, when new buildings and developments are proposed in urban spaces. Current UK planning procedures involve a long and detailed sequence of events, much of which takes place with physical documents being copied, edited and signed. Part of this process requires notices of developments to be placed in the area concerned, as physical paper signs on lampposts or in other prominent places. These signs are very commonly seen, and most people perhaps do not give them a second glance or take the time to read them, unless they are particularly invested in the area concerned. Professor Richard Koeck, an academic working on the project who is not native to the UK, commented that when he first arrived in the country he did not realise these paper signs were official notices but put them in the same category as signs for a lost pet.

The project, whose team also included design company Red Ninja, and Joel Porter and Lara Salinas from the Creative Exchange, proposed two different strategies to increase engagement with the public. The first was a digital app to allow people to easily find out information about developments in a particular physical space. The second was a redesign of the sign itself to more clearly present information and guide readers to further information available online, unlike current signs which require reading dense text to find a specific number to input to a fairly opaque online system. This latter strategy on the surface seems fairly straightforward; present clear details of proposals, perhaps utilising images (drawings or other projections of the proposed changes) and highlighting the most relevant information such as dates for comments and objections to be lodged. But complex issues of bias affect the design of this signage. Firstly, there is the perception bias issue. While those not familiar with the system, such as Professor Koeck, may not pay attention to the sign because they do not perceive it as being relevant to them, those who know what the signs are will immediately recognise the familiar design. Planning notices in the old style are ubiquitous around the UK, and most people will know what they are without stopping to read the details. A redesign of the familiar layout may lead to the public assuming that these are unofficial notices rather than a new design of the formal notification

The second bias issue relates to the content of the signs. While diagrams, drawings and mock ups exist for many proposed building projects, they may vary in content depending on who has provided them. The computer generated image of a new building which has been created by the planners and developers, may look very different to content prepared by local residents who will note aspects such as changes to the view or social make-up of the area. Imagery is powerful, and will affect reactions to the notice. Making decisions about how to display the information is critical, and could be used to influence public opinion. Even if the notices are generated algorithmically from data stored on the council's systems, will the council be held responsible for which of the images are chosen, and held to be in favour of the development if it presents a positive view? The presentation is subjective and therefore can reflect a bias.

With regards to development of the app, implementation of it on a wide scale was limited by the open availability of council data. But this is not the only potential issue. While availability of wider information in the digital space is a worthy goal, there is no denying that the paper notices are effective in that they exist in the space which is directly relevant to the information they are conveying. We are good at noticing information presented in physical space, and at the moment information in digital space has to be presented much more forcefully, or sought out directly, in order to be fully conveyed. It is possible that the extra barrier to get to information online means that it would be less frequently conveyed to those who need to use it, especially if those people are not comfortable as inhabitants of the digital public space.

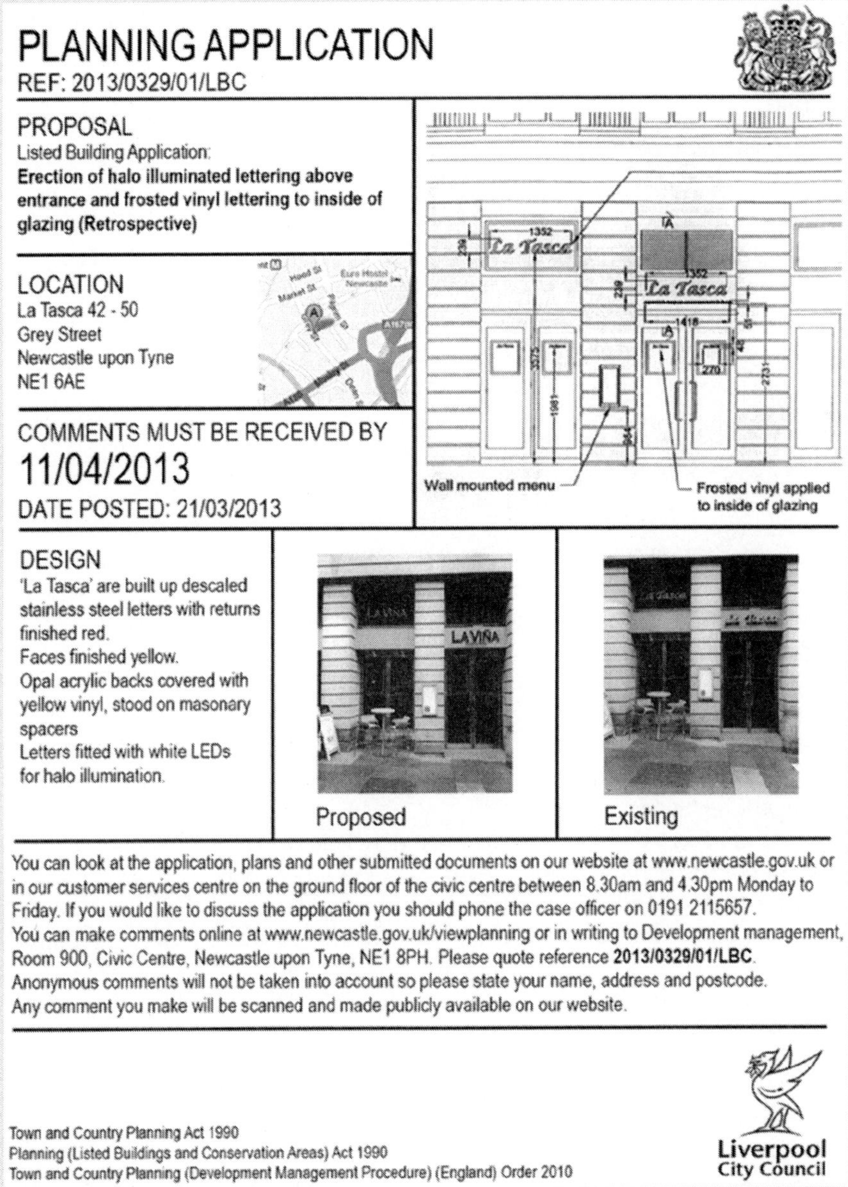

Figure 7.5 Open Planning signage
Source: The Creative Exchange

With this focus on ownership, privacy and bias, we have demonstrated that if thought is not given to the social implications of design decisions in the digital public space, the consequences can be far reaching. When designing new communications technology, and platforms and software which allow us to amend, in effect, our cognitive machinery, we must be aware of these challenges and design sustainably and compassionately to provide solutions. For example, we should be encouraging diversity in collectively compiled information banks in order to provide a more balanced, comprehensive information pool; or developing platforms which allow privacy to be retained, and give the power to construct social personas to the people using it in order to facilitate existence in the digital public space that does not compromise safety for those who may need to conceal their 'real names'.

The final chapters of this book will explore some ways by which design methodologies can approach these problems and suggest some approaches for further development of the digital public space.

Key points

- To allow anything to be found among the large volume of content that makes up the digital public space, it must be curated. A key component of the power of the digital space is this sorting, referencing and cataloguing.
- Perception biases and expectations mean we may not always see what is in front of us. This must be kept in mind when designing for digital public space.
- Maps present a particular view of the world, and no map will ever be completely accurate, because translation to a useable form requires compromise. But making choices can introduce bias.
- Algorithmic curation is not free from bias, because of the choices made in creating algorithms. Such biases can have real-world implications.
- Biases can be introduced intentionally with political or commercial motivations.
- Algorithmic presentation or curation of information can lead to an absence of serendipitous discovery.
- Untrue information can spread easily online, or attribution and context may be lost in transmission. It can be difficult to correct such information once it spreads.
- Systematic biases may occur through missing information, if we assume something is a complete record when it is not, especially if there are reasons particular subsets of information are not included.
- Greater ease of access to more popular work may lead to reinforcement of biases.
- Selection of content can also lead to representation biases, whether a Facebook feed of positive memories, or 'filter bubbles' in the politics of social networks.
- Subtle biases can occur when groups are exclusionary due to their access requirements, which might not be immediately visible to members.
- Bias can be used as a political tool for propaganda or influencing behaviour. The social implications of design of digital technology must be carefully considered.

Notes

1 The 'True Size Map' (http://thetruesize.com) allows you to move countries into different parts of the globe to compare their actual sizes.
2 www.collectionsdivetwmuseums.org.uk/
3 www.snopes.com/info/aboutus.asp

References

Abbruzzese, J., 2014. Amazon tensions with book publisher boil over. [online] *Mashable.* Available at: http://mashable.com/2014/05/23/amazon-blocks-hachette/#9B4f.0KREkq3 [Accessed 20 October 2016].

Bakshy, E., Messing, S. and Adamic, L.A., 2015. Exposure to ideologically diverse news and opinion on Facebook. *Science, 348*(6239), pp.1130–1132.

Best, J., 2004. How eight pixels cost Microsoft millions. [online] *CNET.* Available at: www.cnet.com/uk/news/how-eight-pixels-cost-microsoft-millions/ [Accessed 20 October 2016].

Castells, M., 2011. *The rise of the network society: The information age: economy, society, and culture.* Vol. 1. John Wiley & Sons.

Challis, C., 2015. Computer says no! Gym's sexist system locks female doctor out of women's changing room. [online] *BT.* Available at: http://home.bt.com/lifestyle/computer-says-no-gyms-sexist-system-locks-female-doctor-out-of-womens-changing-room-11363969434794 [Accessed 20 October 2016].

Cohen, N., 2011. Define gender gap? Look up Wikipedia's contributor list. *The New York Times, January, 30*(362), pp.1050–56.

Dodge, M., Kitchin, R. and Zook, M., 2009. How does software make space? Exploring some geographical dimensions of pervasive computing and software studies. *Environment and Planning A, 41*(6), pp.1283–1293.

Donath, J., 2014. *The social machine: designs for living online*. MIT Press.

Garfield, S., 2012. *On the map: why the world looks the way it does.* Profile Books.

Graham, S., 2002. Bridging urban digital divides? Urban polarisation and information and communications technologies (ICTs). *Urban Studies, 39*(1), pp.33–56.

Hardman, I., 2014. The menace of memes: how pictures can paint a thousand lies. [online] *The Spectator.* Available at: http://blogs.spectator.co.uk/2014/11/the-menace-of-memes-how-pictures-can-paint-a-thousand-lies/ [Accessed: 20 October 2016].

Hodson, H., 2013. Smartphone data could be used to influence what you do. *New Scientist, 220*(2942), pp.19–20.

Hodson, H., 2014. Welcome to the oracle. *New Scientist, 223*(2983), pp.18–19.

Hodson, H., 2015. Nothing but the truth. *New Scientist, 225*(3010), p.24.

Jenkins, H., 2006. *Convergence culture: where old and new media collide.* NYU Press, p.208.

Kelly, A.M., 2013. *Google Maps as 21st C. Mercator projection: a map so engrained all other versions simply seem a bit strange.* [Twitter]. 12 December. Available at: https://twitter.com/a_m_kelly/status/411137681364512770 [Accessed: 20 October 2016].

Lerman, K., Yan, X. and Wu, X.Z., 2016. The 'majority illusion' in social networks. *PLOS one, 11*(2), p.e0147617.

Manjoo, F., 2008. *True enough: Learning to live in a post-fact society.* John Wiley & Sons.

Pariser, E., 2011. *The filter bubble: what the internet is hiding from you.* Penguin UK.

Seefeld, B. and Chawathe, Y., 2013. Meet the new Google Maps: a map for every person and place [online] *Google Maps.* Available at: https://maps.googleblog.com/2013/05/meet-new-google-maps-map-for-every.html [Accessed: 21 October 2016].

Seife, C., 2014. *Virtual unreality: just because the internet told you, how do you know it's true?* Penguin.

Seton, M., Williams, S., Zahirovic, S. and Micklethwaite, S., 2013. Obituary: Sandy Island (1876–2012). *Eos, Transactions American Geophysical Union, 94*(15), pp.141–142.

Somebody's Going to Emergency, Somebody's Going to Jail, 2001. *The West Wing.* Season 2, episode 16. NBC.

Streitfeld, D., 2014. Hachette says Amazon is delaying delivery of some books. *The New York Times,* May 9, 2014, B4. Available at: www.nytimes.com/2014/05/09/technology/hachette-says-amazon-is-delaying-delivery-of-some-books.html?_r=0 [Accessed: 20 October 2016].

Swift, J., 1710. The art of political lying. *The Examiner, 14.*

Titcomb, J., 2015. Asking Google 'What happened to the dinosaurs?' leads to a controversial answer. [online] *The Telegraph.* Available at: www.telegraph.co.uk/technology/google/11681955/

Asking-Google-What-happened-to-the-dinosaurs-leads-to-an-controversial-answer.html [Accessed: 20 October 2016].

Weise, E., 2014. Amazon-Hachette: the war of the button. [online] *USA Today.* Available at: www.usatoday.com/story/tech/2014/11/13/amazon-hachette-preorder-publishing/18995643/ [Accessed: 20 October 2016].

Wikipedia, 2016. Wikipedia: systemic bias. [online] *Wikipedia*. Available at: https://en.wikipedia.org/wiki/Wikipedia:Systemic_bias [Accessed: 20 October 2016].

Section 4

How do we design digital futures?

8 Futures of digital public space

In previous chapters, we have explored multiple dimensions of digital public spaces. We have considered what constitutes the archival nature of the digital public space and what and who might exist within it; discussed the importance of physicality, sociality, communication and narrative in the evolution of culture and intelligence; looked at how physical experiences and expectations mesh with the development of digital interaction technology, and examined the 'information space' by way of analogies to higher dimensions. We have also looked in detail at a set of challenges raised by new digital media for communication and interaction.

Many of these discussions have involved describing cutting-edge technology that is newly introduced to the digital public space, or is in development and may be impactful when it is completed and released. What we have not done (and will not do), unlike many authors considering the influence of new technology on human interactions, is to attempt to look into the future with specific predictions of what will happen. Making predictions about the future of technologies and their uptake is risky, and often futile. As stated in the Introduction, we are writing at a particular moment in time, and our writing will be a product of what we know now, and what we cannot foresee. Although some startlingly accurate predictions have been made, sometimes trying to predict the future will make you seem foolish or ignorant of a major development just around the corner.

For example, some remarkably prescient predictions about the year 2000 were made by John Elfreth Watkins Jr, including the switch from horses to automobiles as the primary form of transit, and wireless telephone communication worldwide:

> Wireless telephone and telegraph circuits will span the world. A husband in the middle of the Atlantic will be able to converse with his wife sitting in her boudoir in Chicago. We will be able to telephone to China quite as readily as we now talk from New York to Brooklyn. By an automatic signal they will connect with any circuit in their locality without the intervention of a "hello girl".
>
> (Watkins Jr, 1900)

However his visions of transport technology one hundred years in the future were less accurate:

> Fast electric ships crossing the ocean at more than a mile a minute will go from New York to Liverpool in two days… There will be Air-Ships, but they will not successfully compete with surface cars and water vessels for passenger or freight traffic.
>
> (Watkins Jr, 1900)

The crucial piece of information Watkins was missing, that powered flight would be successfully tested by the Wright brothers in just a few years, meant that he neglected to consider air travel as a critical part of transport infrastructure.

Rather than attempt to predict then, we can look at current trends and think not about what will happen, but what *might* happen: we can explore different avenues and possibilities and think about what the consequences of such advancements might be. In other words, rather than attempting to identify the specific future of the digital public space, we can discuss possible *futures*. Science fiction (and 'speculative fiction' (SF) more generally) is often an excellent avenue for such provocation, exploring ethical issues surrounding theoretical new technologies; these might include issues of personal autonomy, environmental impact and social consequences. Fiction can take the wider, broader, long term view of these questions in a way those developing the actual technologies might not, and provides an alternate lens to traditional research (Dourish and Bell, 2014). By divorcing the question of 'how might we do this' from 'what happens if we do this', we can use the process of storytelling as a form of exploration and risk assessment, much as we suggested in our discussion of narrative as a key tool for innovation in our evolutionary toolkit. Science fiction can sometimes also serve as a road-map to where we want to be: highlighting things that we want from our technology and using stories to imagine desirable systems and technologies. Examples of technology proposed in fiction long before their realisation range from rockets that travel to the moon, to the *Star Trek* 'Tricorder' which can diagnose multiple illnesses simply by running a non-invasive device over the body.[1] Speculative media shapes our desires and expectations, and has a close relationship with innovation (Bassett et al, 2013). The prevalence of computers in fiction that understand our spoken instructions means that we do not find such things outlandish and expect it to be possible, despite the fact that voice-responding systems such as Siri and Google Voice are relatively recent and still imperfect. Touchscreens were portrayed in fiction long before they were a viable technology, and the *Hitchhiker's Guide to the Galaxy* (Adams, 1979) described a portable reference guide not dissimilar to using a tablet computer. Design and development of these technologies is likely to be influenced by the fictional versions as familiar to the designers.

An argument against this is that some technologies common to fictional futures, such as space elevators, are still not realised with current technology despite having been written about for decades. Does this mean that they are not possible? Not necessarily: we might still see these technologies in the future, as the speed of imagination is not mappable to the speed of development. But it does not mean they are inevitable either.

Such literary and fictional imaginings are, as noted, an important avenue for discussing and shaping the future. But such speculation can also be a more direct tool in design. In Chapter 9 we will discuss how digital public space is an emergent phenomenon and an uncertain design space. Speculative design and design fiction are tools which we can use to imagine future objects and explore their implications, leading to insights which can be applied to design in the present. Design fiction, as opposed to traditional design methods, is focused not specifically on problem solving, but on identifying these problems in the first place: creating designed prototypes or 'products' which are not intended to be functional, but rather which prompt discussion and debate. Through this lens, plausible futures can be identified as well as ethical and social dilemmas arising from them. These findings can then inform the design of more implementable outcomes. If this speculative design is done in a participatory fashion and includes insight and input from a wide range of people, it can provide perspectives that a designer coming from a unique background could not alone consider. Speculative design is developing as an important field in current design research. Dunne and Raby are leaders in this field, and many examples can be found in their 2013 book *Speculative Everything*. Fictional design pieces have even been presented at major technology conferences.[2]

Using current advancements and 'blue skies' research areas, and drawing on examples both from speculative fiction and speculative design, the remainder of this chapter will explore a selection of possible futures for digital public spaces and the implications thereof. These will be themed around the different aspects of digital public spaces which we have covered in preceding chapters.

Futures of archives and storage

In Chapter 1, we discussed definitions of the digital public space, and who and what is encompassed within it. This includes focus on the different forms of archives that make up much of the digital public space. Collections of content may be publicly accessible, or contributed to by the public, or contain data pertaining to the public collected in a variety of ways. Archives might also include physical objects which are digitally linked, through the internet of things. Public areas and experiences might also make up part of the digital public space, through digital counterparts which mediate the experiences of these places.

The novel technologies being explored as part of the internet of things may be important to the development of archives as a way to add *tangibility* to multidimensional digital objects and their categorisation. As discussed previously, one down side to the creation of ever more vast archives is that it becomes difficult to both find the content needed, and to navigate in a way which 'feels' comfortable and has physical markers for memory. Browsing a row of books on a shelf is not the same as scrolling down a list of titles on the screen. By renewing tangibility through the use of digital/physical hybrid objects, it may be possible, like the physical playlist, for future archive technologies to renew the physicality of items stored digitally. Future-based fiction offers various examples of how tangibility might provide greater functionality in manipulation of content: for example, three dimensional 'hologrammatic' displays that can be manipulated physically, famously portrayed in *Minority Report* (2002) and in diverse popular culture from *Iron Man* (2008) to *Parks and Recreation* (2015).

We can also consider how potential futures of archives might take advantage of connectedness of stored content in digital archives. This is something that is only just starting to be explored but is, potentially, the ultimate expression of the digital public space archive as conceived by the BBC: that any content which is created is added, in its entirety with linked metadata, to the national public archive and can then be accessed by anyone. This could lead to advances in creation, as well as storage and cataloguing, such as advances in transmedia storytelling, where one thread of narrative can run through many different media (such as television, websites, books and physical spaces for play) and be connected by technology which allows seamless transition and instant recall across contexts. This kind of multi-platform storytelling has been tried in the past (for example in multimedia surrounding the trilogy of films that began with *The Matrix* (1999)) but is generally agreed to have been of limited success to date, perhaps because transition across media is not straightforward.

Advances in distributed information sharing might also provide more direct societal benefits: healthcare is an area of significant focus for these connected information storage technologies. Medical treatment and diagnostic information already creates huge databases of information, which can be spread across many different archives. Your health data may already be shared in a database which lets any general practitioner access your records if you need treatment. A hospital might keep records of all outputs of their MRI scanner. Your insurer may keep detailed records about your family history. But extended linked archives could create significant changes in the way we approach healthcare, if they can surmount ethical challenges and provide data in a form which can be used for the benefit of individual patients and the wider society. Medopad,[3] an example of a company currently trying to position itself as such a database model, is a London-based start-up which has a vision of a distributed medical database, storing patient data so it can be accessed anywhere by doctors. Their current products include a range of apps which gather data from multiple sources in a hospital (such as a heart monitor, notes that doctor takes during an examination, or an X-ray machine). This information is collated and made available to physicians or others via an iPad, or even through a heads-up display: at one point the system featured an app which integrated with Google Glass, before the platform was discontinued (Baraniuk, 2014). It can even perform basic analytical work such as flagging blood test results which might be of concern. If such technologies become a standard part of medical processes there could be significant moves forward in the treatment, diagnosis and understanding of illness and disease, particularly if artificial intelligence agents (discussed later in this chapter) contribute to drawing links from large data sets. As an example of this, one can take IBM's machine 'Watson' which was developed to be able to use natural language processing to answer questions on the quiz show *Jeopardy!*. Watson is now being used to assist in healthcare decisions, using data mining of medical databases and patient information to provide treatment options and recommendations. These analyses might become even more powerful if linked to substantial data sets. Extensive longitudinal health data is being collected by projects such as Biobank,[4] which has recruited 500,000 participants to provide detailed health and lifestyle information. This is intended to form a baseline measure for information on why some people develop a particular disease and others do not.

We can imagine a potential future therefore, where any medical treatment or examination you get, whether it be at home, at a hospital or with your local doctor, is recorded in a central database. This might be cross referenced with information about your lifestyle; perhaps recorded via a quantified self-device that records your heart rate, movements and sleep. As well as flagging health risks, such archives might allow trends in diagnoses and treatments among the general population to be highlighted, and even note a particular cluster of warning signs and alert you to steps that should be taken to prevent acute problems like a heart attack.

But it is important to consider all sides of the impacts that such linked archives might make. With such sensitive data, there are major ethical questions which must be satisfactorily answered. There are already concerns that if proprietary platforms like Medopad are used, it will bind health providers to potentially expensive and perhaps limited forms of archiving. If the system holds or collects data in a certain way, biases may be introduced. This might for example lead to prioritising certain types of conditions, or focusing on the output of machines that are more easily compatible with the system, at the expense of those that are more complex but potentially critical for a subset of patients. These biases might not be intentional, but a consequence of, for example, an all-male executive board making decisions about which data is critical to collect in the initial roll out of software. It is also important to consider what else might be done with the data set once it is collected and stored in such archives: who owns it, what can they do with it and what levels of consent are required? What business models are the companies producing the technology working to, and if this commercial dimension exists, can it truly be a *public* digital space?

As well as the future uses for the content of archives, we can also consider possible future technologies of data storage itself. Currently, archives of data and content such as these are for the most part held on networks of linked computers. But future technology might allow new forms of archiving with even greater capabilities, and this does not have to rest within traditional electronic computers. Karin Ljubič Fister and colleagues have demonstrated proof-of-concept work to store data in DNA base pairs in living plants and seeds. DNA is an excellent way of storing information, given that this is the purpose for which it is evolutionarily suited. They translated a computer programme, 'Hello World' into binary code, and encoded this in the four 'letters' of the nucleotides which make up DNA: A, C, T and G. 00 became A, 10 turned into C, 01 into G and 11 to T (Ljubič Fister, 2016). The DNA for the sequence to encode the programme was synthesised, and inserted into the genome of a tobacco plant (which are often used for plant-based genetic research). In most organisms, the genetic code that makes up the central 'instruction book' for the organism will be present completely in almost every cell that makes it up. Therefore, included in all of these copies of the information, among the instructions for making the tobacco plant, is the 'Hello World' programme. To read it, the DNA is extracted and read using standard sequencing procedures (Ljubič et al, 2014).

Although this is not what many people would consider when they talk about 'digital technology', it is highly digital because it involves translation into the digits of binary code. Preservation in such a living format, harnessing the mechanisms of organic replication and storage, might meet several of the challenges which are faced by silicon-based computing: such as heat generation and energy consumption, storage and durability. Data centres could be replaced by forests, taking carbon dioxide from the air rather than relying on generated electricity which most likely put it there. DNA is a very efficient data storage medium: one gram of DNA can hold over 450 × 1018 bytes, enough to store 'all the archives in the world in one box of seeds' (Ljubič Fister, quoted in O'Neill, 2016). These seeds could be stored in a seed vault, and lie safely dormant for thousands of years, able to be read in the future. Or alternatively, the plants could be allowed to reproduce and pass the information generationally. Ljubič Fister speculates that handheld sequencing readers, a technology already in development, could allow you to scan a leaf and extract the information from a living plant or tree 'with virtually no damage'. It is important to note that this storage would be, for the most part, 'read only' since it is difficult to change the DNA of the cells of a living organism. New gene insertion technologies such as CRISPR (Jinek et al, 2012) would allow changes to be more easily made, but even this would be on a slow timescale more suited to archive storage than data processing. DNA sequences in replicating organisms can persist for millions of years, as we can see from areas of our own genome that can be traced back to our common ancestors with other species.

There are some reasons to be cautious with this approach. Like any method of preservation which relies on copying, be it making handwritten copies of a manuscript or using a photocopier, errors can creep in. Over time, these errors (or mutations, in the case of DNA) might corrupt the original information or make it irretrievable. In the case of important DNA code for living organisms, this is avoided by having redundancy for important functions, and by the fact that so called 'highly conserved regions' are maintained by being necessary; mutations in these regions are likely to be fatal to the organisms. For critical information sequences to be maintained, it might be the case that these are included alongside critical conserved regions that are protected from damage. Technologies which replicate these natural safeguards will be necessary for safe long-term storage in this manner.

A different potential mechanism for different kinds of archive storage lies in the development of novel forms of computation. These include areas such as biocomputing (where living cells are used as components), or chemical computing. Another novel computer technology which is already in practical development is quantum computing. Quantum computers, rather than using 'bits' to store information, (which are a straightforward binary 'on/off' position of switches or electrical current in standard computers), use 'qubits'. Qubits can occur in 'superposition' which means they can be both on and off at the same time, and thus store much larger amounts of information. Some companies already claim to have operational quantum computers, but to date they are limited in operation or require extremely specific conditions (such as very low temperatures) to function. Quantum computing is only one potential such new technology: there may be others which we are currently unable to foresee. But it is likely that some advance will lead to an increase in the capabilities of digital storage and computation. Any such breakthrough would revolutionise computation.

To date in the digital revolution we have seen rapid advancement of computing power in line with Moore's law, which observes that that the number of transistors per square inch on integrated circuits doubles every one to two years, and predicts that this will continue. However, one important limitation which could halt this progress is the physical movement of atoms in the transistors, which maintains an absolute minimum size of such components. If new technologies such as quantum computing overcome this barrier, then we can start to imagine implementing some of the scenarios explored below, where computation is truly embedded in our environment, or even integrated with our bodies themselves.

With such advances, the speed of cataloguing might be much increased, and the cost of storage and analysis minimised. There are many ramifications of this. Such vast computing power could lead to advances in artificial intelligence (see later in this chapter). We must therefore consider in designing the digital public space how we construct and curate our archives, bearing in mind that the technology which holds them may be something currently inconceivable. Any storage medium which is intended to provide long-term archive storage for digital public information must be sustainable, and 'future proofed' to be readable in the future and compatible with new technologies. We must also consider how information included in any such archive is chosen, and the biases which exist when building it.

Case study 8.1: Alicia

Alicia lives in a city surrounded by greenery. She walks down the street, admiring the tall trees which have stood for decades. As she passes, she takes out her personal, solar powered reader to scan what they contain: these are the public archives of history and record which, if scanned, will tell her all about the place where they are planted, and may contain related music, video and other long-term storage information.

She makes her way past the edge of town to the vast fields, segregated so she can see the young plants at one end, getting older as she moves outwards. Mechanical 'bees' buzz around bringing new updates, delivered by means of gene editing vectors which will snip out the old segments and replace them with the new, for the next generation of seeds.

As Alicia walks, she gets an update on her personal device: she's walked further than usual this week, which is good for her heart and earns her savings on her health insurance. But she gets a warning for high pollen counts for the particular plant she is allergic to, and is advised not to walk further into the next field.

If we are able to cognitively integrate access to external digital resources, which increasingly have large capacities and high fidelity, might be we able to get to a stage where we 'remember' everything and have perfect recall? Bruce Sterling (2005) suggests that if you are going to be capturing information from the environment, why not capture as much as you possibly can rather than make decisions about what is needed? This is vital because what is important to you might be different than the priorities of someone else, including a future version of yourself. The BBC for example, thought for many years that there was no reason to keep copies of transmitted television programmes, reusing the tapes and wiping many years of archive programming that is now highly sought after (many episodes of beloved programmes such as *Dad's Army* and *Doctor Who* are lost). If you collect everything, you can be sure that what you need is most likely available. It is not just factual information from databases, or data gathered from our devices that could be captured and connected in this way, but also from our own experience. In Chapter 4 we discussed the practice of 'lifelogging' whereby people wear devices which record everything they experience. If this were added to a digital public space that could be accessed by anyone, we could potentially get to the point where our 'memories' consist not just of everything that is encoded in our biological brains (which can be subjective and unreliable) and not even just everything that we ourselves experience directly (which can be distorted by a particular perspective or attention) but multiple perspectives on multiple experiences collected from the whole of the public.

Will this mean that we no longer have the potential to 'forget' anything? Or will we be overwhelmed with 'memories' and not necessarily be able to identify things when we need them, or perhaps even distinguish the source? It is critical that any extended memory has efficient search functions; a large database of telephone numbers stored on your mobile rather than in your head is only useful if you can find the number you are looking for when you need it. In fact, our biological memories do not function in the same straightforward way as facts stored in a database: links and retrieval criteria are complex and still not fully understood, and the things that bring 'knowledge' or a 'memory' to mind are individual to every person. If a very large volume of memories were stored, it might be that certain types take 'priority' and are more quickly accessed, whereas others would take more effort to retrieve. Certain individuals have been studied who are described as having 'perfect' episodic memories, recalling every detail of everything that has happened to them. But this is usually described as a negative experience with memories becoming intrusive and uncomfortable, to the extent that Jill Price, who has such a memory, describes it as 'agonizing'. 'I don't look back at the past with any distance. It's more like experiencing everything over and over again, and those memories trigger exactly the same emotions in me. It's like an endless, chaotic film that can completely overpower me. And there's no stop button' (Shafy, 2008).

Another hurdle is that for a shared memory to function in a fully integrated way with our own, we would have to be able to replicate the full experience of others including all senses, in the same way that our own experiential memories consist of sight, smell, sound and other sensory inputs. For data coming from other sources to have the same status as biological memories, it would have to replicate this 'thick' experience on multiple channels, including potentially the emotional impact and feelings that were evoked. This might require significant advances in technology, particularly understanding how such complex and deep memories are stored in our brains, and how they could be reproduced artificially. Again, we would also have to look at how memory retrieval on demand functions, and attempt to create a functional system for this. Rather than a technology of virtual reality, we would need one of virtual memory. It might be that this requires creating new biological memories. As portrayed in the Wachowski's film *The Matrix* (1999) you might choose to 'learn' a particular skill by adding the memories of knowing it, directly into your brain. Although this sounds farfetched, it is not such a big step from current technology. In fact, something resembling this has been demonstrated in animal experiments, where memories from one rat were implanted into another, allowing it to solve a task it had not been trained on (Berger et al, 2011). It might end up that rather than having a personal memory, we all have access to a shared public resource, a 'cultural memory' in a much more concrete way than the term is currently applied to archives. However, if such technologies became commonplace, a risk is that we may find ourselves unsure whether something is what we ourselves experienced, or a 'memory' from an alternate source. Questions of ownership bring complexity to such an idea: who owns this data? Nobody, or everybody? This will be discussed further at the end of this chapter. It is also important to note that when we learn, we attach multiple aspects of experience to it; this would not necessarily be gained by 'implanted' memories and may result in a less rounded learning experience.

Having a searchable archive of shared previous experience could potentially vastly improve our creativity and innovation. Being able to instantly 'know' what has already been tried by others, or what has already been achieved by others, may prevent you making the same mistakes or wasting time searching for solutions to problems that have already been solved. Bruce Sterling notes with his discussion of the internet of things that if objects recorded the manner in which they were used, this might cover novel uses that were not conceived of by the designers, but could be potentially adopted by others who might find them useful (Sterling, 2005).

If this could apply to general knowledge and experience, it could significantly progress human development; information is a resource which gets bigger as it is collaboratively shared rather than smaller. By sharing information in directly available digital public space, we would be negating the need for public dissemination and publication through traditional mechanisms. Again though, this would rely on excellent search algorithms, which would need not only to be able to find an answer that was appropriate, but potentially flag something for your attention that you were not aware already existed.

There are several potential issues with connecting all our memories into one vast, public digital store. As mentioned, there are problems of search: how to make sure the 'memories' that you wish to access are available when you need to remember something. If this can be done for memories that are generated externally, could it also be done for those created by our own brains? Could we 'download' our memories to a hard drive, to be retained for later? Could we remove some that we no longer deem important, to save space for other things? And if the edges blur between what we experience ourselves and get from other people, how can we be sure what is 'our' memory? With the potential negative effects to 'remembering' everything, we may choose to integrate forms of forgetting into how we store memories. Our existing memories degrade over time, and it appears that we reinforce them by access; our memories may change over time the more we access them, and can be influenced by our emotions at the time of original encoding (Foster, 2011). With this being the case, we may find that we need a form of forgetting to keep a curated version of our memories that frame our own personality. As described in Chapter 6, Ayalon and Toch (2013) highlighted what appears to be an analogue of this in current aspects of digital public space; that people are less comfortable with sharing personal posts the older the posts become. Some of the newer social media platforms such as Snapchat are specifically time limited, aiming for an 'in the moment' interaction rather than long term retention, and we may find that this, rather than permanent storage, is the desired form of digital public space. As Lothian (2013) describes, much interaction is context dependent, and may suffer if this is absent; 'ephemera' such as posts and comments do not reflect the full reality of the context-dependent interactions. Lothian does not however argue against the archiving of ephemera, as a mode of long term preservation of digital culture. Such choices on what is and is not stored, and how and where, will be critical in the future of digital public spaces.

Futures of human evolution and experience

In Chapter 2 we discussed in detail human evolutionary history, relating it to how culture and technology are intimately intertwined. Many people, though happy to think about our evolutionary past, assume that due to technology, healthcare, and culture, humans are no longer subject to natural selection and change. This is not necessarily the case: there are many ways in which humanity as a species is still developing. Firstly, as long as we are still biological organisms which breed and collect DNA mutations, then we are still subject to natural selection. For an example of this we can look to the very recent change (in evolutionary terms) which has meant that adult humans can digest dairy products. This depends on a single mutation enabling adults to digest the protein lactose, which was still rare even in the Bronze Age (Allentoft et al, 2015). Development of this trait, and its spread through the population in Europe, has been linked to the development of farming. When people began to cultivate large mammals, it made sense to be able to digest the milk, even as adults, as a highly efficient source of food. In this way we can see that the technological revolution of farming led to evolutionary change and the spread of particular inherited traits.

As examined in Chapter 2, direct changes to behaviour and brain function due to use of the internet are likely to be at the individual cognitive rather than species level. However, it is still possible that over the longer term, certain aptitudes and characteristics could become more common in the population due to selection pressures. In considering this, it is important to remember that evolutionary selection works on very specific criteria: how many children you have, and how healthy they are. If a behaviour or trait does not affect this, it is unlikely to be preferentially passed on to a larger proportion of the population. When such changes do occur, although they can be quick in terms of the species lifetime, they are not visible except at the remove of several generations. It may well be hundreds or thousands of years before we can see the full effects of the digital information revolution on our species, similar to how the results of the dairy farming technological revolution took some time. Some changes may be incorporated more quickly, by means of 'epigenetic' change: these are changes not in the content of our DNA code, but which parts of it are switched on and off, and recent research has shown that via this mechanism the environmental influences on one generation can have an effect on the gene expression of the next (Teperek et al, 2016). It is suspected that these epigenetic factors can affect physiology, so that for example height and weight may be affected by stress and nutrition of your parents or even grandparents. As these changes come to be more understood, we might see that technology is also having more subtle effects.

Despite this, the fastest mechanism for change within a group is still culture and technology, which can be passed both within generations and between them. While to this point we have mostly been considering external technology objects and infrastructure, it may soon become possible to fully integrate technology into our bodies themselves, affecting us as organisms. In Chapters 2 and 3, we described how we can already consider our embodied consciousness as extended to include tools that we use to think and remember, which includes 'wearable' technology that reacts to our needs without requiring intentional access. As well as digital technology devices, bodily augmentation could be considered to include clothes (which help regulate our temperature), eye-glasses (that augment our vision to 20-20) and watches and timepieces (that give us an awareness of time). Smartphones, which give us access to the information resources of the internet, might also be considered in this category. But possible futures of the digital public space include ways that we might integrate technology more substantially into our biological bodies. Research in this space has been underway for many years by those who ape the 'cyborgs' of science fiction, by those such as Kevin Warwick[5] (who is well known for having had an RFID array implanted into his arm) and Thad Starner (whose memory recall device was described in Chapter 4). In recent years the integration of these digital experiences is becoming more sophisticated, for example the work of colour-blind artist Neil Harbisson who has provided himself the ability to 'hear' colours and 'see' sounds, converting different frequencies of light into different notes (Jeffries, 2014). The melding of human and machine on a more bodily level has long fascinated SF writers, from *The Six Million Dollar Man* (1973) to the Cybermen of *Doctor Who* (1966).

Technology which affects our bodies might change the way we experience our environment, including physical and digital public spaces. In a similar way to the *Northpaw* system described in Case study 2.1, we may be able to use existing sensory input in new ways to add currently imperceptible information to our experience. As an example, the experimental system called *Phantom Terrains* (an addition to a standard hearing aid) allows the wearer to hear wifi signals. Frank Swain, who trialled this system, described how his already digitally-assisted hearing provides a good opportunity to add additional senses. 'If I have to spend my life listening to an interpretative version of the world, what elements could I add? The data that surrounds me seems a good place to start' (Swain, 2014). This integration of digital information into our sense of experience is one route to making ourselves 'biots' as described by Bruce Sterling; digitally enabled organisms which connect to the digital space. How would being able to listen out for areas of excellent wifi coverage, and notice immediately when you entered or left one, affect your experience of the digital public space? Or, if everyone had this ability, how would it affect infrastructure choices by business and civil planners? Would 'quiet' areas of low signal be sought after as refuges?

More speculatively, we might consider making our bodies and brains themselves part of the technological systems. This might include adding digital storage to our organic brain storage capacity as described above, or utilising our physical form as an aspect of display or function. This could be achieved in a variety of different ways, from silicon-based technology such as implanted computer chips, to more biologically based technologies, such as biocompatible and conductive silk spun by genetically modified silkworms (Ranner, 2013). There are already prototypes in development that, rather than being screen based, use 'digital skin' as a touch sensitive display for digital information (Yokota et al, 2016).

Technology might be allowed to take over certain monitoring and maintenance functions on our behalf. Forsythe et al (2005) write about software agents that monitor cognitive processes (in this case in fighter pilots) via physiological effects, and take over life-critical functions or provide assistance with certain functions as an 'augmented cognition' system when they detect severe human malfunctioning. If this type of technology was connected to others in the digital public space, or to information about the space itself, it could act to protect humanity on a larger scale – for example encouraging you to move away from a particular area if it were dangerous or overcrowded. This might not even have to be through alerts or other consciousness-raising means, but could act through more subtle behaviour-changing nudges. For an example of how this might function, we can look to the work of Pfeiffer et al who created a system whereby a wearable device helped pedestrians navigate by directly stimulating muscles in the leg, causing people to adjust their walking in a specific direction.

They noted that: 'In this way, actuated navigation may free cognitive resources, such that users ideally do not need to attend to the navigation task at all' (Pfeiffer et al, 2015).

These notions of the human body as something we might build on and adapt are an important component of the movement known as transhumanism, which holds this up as an ideal to which we should aspire. The British Institute of Posthuman Studies describes the transhumanist movement as 'the drive to fundamentally revolutionise what it means to be human by way of technological advancements' (Brietbart & Vega, 2014). They describe three central areas of transhumanist thought which they think deserve focus: super longevity (extending life and reducing age-related diseases), super intelligence and super wellbeing. Digital public space related technologies have particular relevance to and implications for the latter two of these, since by extending our minds via a shared public information network we can extend our intelligence through cognitive augmentation (see page 206) and our wellbeing may be improved by technologies which augment and improve our public spaces and our connection to them. In this sense, our bodies and minds would themselves become part of the connected digital information space.

This kind of physical, biological integration may be facilitated by parallel technology developments which are currently underway in the field of synthetic biology. This involves creating artificial versions of biologically based systems using biotechnology and engineering. This might involve genetic engineering, or construction using biological materials on the nanoscale. For example, bacteria could be engineered to digest industrial waste, or produce necessary products. DNA and protein based 'machines' might be constructed which seek out and destroy cancer cells. Future implementations of synthetic biology might even involve growing organically based structural components from first principles. One can imagine futures in which digital public spaces were grown as well as built, but with equally specific design parameters.

As biological engineering technology develops further, it is possible we will be able to build such changes into our very bodies, without the need for silicon chips and metallic implants; this could potentially involve manipulation of our gene expression and giving our bodies the ability to incorporate new systems, processes and structures. New gene editing techniques such as CRISPR (see page 191) may allow editing of specific genes within humans, though there are still technological, not to mention ethical hurdles to its use in this context: the technology is not yet at the stage to be implemented in humans, nor do we currently have sufficient knowledge of gene function to effectively treat more than a handful of very specific diseases. Such techniques allow the genetic code to be specifically changed, and could facilitate incorporation of biologically based 'machines' which connects us to the digital public space. An even more extreme step would be to incorporate designed changes into 'germline' DNA, that is, that which is inherited by our children. It has been proposed that germline gene editing should be banned under worldwide ethical conventions, though this was not, as predicted, agreed at the International Summit on Human Gene Editing in December 2015. They instead stated that it would be irresponsible to proceed with such work until current ethical and safety issues have been resolved, but did not rule it out for the future (Baltimore et al, 2016). Germline editing creates ethical questions even beyond those of changing the genetic code of an individual, because it could cause changes to our species as a whole if such changes spread amongst the population, or cause irrevocable divisions between 'posthumans' who have these new adaptations and connectivity, and those who do not. Distribution of such technologies is important. As discussed in Chapter 1, we are already seeing a divide in the digital public space and access to it. If the digital public space becomes intertwined with our biological bodies, then, we may see the development of two distinct 'species' of humanity that can no longer interbreed and will become more and more separated. This will be particularly deleterious if access to these initially is based on imbalance such as wealth, or privilege.

As hybrid digital/physical organisms, the nature of our interactions in public space, both digital and physical, would change. If digital information about individuals or objects that we encounter 'in public' can be exchanged, sourced or recorded, then the potential anonymity and transience of public space may be changed or lost entirely. Echoing public concerns with the built-in recorders of Google Glass, consider the potential if everything you saw or heard was recorded and could be recalled with perfect clarity. We will return to the possible implications of being able to instantly transfer information between individuals in a connected public space later.

Many proponents of transhumanism are by nature utopian in their visions, describing a world in which humans are better, stronger and more intelligent. But with any such potential technologies, we must also consider the possible negative consequences of futures that take these paths. One important point to consider is that our current existence as biological entities is not inherently reliant on technology, though our culture may be. If we incorporate technology to the extent that our bodies can no longer function without it (like those now who reply on medical implants and regulation such as insulin pumps), then a disastrous collapse of technological civilisation could be fatal not just for us as individuals but also for our species. Integrated high technology such as cyborg style electrical implants, or biological, bacteria based 'nanobots' could be imagined to fail worldwide due to environmental disruption of an asteroid strike or major volcanic eruption, or via a geomagnetic storm from the sun which would severely disrupt electronics based technology.[6] It might even be that malicious attacks could fatally disrupt services, in the same manner that co-ordinated Distributed Denial of Service (DDoS) attacks have brought down internet services worldwide. Even if disconnection is survivable, there might be psychological effects of disconnection if it becomes integrated to our sense of self: 'And when the migration is complete, we shall increasingly feel deprived, excluded, handicapped, or poor to the point of paralysis and psychological trauma whenever we are disconnected from the infosphere, like fish out of water' (Floridi, 2007).

There are also social risks inherent in intrinsic digital public space connection, even if it remains stable. Floridi (2007) refers to such connected humans not as biots, but as 'inforgs' (as opposed to cyborgs). If we digitise our existence as bodies which move through public space, and ourselves become a part of digital public space, then we must consider the privacy, ethical and other implications. As connected biots or inforgs, we may find that like the 'spimes' of the internet of things our entire 'supply chain' from birth to death becomes trackable and searchable. We will return to such challenges in the conclusion of this chapter.

Case study 8.2: Vera

Vera is 120 years old, and lives with her two children. Since she is still fully healthy, as she can tell by the reports from her internal monitoring systems, she is considering having another child before she dies. She goes for a walk while she considers her history so far, consulting her databank of recorded information about her life (and polling her friends and family on which is the most important) to see how she might improve things for any future offspring. As she walks down a particular street she instinctively avoids the left hand side of the road because it feels less comfortable; this is because her implants are providing subconscious cues that the wifi coverage there is less stable.

She goes ahead and decides that another child is a good idea, and that she will include some new extra features for them; they will have additional light receptors in their eyes that can see the ultraviolet that is now often used for extra information on display screens. She sends a signal to her biological control systems: it's time to boot up her fertility systems again. Nanobots in her blood stream flip the necessary gene switches and she's ready to go.

Futures of environments and objects

Chapter 3 covered various different aspects of how physicality is important in the digital public space. Our physical experience is critical in relation to how we learn and interpret the world, and this applies to our interactions with others as well. Being in the same physical space as another person is still, to date, the most effective way to share and collaborate, although digital technology is providing a raft of tools which can allow varying levels of remote interaction. Virtual worlds and virtual reality have been a staple of both science fiction and futurism for many years, but current versions are still limited in what they can achieve in terms of tangibility. Until the physical environment can be seamlessly replicated virtually, it is important to mitigate for loss of physicality through other means when interacting digitally. But it may be the case, as alluded to in Chapter 3, that future technological developments can replicate the feeling of 'really being there' as portrayed by fictional futures such as *Star Trek's* holodecks. These full sensory virtual environments are a key goal for many technology development companies and researchers, who hope that these technologies will improve and play a critical part in futures of digital public space, with a recent resurgence in virtual reality technologies such as the *Oculus Rift*.

But there are other potential futures which we can examine dealing with the physicality of digital interactions. In Chapter 3, we looked at telepresence, and the ways in which current technology is facilitating action at a distance through use of digital proxies. SF writers have often used for inspiration the possibilities of remote operation. An early example of this is the development of 'waldoes', or remotely operated reaching arms. These are used in a variety of industries from space exploration to film making, being popular for example with the Jim Henson company puppeteers. But the term originates from SF and was coined in the eponymous 1942 story 'Waldo' by author Robert Heinlein.

By extending this kind of remote operation to full bodies, and having two-way sensory transmission, limitations of physical locatedness might be entirely overcome. Your digital avatar would no longer be stuck in the digital space, but could also interact in the physical world. This kind of projection or bodily adaptation might be of particular interest to those who feel that their existing bodies have limitations in private or public space, either because of the unsuitability of the human body to particular tasks (like exploring the depths of the ocean) or because their own bodies have physical limitations due to disability. Many SF works have imagined controlling a remote surrogate for one's body. *Lock In* by John Scalzi (2014) is an example which explores this concept, with a vision of a near-future American society in which a significant proportion of the population are entirely paralysed, and interact with the world via 'personal transport' devices; remotely operated humanoid bodies. The technology to realise such feats moves ever closer, with recent advances allowing volunteers with spinal cord injuries to control a robot in simple tasks in near real-time and receive sensory feedback (Thompson, 2016).

The ultimate expression of giving ourselves presence in a physical space through a digitally enabled projection of ourselves might be to experience an entirely different set of senses and thus control a different type of physical body. Again, this is a common conceit of SF, from mechanical exoskeletons often seen in Japanese manga comics, to the 'Jaegers' of the 2013 film *Pacific Rim*: enormous mechanical fighting machines controlled through direct mental control by a pair of operators. A series of SF books by Anne McCaffrey, beginning with *The Ship Who Sang* (2012), explores this idea, set in a future in which newborns with severe and/or life threatening physical disabilities are connected to the sensors and controls of space ships or space stations. These 'shell people' control the ships as if they were their 'bodies', experiencing the world through sensors instead of senses.

> Instead of kicking feet, Helva's neural responses started her wheels; instead of grabbing with hands, she manipulated mechanical extensions. As she matured, more and more neural synapses would be adjusted to operate other mechanisms that went into the maintenance and running of a space ship.
>
> (McCaffrey, 2012)

In Chapter 3, we also discussed the potential of creating hybrid digital physical spaces that bring digital content out into the real world. New forms of play can be facilitated by shared technology, for example physical digital games like those investigated by Garner et al (2013), which require people to act in tandem in physical space. Some games, while remaining screen based, take specific advantage of physicality and co-location to enhance players' experience. *Artemis*, and *Spaceteam*, both of which involve a team of players working together to pilot a space ship, are games which require communication and co-ordination of players physically. *Spaceteam* for example, requires four players to each read instructions on their mobile device (including physical commands such as to 'shake' or 'invert' the phone) which all players must perform together, but which are only displayed to one player. Therefore, the player who has those instructions must relay them verbally to everyone else, who can act upon them.

Another technology strand which might feature in potential futures of digital public space is augmented reality. Although the release of recent games such as *Pokemon Go* and its massive popularity have increased public attention on augmented reality games, the appeal is still niche and limited. In fact the augmented reality aspect is not a critical part of *Pokemon Go*: many people find that it functions more effectively without it, saving on battery life. It seems likely that augmented reality will not find a true foothold as a technology until it is available as an integral part of our experience rather than through screens which distract us from the 'real world'. There are already some forays into this; Google Glass may have been an early attempt but as discussed had many failings. This type of full integration can be described more accurately as 'mixed reality' which incorporates virtual content into the environment in real time (Ceurstemont, 2014). It may also not just be applicable to our visual senses: a fully embodied experience must create a 'new sensoriality' (Verbücken, 2003) to combine with our experience of the world. Various technology companies are exploring, for example, how sound instead of sight might be a suitable medium for augmented reality, developing 'earplugs' that play back an augmented soundscape (Swain,

2016). This might incorporate alerts from your personal devices, or potentially information about your physical environment as you travel through it, without having to divert your attention to a screen.

Devices like these might not need to be worn all the time to have significant impacts. At first they may be adopted in specific circumstances, such as games, or to assist specialist roles such as providing information when it is not easy to get another way. For example, surgeon Steven Horng claimed use of an app on Google Glass saved the life of one of his patients in 2014. While operating, he was able to bring up details of drug interactions (in this case, which blood pressure medications were suitable for a patient with allergies), which would not otherwise have been possible in time (Baraniuk, 2014). Routine use of such apps, especially if they also included access to patient records and databases as described above, or could provide alerts without being prompted, would greatly assist in medical care and circumvent the need to pause operating to use keyboards or touchscreens and therefore 'scrub up' again to be ready for surgery.

Once devices are easy enough to use, and reduce enough in price that they become as ubiquitous as smartphones, why would you not wear one all the time? As technology improves, there may be options which are both comfortable and not obvious to others; 'transparent' design (see Chapter 9) at various levels, which could be 'forgotten about'. Such devices might include an augmented view built into contact lenses and directly 'beamed' into the eye: this is technology which is already in development.[7] The possibilities of continuous mixed reality, allowing you to choose how you view or hear the world, are seductive. Already, you can purchase selectively noise cancelling headphones that 'filter' the real world.[8] If the technology allowed you, you might prefer to 'filter' your view, like services which block certain hashtags on twitter, so that examples of a particular thing that you do not like are covered by something else or blurred out. This personally curated digital public space might sound ideal, enabling you to, for example, use a physical ad-blocker service which covers all billboard advertisements with pictures of puppies. But there are social implications to filtering the real world. What if there was an app which blocked out homeless people? And what if a politician included such an app in their campaign material to improve your view of their city?

So far we have been concentrating mostly on technology which individuals can use to create personalised experiences of digital public spaces, or project themselves elsewhere within real space. But as we saw in Chapter 3, digital public space as a concept also includes augmentation of physical public spaces with digital technology, creating a hybrid environment. One way in which we can consider futures of augmented environments is via the internet of things, which as we have seen is leading to a significant number of connected devices in homes, workspaces and public areas. At a certain saturation point, it is possible that we will achieve a future which has been predicted for a long time by many people: that of ubiquitous computing; that our very environments themselves will become 'smart'. Looking ahead at the consequences of such a fast moving field is dangerous: 'extrapolating from today's rudimentary fragments of embodied virtuality is like trying to predict the publication of *Finnegans Wake* shortly after having inscribed the first clay tablets' (Weiser, 1991). However, we can examine trends in current and upcoming technology, and look at potential consequences.

The goal of much ubiquitous computing is anticipatory computing: that rather than being passive technology which is available if necessary, it works in the background to do what you need it to do before being asked, and potentially before you are even yourself aware that this is something that you need.

> The thing that will happen, I believe, is that the products will be smart enough, or integrated enough, that they will be able to react to us; that the product will know what's going on with us and will be able to do the right thing. I think that's different from the toaster and the blender; the toaster and the blender sit there, not knowing how we are, waiting for commands. I think information appliances, highly technological appliances, will know we're there and anticipate what we want just from the way we act.
>
> (David Kelley, quoted in Moggridge, 2006)

We are already seeing this kind of reactivity in individual devices, for example heated gloves which are designed to maintain a steady temperature for those with circulation issues.[9] To answer more complex needs, devices must incorporate some form of artificial intelligence. Although this sounds firmly within the realm of science fiction, basic artificial intelligence is already a key component of much existing technology, such as responsive digital assistants like Siri, and picture matching algorithms. The important distinction to make is that we are not talking about intelligence in human, broad terms (known as general intelligence, but very specific abilities to perform particular tasks. As technology is able to incorporate electronics into more types of materials and devices, these types of functions will become broader: for example, a recent development in 3D printed manufacturing means that circuitry can be printed directly onto surfaces (demonstrated on phone housing), using conductive ink, allowing huge advances in flexibility and adaptability (Hodson, 2016c). These kinds of innovations could incorporate simple electronics and sensors into products and materials that were not previously connectable.

Individual objects able to react to our needs are one thing, but what if our very environments were adaptable? One route to this is 'smart buildings' as touched upon in Chapter 3, which have features that are designed to meet and anticipate our needs. But these would mostly be in private spaces tailored to individuals or families. What about in public areas? With increasing miniaturisation, some suggest that we could implement what is generally referred to as 'smart dust' (Kahn et al, 1999). Like the RFID tags described in Chapter 3, tiny chips could be embedded in all sorts of objects, but rather than be passive tags which react when read, or even active ones which send out a particular fixed identification message, smart dust could also perform a range of functions including sensing and reporting. Sensors less than $0.1cm^3$ already exist which can transmit information about temperature, humidity and other environmental functions. One can imagine that these functions could be greatly expanded by future technologies, and hundreds or thousands of these devices could be deployed around your environment to provide a constantly updated stream of information.

However, there are many ethical and safety questions which must be addressed before we deploy such technologies. Data being produced constantly throughout the environment in a self-regulated way brings up many issues of surveillance and privacy as touched on in Chapter 6. If such a system is implemented it is important to consider who will have access to the data generated by such devices, and what they can do with it. This is even more the case if they have an active, not just passive role in the environment. There are already cases where connected internet of things devices have been manipulated to act as part of 'botnets' to undertake DDoS attacks on websites and internet infrastructure, causing major outages and security concerns. More direct effects on our environment might be possible, depending on what the devices do: for example, if ubiquitous devices can release pleasant scents or smells to accompany broadcasts, what is to prevent them being adapted, either by terrorists or even an authoritarian government, to emit toxic or behaviour-altering gases? Even in a less extreme scenario, there are waste and disposal issues to consider. If sensors or other devices are distributed as 'dust', they are likely to be unrecoverable. If replacement and recovery is impractical due to their wide distribution and hard to reach locations, they must either be self-powering (drawing energy from themselves or their environment) or powered from an external source that is maintained through the life of the device. Alternatively, they may have a self-limited life-span, in which case they must self-destruct in a way that does not cause polluting waste that will cause clutter and/or impact the environment. An example of the dangers of widely taken up micro technology without proper consideration of the environmental impact can be seen in the common use of so-called 'microbeads' in cosmetics such as facial scrubs, which are now known to cause environmental pollution due to their lack of biodegradability. Many groups are now campaigning to ban such additives, and the UK government has backed this move (BBC News, 2016).

The negative environmental effects of micro devices are even more critical when they are located not just in our environment but in our bodies as well. We must ensure that anything which is going to be introduced to the body, or could potentially be inadvertently taken internally, is not toxic and will not produce any unintended effects, such as immune reactions. Some work is developing even smaller devices, on the nano scale; not smart dust but smart molecules. 'Scientists have started shrinking sensors from millimeters or microns in size to the nanometer scale, small enough to circulate within living bodies and to mix directly into construction materials' (Garcia-Martinez, 2016). These connected nanosensors, as part of the 'internet of nanothings' might be the product of biotechnology, as described above, or might use novel non-biological materials such as graphene that have unusual electrical or mechanical properties. If these sensors were distributed throughout the environment, highly detailed maps of environmental effects, on any scale, could be developed for a range of factors including light, vibration, magnetic fields, or chemical concentrations among many others. Biological nanobots, mentioned in Case study 8.2, could act as both sensors contributing to this network, and pre-emptive treatment mechanisms to implement changes where necessary, while mechanical equivalents could act on the environment, perhaps scrubbing pollution from the air, cleaning graffiti from a wall, producing personally selected and directed music out of any surface, or repainting a wall in moments.

If technology becomes embedded in our environment to the point of invisibility, one potential future of our environments would be to hide it entirely, and return to something more closely resembling non-technological pre-industrial revolution environments. This would be particularly the case if novel forms of manufacture and distribution meant that we did not have to be close to either our places of work or places to purchase goods or services. In Chapter 3 we discussed new forms of distribution that can be facilitated by fabrication at the point of need, via 3D printers and other new forms of manufacturing. New forms of fabrication technology may allow increasingly complex objects to be produced where needed, perhaps from renewable feedstocks with materials which are adaptable on demand, created from a digital copy which can be adapted and made bespoke rather than requiring a physical original. Post-scarcity, post-industrial high technology utopias have been commonly explored in SF, from the replicators of *Star Trek* to 'The Culture' described by Iain M Banks in his eponymous series (Banks, 1987–2012). However, some have also explored the risks of distributed fabrication, for example, *Rule 34* by Charles Stross features criminal activities in an age of widespread 3D printing (Stross, 2011).

It is also important that adaptable, environment affecting technology is responsive not just to a blanket 'everyone', but to specific needs of the individual which may change across time depending on mood, age, activity and so on. It is not enough to just profile people based on stereotypes, but any such systems must intelligently react to anticipate needs, and be able to take into account the diversity of human cultures. Therefore the design of such systems must be very carefully considered to accommodate this reactivity and diversity. As mentioned, intelligent software is a key part of such comprehensive environmental systems, but there are currently limits on the complexity of these. True general artificial intelligence is the focus of much research in a range of disciplines, from computer science to philosophy, though it is unclear if it is something that is possible for us to create. If we were to achieve an intelligent system, there are implications inherent for how we might use it. If such a system with general intelligence at the level of a human were created, could it be said to have sentience? And if it did, would we need to consider the ethical considerations of its use, and what a building might 'feel' about the demands we place upon it?

Case study 8.3: Phil

Every day, Phil goes to work in a remote mining facility, by accessing his external body which is located there. He can lift heavy weights but also feel the texture of the rock samples to tell if they are suitable for what he needs, and have 'face to face' meetings with his colleagues.

When he has finished work for the day, Phil prepares dinner, using ingredients that were selected and transported for him by his smart kitchen when he chose the recipe yesterday. The recipe also needs a piece of specialist cookware, which was fabricated on his home printer and will be recycled when he has finished using it. As he cooks, instructions for which ingredients to add at which stages are visible via his contact lenses, with arrows pointing to the correct items, and visible measurements for the liquids.

After dinner he will meet his friends at the local pub, where his lenses will highlight his favourite drinks but block out the football match which he wants to watch at home later at his leisure.

It may not just be artificial intelligences that we need to consider in terms of their exposure to input and rights in the connected digital public space. Combining archives, augmentation of our minds and bodies, and augmentation of our environment with ubiquitous computing could make our minds part of these systems, and an integral part of the digital public space. We will now look at potential directions and implications of this.

Futures of connectedness

In Chapter 4, we laid out the principles of the digital information space, and how our access to knowledge (and in turn how we use it) has changed in its form due to digital connectivity. This connectivity is critical for forming a true digital public space. In writing about the internet of things, Bruce Sterling identified connectedness and information access as critical factors in the difference that such systems make:

> The primary advantage of an INTERNET OF THINGS is that I no longer inventory my possessions inside my own head. They're inventoried through an automagical inventory voodoo, work done far beneath my notice by a host of machines. I no longer bother to remember where I put things. Or where I found them. Or how much they

> cost. And so forth. I just ask. Then I am told with instant real-time accuracy. I have an INTERNET OF THINGS with a search engine. So I no longer hunt anxiously for my missing shoes in the morning. I just Google them. As long as machines can crunch the complexities, their interfaces make my relationship to objects feel much simpler and more immediate.
>
> (Sterling, 2005)

This idea of not needing to bother remembering where things are harks back to discussions earlier in this chapter about using external sources as a component of 'your memory'. SF writers have long been interested in the possibilities inherent when you link access and augmented reality not just to specific functions, but to general access archives, databases and artificial intelligence so that you can harness the power of the internet while doing other things to create a fully augmented mind. Many of these scenarios use the idea of in-sight displays which can provide additional information about the scene or people you are interacting with, for example the names and biographies of the people you meet. In Marvel's *Iron Man* film (2008), Tony Stark's high-tech suit does not just provide him with superhuman strength and the ability to fly, but also connects to his artificial intelligence programme JARVIS which can provide important information into his visual display; either when asked or in anticipation of his needs. The *Commonwealth Saga* books by Peter F Hamilton (2004–2005) portray an intriguing version of connected digital public space which is almost ubiquitous in the population. The majority of people, except those who eschew technology, have implanted 'transceivers' which connect them to the web of information known as the unisphere and personal assistants called 'e-butlers'. Many also possess 'organic circuitry tattoos' (OCtattoos) which may facilitate this connection or have other features such as sensors. This connectedness gives instant access to many different technologies, from augmented vision to sending email-style messages to controlling household lights and heating. One character uses her implanted technology to control a hand-glider style vehicle:

> She put her hands down on the console's i-spots, fingers curling round the grip bars; plyplastic flowed round them, securing them for what promised to be a turbulent flight. The OCtattoos on her wrists completed the link between the i-spots and her main nerve cords, interfacing her directly with the on-board array. Virtual hands appeared inside her virtual vision. Her customization had given them long slender fingers with green nails and glowing blue neon rings on every finger. A joystick materialized amid the icons, and she moved her virtual hand to grasp it. Her other hand started tapping icons, initiating one final systems check. With everything coming up green, she ordered the on-board array to deploy the wings.
>
> (Hamilton, 2004)

As technology develops, these futuristic scenarios become less implausible. Already, gold electrode 'tattoos' have been developed as prototypes which can function to provide EEG (electroencephalogram) readings, giving indication of brain wave patterns (Ma et al, 2010). This is the first step towards such thought-controlled body mounted technology. In our discussion of digital archives functioning as memories, above, we discussed the potential of being able to remember things which we might currently forget, or which originate outside of our own experience. With two-way interaction via EEG reading (or an equivalent technology) our 'thoughts' and 'senses' could act like the internet, being able to draw on multiple sources of information and having it to hand whenever we need it.

Even if technology does not enable this direct connectivity to digital archives, information, and other people, learning in the digital public space could still function in a different way than currently. For example, many academic institutions and organisations are experimenting with the idea of 'MOOCs' (Massively Open Online Courses) which allow anyone with access to the internet to learn via online lectures and teaching materials. This flattening of training and information could lead to more democracy in education; anyone could take any training they wished, which could be delivered by world experts, rather than those who happen to be geographically or linguistically available. This still relies on openness and availability in an unbiased manner: many of the people taking these courses are still from privileged backgrounds, if only because they are the ones with available time to invest.

Rather than sharing memories and thoughts, we could instead consider sharing our mental process: distributed cognition. In Chapter 4 we discussed the significant benefits associated with sharing 'cognitive surplus' and the positive outcomes available through collaborative problem solving, computation across large networks, and distribution of workloads. In terms of digital public spaces for management of cities and large scale data networks, the benefits are evident: in healthcare, traffic management and other complex logistical problems, distributed information and large scale data processing can uncover emergent patterns that would not be visible with a much smaller amount of data. Fluid, changing networks can emerge and coalesce for short timescales, but through digital connectivity can wield real power; as for example the changing needs of tourists in a busy city being analysed and accommodated. But if we connect our own minds we may see specific individual benefits to this kind of data crunching. Featherstone (2009) suggests: 'If we understand human consciousness as emergent from lower-level distributed cognitive processes, then human cognitive and sub-cognitive processes can be connected to distributed mechanical cognition'.

We may find that we become part of an augmented cognition system in which both human and machine are components. This links back to the idea of *extelligence* which was discussed in Chapter 2; except that rather than the sum total of human knowledge being stored externally in archives, it could be directly linked to individuals and potentially form part of a greater unified intelligence. We noted in Chapter 4 that individual human minds do not have the capacity to process the large amounts of data that we now have access to, and that this is something requiring connected digital data spaces. It also requires algorithmic processing, and potentially artificial intelligence software which can tease out the critical implications of the data. Future technologies could potentially involve artificial intelligences operating independently, in order to be able to react more quickly to emergent information than humans ever could, and then feed the processed results back to us directly. As highlighted in Chapter 7, we must be able to trust that any such decisions made on our behalf are made based on reasoning that is in our best interests, and as unbiased as possible.

Chapter 4 also discussed our lives as inhabitants of a digital information space, and how this shapes and is shaped by social constructions. If our level of interconnectedness increases further due to the introduction of new technologies and the further integration of digital public spaces into our lives, we may find that the way we conduct our sociality also changes. At the beginning of this chapter we touched on the idea that creative collaboration is affected by digital connectivity, with the example of transmedia storytelling. It is currently possible to tell broad, deep stories across many platforms, but professional media production companies often still follow a broadly traditional model of film or television as a primary channel of delivery, and other media being used for supplementary material. Paul Booth (2010) describes how a blog, even in the form of an individual post, may not act as a fixed text but rather an ever evolving shared discourse between the original entry and the comments which are appended to it. More and more, we may see these kinds of fluid 'intratextual' objects becoming the norm. There are now emerging instances of both production and consumption that rely on a more distributed media, and this seems likely to continue in future.

Henry Jenkins (2006) highlights how collaborative information sharing in online communities can act as a higher cognitive efficiency model. He examines communities which undertake detailed analyses of the show *Survivors* to uncover 'spoilers' and connect information to uncover facts about the production. He notes that a group working together can be far more productive than any one individual can be, and this collaborative discussion has led to a different kind of culture and expectation around such shows. A group of minds contributing to a shared problem can provide more details and connected thinking than any one individual could gather on their own. This is something that is becoming evident everywhere online, where detailed analysis can be conducted by a group extremely quickly, whether for journalistic investigation of a political voting scandal (Benkler, 2006) or unpicking of complex plotlines and hinted secrets by fans of a particular television show.[10] Collective analysis and discussion online is a much speeded-up version of traditional scholarly sharing of knowledge through publication, retort and revision. The examples here show how connected behaviour can facilitate this collective thinking in a very specific set of challenges, but we can extend this into possible futures where such collective intelligence is brought to many domains and problems. If you are interested in a particular area, why not connect instantly with others, locally, nationally or globally, who may be coming up against similar questions or issues, and may have expertise that can be shared and built upon collectively? The nature of how this collaboration can occur will differ depending on whether we are looking at near futures (where it could be argued to be happening already) or more distant, speculative ones where people are linked more directly when needed, such as the direct linkage between a group of minds in the television show *Sense8* (2015). Being exposed in such a way to a variety of different collaborators could also potentially be a source of greater empathy, giving us insight in to how different people think and what their concerns are.

Case study 8.4: Alison

Alison is trying to find the most effective way to improve the irrigation system in her garden. She remembers that out of 14 years she has lived in her house, 8 years have seen below average levels of rainfall, and she recalls the different percentage of flowers that have survived. She is able to bring these numbers to mind precisely, because they are stored in her cloud based memory storage, and she has a wireless link through which she can access them directly by thinking about it.

She gives the command to widen her search, and 'remembers' solutions that other people with similar gardens have tried. Nothing comes to mind quickly which seems to exactly match what she needs, so she posts her problem on a gardening board. The people there work together to solve her problem and between them come up with a new solution that saves water and lets her grow more plants.

In light of sharing our intelligence and pooling resources with others through networked communication, we might also consider there may be consequences with regards to the notion of self. Through extended self and bodily telepresence, as well as existence as a networked individual in a digital public space, we may find that the edges of what we currently perceive as individuals begin to blur. If we can extend our 'self' out into another physical space where it can act in real time with others, we may find that our connection to our individual bodies becomes less relevant in interactions. What consists of 'me': is it the biological body that I was born with? The prosthetic leg that responds to my mental commands? Or also the secondary body that I control via a digital connection and is located somewhere else?

Soraj Hongladarom discusses not only the extension of self that comes from being able to connect with non-biological technological extensors, but also with other people, suggesting that:

> ubiquitous computing provides support to the idea that the self lacks a core identity in such a way that there is no actual mental or physical entity that functions directly as someone's self. Furthermore, as many selves are able to be connected through the network, they can directly communicate with one another so that real empathy can result.
>
> (Hongladarom, 2013)

This proposes that if ubiquitous computing technology extends the sense of self, and this technology is shared between different people, then the 'self' will overlap with those of others enabling this enhanced empathy. Such an all-encompassing mind to mind network, might enable a 'global consciousness' or 'noosphere'[11] and one can envisage that in a society where you can feel and experience what happens to any other person, there would be increased motivation for peace and harmony. But purely optimistic viewpoints such as this can be risky, and other scenarios are also possible.

As we have seen several times in this chapter with descriptions of futures of digital public space technologies, we can conceive of either positive or negative versions of these scenarios depending on both their design and usage. Stupid or dangerous ideas could spread through a connected consciousness as quickly as good ones. While the idea of a distributed self, and a form of immortality, might sound beguiling, negative aspects have also been explored in fiction. An extreme scenario which could be envisaged as a future endpoint of this type of technology is described in the *Imperial Radch* books by Ann Leckie, which feature a form of 'distributed self' via networking to different physical bodies. The ruler of the empire, Anaander Mianaai, has a distributed consciousness (otherwise used only by artificial intelligences) that allows her to control multiple distant parts of her empire simultaneously, having many 'ancillary' clone bodies that contribute to her combined experience. The books question what might happen to individual instances of these ancillaries if they are separated from the main governing intelligence, be that human or artificial. Mianaai has an 'argument with herself', effectively leading to civil war. Many stories of distributed selves, or combining many minds into a whole, deal with issues of body horror associated with losing individuality and joining a 'gestalt'. A well-known example is the Borg of *Star Trek*, an alien race who lose autonomy as part of the group mind. When the boundaries between individual selves blur, and one mind knows what all the others also know, can we still remain as individuals? A total digital publicness which works by removing individuality has repercussions which must be considered.

So far we have discussed shared human cognition due to connectedness, and briefly mentioned the possibilities of artificial intelligence and its effect in the digital public space. These topics are both important to a concept common to futurism, speculative design and speculative fiction, which has been called the 'singularity'.

The 'technological singularity' is an idea that was initially discussed in the 1950s. First coined by Jon von Neumann, it was described by Stanislaw Ulam as a point: 'beyond which human affairs, as we know them, could not continue' (Vinge, 1993). The name comes from a concept in physics, singularities being an infinitely compact point in space in which the laws of the universe cease to function, and the true nature of which cannot be described or understood; this is the hypothesised nature of a black hole. The technological singularity therefore, is a point in time where models of the future fail to give reliable answers. The idea of the technological singularity was popularised by Verner Vinge in the 1990s, and by Ray Kurzweil (2005) who linked it specifically to the idea of the emergence of superintelligence. This could be achieved by the creation of a 'strong' artificial intelligence smarter than humans in a range of different domains (rather than one specific task) possibly with the ability to make improvements to itself. General intelligence, which would be the necessary first step for this, is a goal which artificial intelligence researchers have sought for many years, but so far with limited success. An AI research boom in the 1970s led to predictions by leaders in the field that the development of general intelligence on a par with humans was imminent, however it became apparent soon afterwards that the leading research direction, rule based learning, was not going to achieve this. Current AI research is seeing a new resurgence based on 'deep learning' and neural networks, and some think it is more likely to be successful, while others predict a similar disappointment to the previous optimism (Adee, 2016). However, if an artificial intelligence model is successful, and achieves 'human' aspects such as creativity and innovation along with rapid processing speeds and vast data processing abilities, we may see it outstrip us as it learns to improve itself.

A common theory of what might happen in such a scenario is demonstrated in the 2013 film *Her*, which features a digital assistant similar to Google's Siri, which is self-aware and appears to display sentience, and sapience. Through the course of the film, the computer intelligences develop as entities, learn about experience and love, but also work together to progress technologically to the point at which superintelligence is reached. At the end of the film, the AI programmes, having transcended human timescales of thought, join together and leave humanity behind. Such scenarios often end with abandonment by the AIs, or their revolt against us. An example of the latter is seen in the television programme *Person of Interest*, which features superintelligent artificial intelligences built initially to monitor behaviour across public and private spaces (via all connected devices) and prevent acts of terrorism.

An alternate version of the singularity is linked not to a wholly artificial intelligence, but the ability to enhance human cognition to a significantly higher level – intelligence amplification as opposed to artificial intelligence (Balsamo, 2011). This is a possible conclusion to the progress of connected digital space and a ubiquitous public: that our cognitive power becomes linked to the extent at which superintelligence is achieved synergistically. Such transcendence is again a popular SF topic. It is possible that such an outcome would lead to godlike attributes, which may be seen as the goal by some, for example Stefano Marzano: 'the ultimate goal of our species is omnipresence, omnipotence, omniscience: the ability to be everywhere, to do everything and to know everything' (quoted in Stoop, 2003). This would be the truest expression of digital public space, but would irrevocably change our society.

Future challenges (ownership, privacy and bias)

As we have seen, there are many possibilities in the future of the digital public space, and therefore there are many potential issues that we must consider when designing technology which will lead us to or through these futures. In Chapters 5, 6 and 7 we looked in detail at challenges of ownership, privacy and bias which are affecting current digital public space technologies. These challenges, and others, will only become more important as we move forward into this technological revolution.

We must, for example, carefully consider the ownership implications of technologies such as the distributed fabrication described above, and connected archives that are available widely for anyone, anywhere to use. While authors such as Lawrence Lessig (2008) and Yochai Benkler (2006) have looked in detail at the legal implications of digital commons and decentralised production, and at new forms of ownership such as Creative Commons licensing, we should also consider how communities of work and production will generate information that is not only not owned by anyone, but cannot be. We are already seeing much knowledge production being done for free, from Wikipedia, to the content of our Facebook profiles, to artwork drawn and posted on Tumblr. We must consider the implications of this economy of immateriality, and create new structures for goods that are not 'used up' but need protections in place for compensation to the contributors. In many cases this may be as simple as giving credit for creation, and we may need to develop new forms of acknowledgment and attribution, perhaps built into the nature of digital objects using technologies like the blockchain.

Ownership in the multidimensional digital/physical hybrid space of the future must also be considered. We can see the first inklings of possible future disputes in this area demonstrated by reactions to the *Pokemon Go* game, where businesses and in some cases individuals are having to choose how they react to virtual monsters roaming the digital spaces which correlate to their physical property. In one reported case, a man in Massachusetts (Sheridan, 2016) noted many people visiting his house, which as a converted church was listed as a 'place of interest' in digital maps and thus a hub of activity in the game. He questioned whether the status of his home as a 'gym' would have a positive or a negative effect on property values. When ubiquitous digital space becomes an integral part of everyone's experience of the world, will you have a say over the digital counterpart of your house or business? Or will you have to contend with businesses filling your space with advertising, or 'digital graffiti' making your home an eyesore? Designers who develop technology for these digital spaces must consider the ramifications for interaction in all dimensions which they cross, including that of the physical.

It is also important to consider the implications for self-ownership of some of the speculative technologies described above. For example, if wearable technologies and other cognitive extension equipment (such as the external memory described above) become integrated into a person's sense of self, would malicious harm to such items be considered damage to property, or bodily harm? People may already feel personally violated or experience a sense of loss if their laptop is stolen or files irretrievably lost, how much worse will this feel if these are more fundamentally integrated to one's sense of self? This is also interesting to consider in the context of proprietary systems such as social media platforms, which may feel like they are 'yours' and part of 'yourself' but are ultimately owned by corporations. If content may be removed at any time for violations of terms of service, who really owns it? If we are developing infrastructure that will come to be relied upon by many people living in digital public spaces, we must be socially responsible in the design of these, and take inspiration from examples of such spaces which have been co-designed by users for their needs.[12]

Issues of intrusion into what is 'yours' are also a factor in privacy concerns with regard to the future of the digital public space. This is of particular concern in the more extreme scenarios described in 'Futures of connectedness' on page 204, if the boundaries between individuals begin to break down. Some people conceive of the singularity as a utopian ideal, but just as often it is a worst-case scenario, a potential end of our species, not necessarily just as a stepping stone to something else, but an unmanageable state for humans to exist in. Being able to experience and know anything occurring means that anyone else can know and experience anything of yours – true unified digital experience would be the end of privacy, and most likely the end of individuality.

As discussed in Chapter 6, privacy is important, and part of basic human rights. A communal mindspace where information is freely shared between many people is likely to reduce individual privacy and mean that your personal thoughts and actions cannot be protected from others. We already share much of our thoughts and daily processes in online spaces via social media, but this, as discussed, is highly curated and we choose much of what makes it into our digital networks and profiles. The speculation we have been doing in this chapter involves direct sharing of all experiences and thoughts in a shared public space: what could in effect be called telepathy or mind reading, and of which it may be more difficult to control the boundaries. This has been explored as much in fantasy fiction as it has in science fiction, and the implications thereof discussed. One example is the practice of 'Legilimency' in the *Harry Potter* novels, which allows one person to look into the thoughts and memories of another. This is portrayed as dangerous skill, often used for harm, which is invasive and must be protected against because it reveals things that people would rather not be known. By contrast, the protagonists in the *Young Wizards* series by Diane Duane share thoughts freely and speak with each other mind to mind. But the benefits of privacy are also highlighted as important. One of the characters is given justification as to why she is losing the ability to directly hear the thoughts of her friend as they grow closer: 'Intimacy is meaningless without barriers to overcome and to lower' (Duane, 2001).

We can also consider how misadventure or malicious intent might damage such a system. Computer viruses and malware can infect connected computers and be shared via communication methods such as email, and also public shared spaces such as social media. If our 'minds', be that our biological brains or simply the digital stores that we choose to use as our memories, can become infected by such viruses, will we find that these might spread and affect our brain? One can also imagine a scenario where a software virus is constructed which removes or adds particular memories to the shared archive, or to individuals (which in some scenarios might be the same thing).

As mentioned above, individuality, and privacy, require barriers. Thus even if we were to upload ourselves to computers we must, if we value these things, remain as individuals. To reference Diane Duane's work again, such an argument is given to a superintelligence on why it should split itself into discrete individual personalities: 'But again and again and again. A thousand of you to share every memory with, and each one able to see it differently… and everyone else'll see it better when the one who sees it differently tells all the others about it' (Duane, 2001). Without the ability to discuss and compare, our experience will be diminished. However, it is possible to take advantage of closeness and connectivity while still retaining individuality and privacy, for example one might conceive of augmented environments which can adjust your level of information sharing dependent on the level of familiarity and trust, perhaps based on information from the connected devices of others.

The degree to which our words and actions persist may, as described in Chapter 6, affect how private they are. Viktor Mayer-Schonberger (2007) suggests that in a world where 'our words and actions may be perceived years later and taken out of context, the lack of forgetting may prompt us to speak less freely and openly'. However, anecdotally at least, this does not appear to be the case as yet. The persistence of online discourse still results in people being embarrassed or discomfited by things they posted earlier, even when they at the time had full realisation that these were 'permanent' records. Perhaps this is because their attitudes and levels of comfort with such things have changed, or because of an assumption that 'privacy in obscurity' means that their words will never be found, and will never become relevant again in the mass of content that is produced constantly. As discussed in Chapter 6, there are also issues of awareness in terms of who has access when the presence of 'invisible' viewers may be forgotten. Two possible future outcomes of this new reality may result: we may see more of a shift in future to such guarding of speech online; or alternatively technology may be designed to include safeguards to more closely replicate the real-world 'privacy' of being able to speak 'off the record'. Such safeguards may provide more awareness of who is reading our words (or be able to in future). If trends of persistence and lack of security continue, this will be necessary to retain our ability to negotiate, collaborate and create.

Case study 8.5: Joseph

As Joseph arrives home, he hears (through his augmented reality earpiece) promotional jingles follow him along the street. These have been placed by advertisers, but stop when he crosses into his front garden, which is his property.

He gets to work creating a video for some music which has been shared in the digital public space; he has taken some footage that he created, and some by other people, and will put it together with the music. He knows that the music has been released with this kind of re-use in mind, and that when he posts his video, the file will include (for anyone who is interested) both the information that he made it, and where every component came from and who made that. He hopes that if people like his work they will ask him to do other things, or contribute to his funding.

Joseph's doorbell rings: it is someone delivering a package, therefore their privacy settings are both set very high: he can access the name of the person, which company they work for and details of what the package is and its delivery route: but he does not have access to their more personal details. Similarly, the delivery woman does not know his personal details, but sees him glowing faintly: she knows that anything they say to each other is being recorded by him and might be shared with her employers.

Joseph's neighbour Manuel comes to say good afternoon, and the glow that the delivery woman sees increases, and also turns blue: this makes her aware that Manuel has a particularly broad social network and things that she says to him might be shared more widely.

Many writers considering the future use extremes of privacy; either utopian ideals (that privacy is no longer necessary because we are all part of the shared noosphere), or dystopian cautionary tales. Current political momentum, from the UK's 'Snooper's Charter' IP bill, to increasing surveillance and governmental control, means that the latter can often be set in the near future. In Chapter 6 we highlighted examples such as Cory Doctorow's *Little Brother*, in which gait recognition cameras are common, yet commonly circumvented. The novel also features heavy use of 'arphids' (RFID embedded items) which are used for tracking movement of objects, and by extension the people that carry them. These too are foiled in their intended use by, for example, people exchanging train passes. (Doctorow, 2008). We must consider similar ways in which aspects of digital public spaces may be circumvented if they are not providing the services or provisions that particular sections of society want. If this pushback is by only a limited proportion of the population, this may be challenging and we must consider the reasons, which may be variously virtuous or harmful to society in general. If the majority of people are positive or apathetic about a technology, it may be because the minority who oppose it have negative intentions in mind, or it may be that they are an oppressed minority who have different needs from the majority. Or they may even be campaigning for freedoms which the majority do not realise they are losing because it does not directly impact their daily lives.

Another example of this is developments using predictive technology to prevent crime, which is familiar from speculative fiction scenarios such as *Minority Report*, but is being developed as a real consequence of connected technologies. Data sources such as emails, texts, chat files and CCTV can be used to predict likely sites for crime and deploy increased policing accordingly. '*PredPol*' is a system which is being adopted in the US and UK and in a similar fashion generates suggestions of areas that should be patrolled based on analysis of recorded crimes (Baraniuk, 2015). While this sounds like a legitimate use of such data for the purposes of crime reduction, one must consider not only whether the loss of privacy is worth this benefit, but also the efficacy of such policing when biases are taken into consideration. The level of false positives (crimes predicted where none occur) and false negatives (crimes which are not predicted but do occur) must be considered with any such technology, and whether conclusions drawn from this data could harm minority communities if biases in the data lead, for example, to a vicious cycle of depravation and harsh punitive measures.

As already hinted at, bias might also lead to unforeseen issues with regards to connected environments and the intelligent software which will need to manage them. If we come to expect our environments to react in anticipation of our needs, we must ensure that their reactions are appropriate, and designed in such a way as to provide equal support no matter who is using them. We might also consider that 'intelligence' is not necessarily a desirable quality in systems that we are expecting to serve us: we should not design systems that can get bored with their roles!

We spoke earlier about the probable necessity of implementing some kind of 'digital forgetting' if we are to build substantial digital memory archives, particularly those that are highly integrated with our biological memories. This also leads to questions of bias, with regard to what gets 'remembered' or 'forgotten'. We may choose to curate our own memories, but what if we also have the power to shape those of others, with features such as the 'right to be forgotten'? We must consider bias as a critical factor when implementing such features in the design of these technologies. We have already established that our biological memories are subjective. If we start using technological extensions to our memory, this will no less be the case, but it may falsely appear that because 'everything is recorded', that it is more objective.

If we are influenced in our reaction to places by digital overlays, for example augmented reality that lets us see and hear the world differently, we may find that this filtered version of the world is extremely subject to bias: both our own, and that imposed externally. This may be benign, unintended, or perhaps used for criminal or unethical purposes. This might be because a hacker removes your ability to see them robbing your home, or because a government prevents you from seeing negative effects of their policies. We must also consider similar effects resulting from self-imposed social biases, the filter bubble discussed in Chapter 7.

Careful design processes are needed to safeguard against negative effects of such invisibility of technological implications. We must be careful that, if we start to rely on the digital public space, and trust records that it includes more highly than we do that of individuals and their memories, that this trust is warranted and hidden biases do not lead to incorrect conclusions. This necessity to consider multiple implications of technology applies across the board, and the final chapter will discuss how we might use a design-led approach considerately in this context.

Key points

- It is almost impossible to predict the future of technology, but examining trends and looking at possible futures can give important insights.
- Using speculation, either through fiction or design, can provoke examination of important ramifications of technological progress, and shape our expectations and hopes.
- Tangibility, connectedness and analytical capabilities are important factors to consider with regard to futures of archives.
- Linked archives could deliver societal benefits but have risks in terms of bias, privacy and consent.
- New storage technologies could change the nature of archives, include novel forms of computing, such as DNA based storage, chemical or quantum computing.
- Archives which can be accessed in a cognitively integrated way can act as digital 'memories', and future developments might give us 'perfect recall'.
- For extended memory to be effective it must be rapidly searchable, and not create overload.
- True shared memory of experience must be 'thick' and include multiple senses, associations and emotions.
- Perfecting such shared or artificial memories might make it difficult to discern the source of such memories as external or internal, and create questions around ownership.
- Being able to directly access experience and knowledge gained from others may improve creativity and innovation, but again relies on effective search.
- With retention of large amounts of digital content, it may be necessary to build in forms of 'forgetting'.
- Humanity is still subject to biological evolution, which selects only for traits which increase reproductive success.
- Shorter term heritable traits may come about through epigenetic change, and via cultural transmission.
- Technology to augment our bodies is the subject of continual research and development, and may change the way we experience our environment.
- Future technology may also involve incorporating our body as part of technological systems.
- These biologically incorporated changes represent the notion of transhumanism, and may culminate in engineering of our species genome. This risks division between those who can and cannot access such technologies, and precariousness if our continued survival is reliant on technology.
- Future technologies may be able to achieve full sensory virtual environments.
- Telepresence, remote operation and body surrogates may allow new ways of interaction with physical spaces.
- Augmented reality may require integrated experience, perhaps utilising multiple senses, to be more widely exploited.
- Experience of the real world mediated by augmented reality technology may be subject to bias.
- Reactivity and intelligence in devices in the physical public space is a step towards ubiquitous computing.
- Reactive environments might employ smart buildings, 'smart dust' or technology on the nano scale.
- The safety of such systems must be considered, in terms of pollution and environmental effects, and how they might be subject to hacking and malicious intent.

- Intelligent anticipatory computing must accommodate diversity.
- Futures of connectivity may mean instant access to information as needed.
- Sharing of cognitive processes and capacity could enable distributed media production, distributed cognition, and collective intelligence brought to bear on problem solving.
- Direct linkage between individuals may have consequences for 'selfhood' and individuality.
- Some have predicted that the technological singularity is inevitable, with consequences for individuality, society, and humanity.
- Questions of ownership become more complex with these projected futures, including those over intangible goods and virtual spaces, and self-ownership of augmented selves.
- Risks to privacy may be introduced by technologies that make it more difficult to control and identify information boundaries.
- Future scenarios involving privacy are often utopian or dystopian extremes.
- Bias is an important factor to consider in future technologies which use algorithmic or other intelligence systems to take actions that affect society, or our perception or experience of the world.

Notes

1 Such a device is the subject of the Qualcomm Tricorder X-Prize, a $10 million competition to develop such a technology.
2 For example, the paper by Sturdee et al (2016) on 'How to build a Voight-Kampff machine' at the CHI conference on human factors in computing systems.
3 www.medopad.com/
4 www.ukbiobank.ac.uk/
5 www.kevinwarwick.com/
6 Such as the 'Carrington Event' which occurred in 1859. If such a storm happened now, it would likely cause major disruption to power grids, satellites and radio communications (Baker et al, 2008).
7 http://innovega-inc.com/new-architecture.php
8 https://hereplus.me/
9 In development by 'Made with Glove' and supported by the Creative Exchange. www.michellehua.co.uk/made-with-glove/
10 The change in the ways that fans interact with televisual media may also be in part attributable to the fact that content is now often available in the digital public space for evaluation and re-evaluation, when previously it might only have been available to view once on original broadcast, or to those who made a purchase of physical media.
11 This term was coined by Pierre Teilhard de Chardin, though arguably with a somewhat different original context and meaning (Fuchs-Kittowski and Krüger, 1997).
12 For example, the 'Archive of Our Own', developed by communities of, mainly women, fanfiction writers (see Fiesler et al, 2016).

References

Adams, D., 1979. *The Hitchhiker's Guide to the Galaxy*. Pan Macmillan.

Adee, S., 2016. Will AI's bubble pop? *New Scientist*, *231*(3082), 16–17.

Allentoft, M.E., Sikora, M., Sjögren, K.G., Rasmussen, S., Rasmussen, M., Stenderup, J., Damgaard, P.B., Schroeder, H., Ahlström, T., Vinner, L. and Malaspinas, A.S., 2015. Population genomics of Bronze Age Eurasia. *Nature*, *522*(7555), 167–172.

Ayalon, O. and Toch, E., 2013, July. Retrospective privacy: managing longitudinal privacy in online social networks. *Proceedings of the Ninth Symposium on Usable Privacy and Security*. ACM, p.4.

Baker, D.N., Balstad, R., Bodeau, J.M., Cameron, E., Fennel, J.F., Fisher, G.M., Forbes, K.F., Kintner, P.M., Leffler, L.G., Lewis, W.S. and Strachan, L., 2008. *Severe Space weather events–understanding societal and economic impacts: a workshop report*. National Academies Press.

Balsamo, A., 2011. *Designing culture: the technological imagination at work*. Duke University Press.

Baltimore, D., Baylis, F., Berg, P., Daley, G.Q., Doudna, J.A., Lander, E.S., Lovell-Badge, R., Ossorio, P., Pei, D., Thrasher, A. and Winnacker, E.L., 2016. On human gene editing: International Summit Statement by the Organizing Committee. *Issues in Science and Technology*, *32*(3), p.55.

Banks, I. M., 1987–2012. *The Culture* (series). Macmillan.

Baraniuk, C., 2014. Digital doctors. *New Scientist*, *222*(2973), p.21.

Baraniuk, C., 2015. Caught before the act. *New Scientist*, *225*(3012), 18–19.

Bassett, C., Steinmueller, E. and Voss, G., 2013. Better made up: the mutual influence of science fiction and innovation. *Nesta Work. Pap*, *13*(07).

Brietbart, P. and Vega, M., 2014. *Posthuman: introduction to transhumanism.* [video] Available at: www.youtube.com/watch?v=bTMS9y8OVuY Transcript available at: https://peterbrietbart.wordpress.com/2014/09/09/posthuman-introduction-to-transhumanism-script/ [Accessed 26 October 2016].

BBC News, 2016. *Plastic microbeads to be banned by 2017, UK government pledges*. [online] Available at: www.bbc.co.uk/news/uk-37263087 [Accessed 28 October 2016].

Benkler, Y., 2006. *The wealth of networks: how social production transforms markets and freedom*. Yale University Press.

Berger, T.W., Hampson, R.E., Song, D., Goonawardena, A., Marmarelis, V.Z. and Deadwyler, S.A., 2011. A cortical neural prosthesis for restoring and enhancing memory. *Journal of Neural Engineering*, *8*(4), p.046017.

Booth, P., 2010. *Digital fandom: new media studies*. Vol. 68. Peter Lang.

Ceurstemont, S., 2014. The virtual, in reality. *New Scientist*, *221*(2950), p.17.

Doctorow, C., 2008. *Little Brother*. Tor Books.

Doctor Who, 1966. (First appearance of the Cybermen: *The Tenth Planet*). TV series. BBC Television.

Dourish, P. and Bell, G., 2014. 'Resistance is futile': reading science fiction alongside ubiquitous computing. *Personal and Ubiquitous Computing*, *18*(4), 769–778.

Duane, D., 2001. *High wizardry*. Houghton Mifflin Harcourt, p.179.

Dunne, A. and Raby, F., 2013. *Speculative everything: design, fiction, and social dreaming*. MIT Press.

Featherstone, M., 2009. Ubiquitous media: an introduction. *Theory, Culture & Society*, *26*(2-3), pp.1–22.

Fiesler, C., Morrison, S. and Bruckman, A.S., 2016, May. An archive of their own: a case study of feminist HCI and values in design. *Proceedings of the 2016 CHI Conference on Human Factors in Computing Systems*. ACM, pp. 2574–2585.

Floridi, L., 2007. A look into the future impact of ICT on our lives. *The Information Society*, *23*(1), 59–64.

Forsythe, C., Kruse, A. and Schmorrow, D. 2005. Augmented cognition. In Forsythe, C., Bernard, M.L. and Goldsmith, T.E. eds. 2006. *Cognitive systems: human cognitive models in systems design*. Psychology Press.

Foster, J.K., 2011. Memory: how memories are made. *New Scientist*, *2841*.

Fuchs-Kittowski, K. and Krüger, P., 1997. The noosphere vision of Pierre Teilhard de Chardin and Vladimir I. Vernadsky in the perspective of information and of world-wide communication 1. *World Futures: Journal of General Evolution*, *50*(1-4), 757–784.

Garcia-Martinez, J., 2016. Here's what will happen when 30 billion devices are connected to the internet. [online] *World Economic Forum.* Available at: www.weforum.org/agenda/2016/06/nanosensors-and-the-internet-of-nano-things [Accessed 28 October 2016].

Garner, J., Wood, G., Pijnappel, S., Murer, M. and Mueller, F.F., 2013, September. Combining moving bodies with digital elements: design space between players and screens. *Proceedings of the 9th Australasian Conference on Interactive Entertainment: Matters of Life and Death*. ACM, p.17.

Hamilton, P.F., 2004–05. *The Commonwealth Saga* (series). Macmillan.

Hamilton, P.F., 2004. *Pandora's Star.* Macmillan.

Heinlein, R., 1942. Waldo. *Astounding Science Fiction.*

Her, 2013. Film. Directed by Spike Jonze. USA: Warner Bros Pictures.

Hodson, H., 2016. Printable electronics for all. *New Scientist, 230*(3079), p.22.

Hongladarom, S., 2013. Ubiquitous computing, empathy and the self. *AI & Society, 28*(2), 227–236.

Iron Man, 2008. Film. Directed by Jon Favreau. USA: Paramount Pictures.

Jeffries, S., 2014. Neil Harbisson: the world's first cyborg artist. [online] *The Guardian.* www.theguardian.com/artanddesign/2014/may/06/neil-harbisson-worlds-first-cyborg-artist [Accessed 26 October 2016].

Jenkins, H., 2006. *Convergence culture: where old and new media collide.* NYU Press.

Jinek, M., Chylinski, K., Fonfara, I., Hauer, M., Doudna, J.A. and Charpentier, E., 2012. A programmable dual-RNA-guided DNA endonuclease in adaptive bacterial immunity. *Science, 337*(6096), 816–821.

Kahn, J.M., Katz, R.H. and Pister, K.S., 1999, August. Next century challenges: mobile networking for 'Smart Dust'. *Proceedings of the 5th Annual ACM/IEEE International Conference on Mobile Computing and Networking.* ACM, 271–278.

Kurzweil, R., 2005. *The singularity is near: when humans transcend biology.* Penguin.

Leckie, A., 2013–15. *Imperial Radch (*series). Orbit.

Lessig, L., 2008. *Remix: making art and commerce thrive in the hybrid economy.* Penguin.

Ljubič, K. and Fister Jr, I., 2014. How to store Wikipedia into a forest tree: initial idea. *Presented at First International Conference on Multimedia, Scientific Information and Visualization for Information Systems and Metrics.* Available at: www.iztok-jr-fister.eu/static/publications/35.pdf [Accessed 26 October 2016].

Ljubič Fister, K., 2016. Breaking the wall of data storage. *Presented at Falling Walls Lab Berlin 2015.* Available at: https://vimeo.com/148192811 [Accessed 26 October 2016].

Lothian, A., 2013. Archival anarchies: online fandom, subcultural conservation, and the transformative work of digital ephemera. *International Journal of Cultural Studies, 16*(6), 541–556.

Ma, R., Kim, D.H., McCormick, M., Coleman, T. and Rogers, J., 2010, August. A stretchable electrode array for non-invasive, skin-mounted measurement of electrocardiography (ECG), electromyography (EMG) and electroencephalography (EEG). In *2010 Annual International Conference of the IEEE Engineering in Medicine and Biology.* IEEE, pp.6405–6408.

The Matrix, 1999. Film. Directed by The Wachowskis. USA: Warner Bros Pictures.

Mayer-Schoenberger, V., 2007. *Useful void: the art of forgetting in the age of ubiquitous computing.* KSG Working Paper No. RWP07-022. Available at: SSRN: https://ssrn.com/abstract=976541. [Accessed 7 November 2016].

McCaffrey, A., 2012. *The ship who sang.* Random House, p.8.

Minority Report, 2002. Film. Directed by Steven Spielberg. USA: 20th Century Fox.

Moggridge, B., 2006. *Designing interactions.* MIT Press Books.

O'Neill, S., 2016. I plant memories in seeds. *New Scientist, 229*(3056), p.27.

Pacific Rim, 2013. Film. Directed by Guillermo del Toro. USA: Warner Bros Pictures.

Parks and Recreation (Season 7), 2015. TV series. NBC.

Person of Interest, 2011–2016. TV series. CBS.

Pfeiffer, M., Dünte, T., Schneegass, S., Alt, F. and Rohs, M., 2015, April. Cruise control for pedestrians: controlling walking direction using electrical muscle stimulation. *Proceedings of the 33rd Annual ACM Conference on Human Factors in Computing Systems.* ACM, pp.2505–2514.

Ranner, V., 2013, December. UISilk: towards interfacing the body. *Proceedings of the Second International Workshop on Smart Material Interfaces: Another Step to a Material Future.* ACM, pp. 13-18.

Scalzi, J., 2014. *Lock-In.* Tor Books.

Sense8, 2015. TV series. Netflix.

Shafy, S. 2008. An infinite loop in the brain. [online] *Spiegel International.* Available at: www.spiegel.de/international/world/the-science-of-memory-an-infinite-loop-in-the-brain-a-591972.html [Accessed 3 November 2016].

Sheridan, B., 2016. Living in an old church means many things. Today it means my house is a Pokémon Go gym. This should be fascinating. [Twitter thread] Available at: https://twitter.com/boonerang/status/751849519407595520 [Accessed 27 October 2016].

The Six Million Dollar Man, 1973. TV series. ABC.

Sterling, B., 2005. *Shaping things*. Mediaworks Pamphlets, MIT Press.

Stoop, E., 2003. Mobility: freedom of body and mind. *In*: Aarts, E. and Marzano, S., eds. *The new everyday: views on ambient intelligence*. 010 publishers, p.140.

Stross, C., 2011. *Rule 34*. Hachette UK.

Sturdee, M., Coulton, P., Lindley, J.G., Stead, M., Ali, H. and Hudson-Smith, A., 2016, May. Design fiction: how to build a Voight-Kampff machine. *Proceedings of the 2016 CHI Conference Extended Abstracts on Human Factors in Computing Systems*. ACM, 375–386.

Swain, F., 2014. I can hear Wi-Fi. *New Scientist, 224*(2995), p.20.

Swain, F., 2016. Heard but not seen. *New Scientist, 231*(3080), p.20.

Teperek, M., Simeone, A., Gaggioli, V., Miyamoto, K., Allen, G., Erkek, S., Peters, A., Kwon, T., Marcotte, E., Zegerman, P. and Bradshaw, C., 2016. Sperm is epigenetically programmed to regulate gene transcription in embryos. *Genome Research*, pp.gr-201541.

Thompson, H. 2016. Being in a robot's shoes. *New Scientist, 3097*.

Verbücken, M., 2003. Towards a new sensoriality. *In:* Aarts, E. and Marzano, S., eds. *The new everyday: views on ambient intelligence*. 010 publishers.

Vinge, V., 1993. The coming technological singularity: how to survive in the post-human era. *Vision 21: Interdisciplinary Science and Engineering in the Era of Cyberspace*. Vol. 1. 11–22.

Watkins Jr, J.E., 1900. What may happen in the next hundred years. *Ladies Home Journal, 8*.

Weiser, M., 1991. The computer for the 21st century. *Scientific American*, 265(3), 94–104.

Yokota, T., Zalar, P., Kaltenbrunner, M., Jinno, H., Matsuhisa, N., Kitanosako, H., Tachibana, Y.,
Yukita, W., Koizumi, M. and Someya, T., 2016. Ultraflexible organic photonic skin. *Science Advances*, *2*(4), p.e1501856.

9 Design processes and management in digital public space

Throughout this book we have illustrated how digital tools and technology are having a major impact on society, and how humans have developed and responded to these technologies. We have also looked at challenges that are arising through implementations of technology when the full implications may not have been fully considered by those who create such systems and those who use them. We have looked into the future and considered the possibility of both near and far futures that illustrate how fundamentally digital technology is influencing almost every aspect of human life on the planet.

Clearly, how we develop and design future technologies now can influence how we help people move between the digital and the physical worlds and combine them into a cohesive whole for themselves and for others. The design of digital public space/s and associated technologies goes beyond individuals and communities to embrace both the material and immaterial world, and will affect all aspects of people's lives because the technology that is being designed now will itself have these wider emergent and unknown properties. In this chapter we will therefore examine the role of design in the development of digital public space, and why it is important going forward.

Designing futures

Designing the future is challenging; professional designers are trained to create futures most often through incremental changes to existing products, places and systems. However, occasionally they deliver radical changes through thinking beyond the short term, fundamentally rethinking the user experience and applying emerging technologies. For examples of this in everyday items, think about Dyson's engineering of vacuum cleaners, hand dryers and other products, Apple's development of the Smart phone, or indeed the world wide web and wifi as the facilitator of numerous digital services.

Digital public spaces will result from both incremental and radical design. However, if we are to consider the raft of new issues that arise from this new domain (privacy, security, trust, bias and other issues discussed in previous chapters), a new era of digital public space design is needed. This will mean understanding the product service system; it will mean understanding the human physical/digital experience; and it means incorporating service design, policy design, participatory design and co-creation methods to take a whole systems approach rather than just creating products that 'do' something.

Designing any type of future product or service always comes with responsibility. In the 60s and 70s environmental design and inclusive design philosophies emerged, followed over the years with a broader perspective of socially responsible design that covers multiple perspectives; from design against crime, or design for health and wellbeing to latterly design for social innovation. It follows then that this carries forward into the digital design. Surely because digital technology and digital public spaces are so flexible, it is up to us to design our spaces in a way that enables us to live within them better, as spaces that have qualities that can enhance the human condition, our culture, and allow people to have a more meaningful engagement within them. Socially responsible design means incorporating not just young and experienced users that have the enthusiasm and drive to create new things, but also older people who might struggle with the transition to new forms of culture and society that will inevitably result when we have such fast-moving development.

Both in the near and far future it is clear some platforms and services that we now consider the cutting edge, or intrinsic to how parts of digital public space now function, will not survive. This might be due to them becoming obsolete through the introduction of new, superior forms. Or it might be because their utility is transient and reliant on a specific set of circumstances that passes, or is part of a life cycle.[1] We are already seeing issues in digital public space where spaces and services that are created for the benefit of all are being taken advantage of by a small number and thus risking the whole system. Rheingold (2002) points out that this scenario is familiar from other domains:

> The presence of flamers, bullies, bigots, charlatans, know-nothings and nuts in online discourse poses a classic tragedy of the commons dilemma. If too many people take advantage of open access to seek other people's attention, the excesses of the free riders drive away the people who make the conversation valuable.
>
> (Rheingold, 2002)

On the other hand we must also, as referenced above, think about contingencies for systems which become so embedded in our way of life that their collapse would be disastrous. As we currently find it difficult to imagine a world without the written word, new generations may be unable to remember or conceive of a world before ubiquitous anticipatory computing. We used to talk about surfing the web, but this is not a phrase that gets used recently; not because we are no longer going online, but that we are always online and no longer need a phrase to distinguish entering that space. Modern teens find the idea of having to wait to connect to the internet almost unimaginable, and cannot remember a time when they did not have access instantly to online information.[2] Floridi (2007) predicts that 'in the near future, the very distinction between online and offline will become blurred and then disappear. To put it dramatically, the infosphere is progressively absorbing any other space'. Some would argue that this has already happened.

Writing is 'low technology' in that it utilises what we were in Chapter 2 referring to as 'tools' rather than 'technology'. You can write and be understood just as easily with your finger and a patch of sand as you can with a typewriter, or a digital pad and stylus. The only thing that changes across these different media is the permanence. In contrast, digital public space relies on a whole suite of technologies that are complex and distributed. Much of it relies on electricity, which currently is still mostly generated by burning unsustainable fossil fuels. Smartphones rely on rare metals that are in finite supply. We must be cautious of building our society and culture on foundations that are ultimately unsustainable, unless we are prepared to put resources into finding ways to make them so.

For a small scale exemplar of this, we can take the example of the authors' personal experiences during floods which took place in Lancaster in 2015. The flooding affected the power station, and meant that electricity, mobile telephone and data signals were not available to the city for several days. Suddenly, undergraduate students who had never had to use a telephone box before queued to use the landline and connect with the outside world. While nationally, social media discussed the floods and their results at length, people in the city were cut off from this discussion and could neither see it nor contribute due to lack of ability to connect. Much information sharing was done by reverting to a backup technology – that of radio, available to those with battery operated and wind-up radios, which provided a crucial service.

Future planning and design cannot plan for all eventualities, in part because if digital public spaces are truly open, shared and democratic, then they will also be open to appropriation and use in ways which their designers never imagined. As we have seen, digital public spaces, and the information space, is already emerging as a consequence of the connected technologies which are being rapidly developed. The question is, how can we design digital public spaces with forethought, so that what emerges is designed considering some of the issues discussed in this book. If we do not create these spaces, they will be created for us, and we might not like the results.

Being a designer in digital public space

Norman Potter in 1969 stated 'Every human being is a designer. Many earn their living by design – in every field that warrants pause, and careful consideration, between the conceiving of an action and a fashioning of the means to carry it out, and an estimation of its effects'. In the context of digital public space, we should perhaps realise that every human being will in one sense be a designer, whether that is accessing content, repurposing it, re-using it and reissuing it, or whether it is a journalist, publisher, broadcaster or other professional creating content, through access to data and information. Indeed, the question of who is creating digital public space is not insignificant; the answer is everyone incrementally through interaction with the digital technologies that form it. However, for the purpose of this chapter we will focus on those people who are currently engineering and designing new software, hardware and interfaces that will radically change the way in which we interact with digital public space. In the Creative Exchange project that informs the case studies in this book, the research team that worked on the products and services comprised people with backgrounds in art, design, computing, engineering, architecture, biology, psychology and social science. These people were the designers, a multidisciplinary group working together to develop the products and service. In many cases they also co-designed with users, consumers, manufacturers and all the other stakeholders in the production, adoption and use of the products and services. In this chapter, we are in the first instance addressing the group of professionals that will choreograph the development of new digital public space technologies and tools: these are digital public space designers of the future.

Traditionally, design as a discipline and profession has been practiced in silos, for example, of product and industrial design, graphic and information design, fashion and textiles design, interior design and architecture. The focus for these disciplines has been on discrete problems, such as developing a new product, article of clothing or building and interior, print media and more recently web based communication. As the digital products emerged, designers of both hardware and software worked alongside interface designers, and engineers of many specialisms, and now as we see with the Creative Exchange team and in companies such as Google and Facebook, new configurations of skills and competencies are required.

Designing digital public spaces, as we saw in our Creative Exchange project, will require multidisciplinary teams, including the 'owners of the space', the creators of the architecture, digital and physical, designers of the business models and revenue streams, licensees and the inhabitants/users of the space. Accepting the principle that skills and competencies required for designing digital public spaces will change and develop, it is necessary to consider what critical concerns need to be taken into consideration and how the process of creating the form and function could take place.

Designing interactions

Designers have always attempted to learn about customer experience in order to apply that insight to the design of the product and service, therefore how users interact with the digital and online experience resulted in interaction design as a design practice and discipline that has emerged over the past 10-20 years. Its origins may be found in human-computer interaction, ergonomics and user centred design, where the underpinning scientific contribution comes from psychology and behavioural science on the one hand, and physiological and anthropomorphic understanding on the other. It is primarily concerned with designing how the human interacts with an aspect of the physical or digital world in order to achieve an objective. Ehn (2006) suggests that:

> Interaction design is not computer science, or even human-computer interaction (HCI), even if it deals with humans, computers, and interaction. Interaction design is not graphic design, even if it is both visual and communicative. Neither is interaction design another computer study like computer-supported cooperative work (CSCW) or participatory design (PD), nor another design discipline like product design, architecture or media studies, even if all of these disciplines and practices and many others in the background give shape to interaction design.

This illustrates the point that we develop hybrid design disciplines to deal with the complexity when the digital and physical worlds collide. Ehn also quotes De Michelis (2003), saying that 'interaction design deals with a new kind of combined interactive narrative and architecture, a kind of mixed object' and indeed places.

Paul Dourish (2001) in 'Embodied Interaction' looks at the philosophical basis of human-computer interaction reflecting phenomenological approaches of, for example, Martin Heidegger, Ludwig Wittgenstein, and other twentieth-century philosophers. He takes an approach to interacting with software systems that emphasises skilled natural practice over 'disembodied rationality'. He suggests that tangible and social approaches to interaction are related, and that by analysing them, we can understand embodied interaction and thus it helps us to design future systems.

This focus on embodied interaction is critical when working in digital public space, because as we have seen in previous chapters, a fundamental aspect of the way that digital information integrates with our experience is the embodied nature of the interaction. From the way we incorporate avatars into our sense of self, to the 'ambient awareness' of connected technology, how we interact with the technology will be affected by our behaviour as evolved human beings. While interactions are complex and often unpredictable, we must not forget to allow for the ways in which humans naturally react to certain stimuli that may not have been intentionally incorporated into how technology works, but emerge as a function of its form, or influences of those who design it.

We must also take into account that digital public space functions as a means of making connections between people. Therefore, the interactions between people, mediated by the technology, are as important as any one individual's interactions with the product or service that is being designed. As multiple people interact, the outcomes become more complex.

Working in digital public space means radically reconsidering the notion of interaction design, because it is not simply the interaction between known elements that can be designed and delivered. It is also understanding how emergent qualities will be generated in any system built upon the interaction between people and products, people and places, products and places, people and physical and digital spaces/places and any further combination. You are not simply designing the specific interaction, but the potential interaction and indeed the whole journey.

What is also important to consider in this context is that we are designing interactions that may take place in the material world, or in the digital world, or in the hybrid digital physical spaces which we have been describing. Design decisions taken for one of these particular contexts may have repercussions in the others, for example as discussed in Chapter 8: Google Maps emphasising particular features of landscapes or cities might affect their usage by people who use the maps for navigation.

Designing interactions means understanding the key touch points of the journey and designing both the route and the place and how the user/ 'journeyman' can navigate the space (whether this be a physical, virtual or hybrid space) to achieve the desired objectives without (adding to their) stress whilst enhancing their (joy of the) experience, making the emergent process intuitive and ***clear***. This is the great design challenge; we will discuss what this might mean in more detail below.

Design using cognitive/behavioural information

Moving forward with user centred design and designing experiences, we can explore further how this relates to earlier chapters in relation to specific biological, behavioural and cognitive aspects of human development. Two principles are relevant here, first the adoption and use of higher order tools (technology developed through cumulative culture), and secondly the behavioural and cultural tendency towards sharing and collaboration.

Use and adoption of tools and technology

We can look first at our use and adoption of tools and technology. In Chapter 2 we explained that while use of tools is not itself unique to humans, the incorporation of invention and technological development into our culture is. Clark in *Natural-Born Cyborgs* (2004) similarly argues that what makes humans different from other species is our capacity to fully incorporate tools and supporting cultural practices into our existence; that all forms of technology from the pencil to a neural implant can be exploited by our adaptable and plastic brains. We have adapted our lives though the way we use technology, for example the measurement of time remaking our social interactions, and digital information storage remaking our memories.

One critical question we might pose in designing these tools, is how much we are aware of them as an extension to our bodies and cognition. This can be thought of in terms of 'transparency'. Norman (1998) argues that the focus needs to be on the user and to make, for example, a computer more convenient and pleasurable, the technology needs to be 'invisible and hidden from sight'. This is often referred to as transparent, that is, technologies which are so well adapted to our lives that they become invisible and fully integrated. When this occurs, the fluidity with which technology is used means that no conscious thought is required for its implementation. An example of this might be a highly skilled typist touch-typing on a keyboard, for whom it is as rapid and unthinking as writing or speaking. In contrast, opaque technology requires skill to use, is in conscious awareness, and can be separated from the user. Using opaque technology requires constant focus and cannot be forgotten about without interrupting its use. *Therefore, designers must consider: whether technologies should be opaque or transparent, or indeed when is it appropriate for them to be transparent or opaque?* Perhaps invisible versus visible is another and better terminology for reasons described below.

Transparent technologies that are not in conscious awareness become integrated; they become a part of our 'extended cognition' as described in Chapters 4 and 8. 'What matters is that as far as our conscious awareness is concerned, the tool itself fades into the background becoming transparent in skilled use. In this respect the technology becomes, to coin a phrase "pseudo-neural"'(Clark, 2004). This can occur when the technology is additional to the body but is no longer noticeable as something outside of our 'selves'. Clark describes an example of this with the non-digital technology of a watch: you might be asked on the street if you know what the time is. The answer to this would normally be given as 'yes'; even though the time information is external to yourself, and needs consultation of your augmented, external cognition (the watch), it is part of the information that is 'known' to you and treated as internal to your sense of self. In contrast, if you are asked 'do you know how many films Tom Cruise has made?' you would not necessarily say 'yes' if you needed to consult the internet via your smartphone to answer the question. If digital public space content becomes accessible and as second nature as consulting a watch, this may change. If we are asked a piece of information that is stored in our 'external' knowledge bank of digital public space, we may say we know it because it is immediately available to us even if it is outside the body. Thus the boundaries of the thinking system are fluid and related to ease of access.

This transparency might be a good thing if it is the intended outcome. The touch typist described above does not have to think about which keys to hit; it is almost as easy as speaking because the keyboard is integrated enough to be an output of their expression. *If designers wish to augment our cognitive capacities with digital public space, then transparency is necessary for our plastic brains to integrate them fully. Technology does not need to be simple to be transparent. Indeed it may have to be extremely complex to be truly transparent.*

Smartphones can be good examples of transparent technology. They are often designed to be a natural fit to biological profiles, fitting smoothly into the hand and with screens that draw the eye to the designers' intended areas. Apple technologies in particular are designed for 'out the box' use, where there is no need for a manual or lots of training. These human centred products rather than technology centred products are designed for intuitive use, and as a result can be used by children before they walk or speak. A study (mashable.com 2016) in the US showed in 2016 that 38% of US children under the age of two are using mobile devices. Smartphones are now so ubiquitous that they may become something as second nature to using a pencil – which does not require cognitive effort to use as a tool, and you can 'switch off' while using. Many other companies, from vacuum manufacturers to those that make washing machines are trying to mimic this aspect of using technology and make the cognitive effort required for use as low as possible. This extends to accessing knowledge from the internet directly via asking, due to emerging voice recognition services such as Amazon Alexa or Apple Siri. *The cognitive load is being placed on the technology, and its designers, rather than the user.*

This notion of transparency or technology as second nature was what Norman (1998) calls the need for information appliances. An information appliance is geared to support a specific activity, and to do so via the storage, reception, processing and transmission of information. Information appliances form an intercommunicating web; they can 'talk' to each other, and are designed to be easy to use and to fade into the background. These systems are poised to be taken for granted, and if they do so may form what has been described as 'pervasive computing'. With the emergence of the internet of things (and the internet of nano-things, as described in Chapter 8) plus the availability of huge quantities of data, digital public space has the potential to be that transparent technology on a grand scale, that permeates our environment but can be entirely forgotten about.

While we can attribute advantages to tools that are complementary to our biological brains, biological brains are good at certain types of things, and not others. They are very good for example at pattern matching and simple associations, while not good at recalling long arbitrary lists of instructions and numbers. Good technology is complementary; thus it might be that we prefer to outsource some of the computational work to external sources rather than integrate it into our cognition.

There are a great many perils in the development of tools to navigate the digital and the physical public space, as described in earlier chapters. Significantly we are often designing artifacts and affordances that change us forever. We can see that once these aids are transparent, and successful, they can become addictive. For instance, head-up augmented reality such as Google Glass and similar technologies make more apparent the sea of information in which we live; if these are fully deployed they may make it even easier for constant access. It is possible that people will in consequence feel 'lost' when they take it off, as many people now feel bereft without a mobile phone, with loss of the internet. Already we do not remember phone numbers as we used to because we do not need to; because they are stored in our phones rather than our biological brains. It may be that we prefer not to make vital functions transparent because of the risks of forgetting that we are vulnerable to their loss.

Transparency versus opacity is not the only design choice that is related to our behavioural systems.

For example, the vast amounts of data now available via pervasive data monitoring and devices presents an opportunity to understand and manage the dynamics of human behaviour. As discussed in Chapter 7, the power of bias in digital information can hide information or present it in a particular and purposeful way, creating particular stories or messages targeted at reinforcing, influencing and indeed changing behaviours. A dystopian extension of this would be that systems could be designed to control behaviour. Weise et al (2012) talk about design choices that can potentially promote or inhibit democratic participation in the design, control and use of the systems and data.

More sinister perils lurk in transparency that might mean ease of use and also unconsidered use; who for instance is in the space, and are they operating with good intentions or will their actions have negative consequences for individuals, communities and society at large? We have discussed in Chapters 6 and 7 the danger of invasion of privacy, of misuse of data, of cyber-attacks and of media and information bias. Indeed there are as many opportunities to do harm as good in digital spaces, and as many as there are in physical spaces, often with less ability of the ordinary citizen to observe and identify nefarious activity. Therefore, in the development of digital public space and the affordances it offers, it is critical that designers take care to consider the value of transparency/invisibility of the technology. Technology may need revisiting to make visible attributes that illustrate vulnerabilities in terms of identity, trust, and security of individuals and communities, ensuring that as we adapt as a species, we do so with a positive benefit overall.

Towards sharing and collaboration

We now turn to the notion that human behaviour and culture is intrinsically linked to *sharing and collaboration* (something we naturally tend towards) especially in public spaces which are in fact domains of sharing where we have the natural tendencies to adopt social and sharing technologies and practices. Digital sharing can be of 'non-rival' or 'rival' goods that are multiplied or destroyed by their consumption; sharing knowledge through sites such as Wikipedia or using the internet to share services and goods. As discussed, we also now share the design of products, services and systems through participatory and co-design. *If sharing is a key factor in digital public space, we need to consider how to design sharing mechanisms and what forms of technologies are appropriate, that is, transparent, opaque or visible.* Because as we start to think about the design of systems, how easy they are to use may have an influence on the degree to which we can protect our privacy and identity and the measure of trust we apply to such systems.

That sharing is an intrinsic property of interaction in digital public space also leads to another key property: that the functions and aspects of digital public space are *emergent.* Paul Booth (2010) in describing online spaces including collectively updated wikis and blogs that have spaces for commentary and discussion, notes that the content is never finished, always evolving and part of an ongoing conversation that he calls intra-textuality; 'meaning that occurs *inside* the trans-mediated text itself' (his emphasis). This fluid, combinatorial aspect of meaning arising from the combined effort of multiple contributors is something that reflects the nature of digital public space in general: it exists not as a collection of content on a computer or collection of computers, but of the interactions that people have within and through digitally mediated spaces.

As described above, interdependency is a key factor when it comes to these kinds of interactions, and as examined in Chapter 4, this stretches out in many dimensions. What we do in a virtual world may impact on our interactions with those whom we see in physical spaces, and vice versa; and information sharing can lead to outcomes which no single individual or group that was functioning without the support of digital technology could have achieved. The form that this mediated interaction takes will therefore affect these multidimensional outcomes, and we can see many phenomena in our current digital public spaces that have arisen because of quirks and choices made in the design of these systems. Because digital public spaces are often not created through the intentions of one person or group, it is not intentionally designed and therefore the outcomes are not predictable. Sometimes features that were created for one purpose can be appropriated and used for something different, or cause 'side effects' that have repercussions for the wider system.

In terms of sharing, collaboration and digital public space in organisations, there is a need to understand the organisational strategy, cultural change and investment necessary to create digital public space as it applies to them. Tony Ageh (2012) sees digital public space as something you design in an organisation; it does not just grow: 'Designing something that serves a purpose is very different from just letting the medium grow without purpose. It's the difference between the BBC and broadcasting, between Radio 4 and radio.' An organisation like the BBC whose mission was to 'Inform, Educate and Entertain Everyone equally and without systemic privilege or favour' through a licence fee, that also needs to generate further funds, must consider how it turns its massive archives into a digital public space: should it be for free or must designers put in systems for generating income. As Ageh says:

> The UK's public sector owns or controls a wealth of assets which have been built up over many decades, centuries even, going all the way back to the Domesday Book.... In the UK we have over 2,500 museums and galleries – many of which maintain unique and irreplaceable archives and libraries. Yet very few have the people with the digital skills or have the resources to liberate their value – either for paid or unpaid access. And apart from our 6 National Libraries, there are a further 1,000 academic libraries and of

> course over 4,000 public libraries. They are largely still mainly in analogue form. The overwhelming majority of the UK's assets have yet to be digitised. And it is pretty much unimaginable that there will ever be enough public money to digitise them all or even a significant proportion of them. They incorporate a remarkable diversity of material – ranging from documents to 3D objects, from books to films, from paintings to microfiche. From century to century, from nation to nation, from prince and from pauper.
>
> (Ageh, 2012)

It is clear that most of the UK public sector does not have a clear policy on how it responds to the notion of digital public space, though in a sense they have a duty to make their archives public and have to develop a strategy for the design of their public spaces. Other public sector bodies like the National Health Service or local government must define how they engage with digital public space while ensuring they understand the dimensions of privacy ethics and security. At the same time in the commercial sector, assets related to digital public space are being designed and released onto the market rapidly, changing the nature of the space every day. The pace of new development is in some senses outpacing larger organisations' ability to respond. Organisations like the BBC must adapt to the changing technological landscape and come up with new paradigms outside the commercial space otherwise we will lose freedoms that we did not even know we had.

Design fundamentals

At this point, having considered some specific factors in relation to digital public space as it has developed organically to date, we might consider some fundamental issues that relate to the design of anything and how they apply to designing digital public space.

Designing for the user

Obviously the user is still a key focus of the design system, and Balsamo (2011) argues that designers should pay attention to the technological literacy of the intended users of the technology under development. It is critical to consider their cognitive capabilities and how far they can take or be taken by the technology. Balsamo states:

> How one does this – or the timing of such considerations – is not an issue of methodology or design protocol; rather it is an insight of the technological imagination, exercised in the process of technological development. To address this question consciously – of the technological literacy of the intended users – raises a set of related questions: how does one design for audiences (and their literacies) that don't yet exist? How does a designer design new technologies that require skills (literacies) that are not yet common? How do new technologies reconfigure the literacies that will be common in the future?
>
> (Balsamo, 2011)

Here the design team must not just understand the needs, behaviours and desires today, but imagine the desires of the future, when technologies of today are as incorporated into our cultural literacy as the idea of reading and writing.

Using principles of cognitive engineering, that is, using theories and models from cognitive psychology to support the design of digital public space, is not enough, since often these approaches are potentially superior but are difficult to transfer into practice with limitations of ecological validity. Therefore, they are limited in applicability to real-world problems. Applied psychology attempts to address this gap, by user modelling, combining current theories of human cognition with real-world systems.

Despite the fact that designers have to some extent always thought they were designing for a customer, it was Donald Norman who popularised the subject (Abras et al, 2004) through his publications *User-Centred System Design: New Perspectives on Human Computer Interaction* (Norman & Draper, 1986) and his seminal book *The Psychology of Everyday Things* (Norman, 1988). Interestingly the computing world has often led the way on this perspective, but as Norman expanded it, user centred design became a central feature of most design disciplines. This means that the user is at the centre of the design process; the role of the designer is to ascertain the needs of the user, to incorporate that understanding in the design and to test it with the user throughout development. Norman advocated 'user centred design, a philosophy based on the needs and interests of the user with the emphasis on making products usable and understandable', and suggested that:

> Design should:
>
> - Make it easy to determine what actions are possible at any moment (make use of constraints).
> - Make things visible, including the conceptual model of the system, the alternative actions, and the results of actions.
> - Make it easy to evaluate the current state of the system.
> - Follow natural mappings between intentions and the required actions; between actions and the resulting effect; and between the information that is visible and the interpretation of the system state.
>
> (Norman, 2002)

In other words, make sure that 1) the user can figure out what to do and 2) the user can tell what is going on as well as, making the technology transparent for users.

This focus on users has led to the introduction of various methods to ascertain the users' needs, so for instance, social science methods of surveys, focus groups, interviews and so on. Furthermore sociologists, psychologists and anthropologists have been employed to apply such methods, rather than designers who may not have the knowledge or skills to apply the techniques with any rigour. This approach has evolved further to include the user in the design process resulting in participatory design and more recently co-design, whereby the users are brought into the design process.

As discussed, digital public space is not solely related to an understanding of the human-computer interaction: the technology and tools are truly ubiquitous and embodied in both the human and the physical environment, so as in the HCI world (Harrison et al, 2007) we are in an epoch of embodied and situated technology. Understanding the users' cognitive and physical biology is critical to designing imaginative and emergent attributes of a digital public space. But possibly understanding the user as part of the whole system is also important.

Designing for the social/cultural context – inclusive design

In designing in and for digital public space, we are designing for a shared space, and we are designing a social space. It is therefore critical to understand not only the nature of individual users but also group behaviour. This group behaviour can cause positive emergent outcomes but also negative ones, which can be subject to debate as we have seen with the appropriation, use and misuse of social media by various activists, especially in the political arena. Donath (2014) observes that the design of social network services influences the structure of the network that arises with them, the communities that use them and the trustworthiness or other factors of people using them. Design decisions therefore affect the reliability of digital public space, and the networks, and therefore influence behaviours and cultures. The question is, can designers foresee a social purpose? In some cases they might, when they identify and design to address a specific social problem, but it is difficult if not impossible to predict how an asset of a digital public space might be appropriated by a community. Despite this, there are mechanisms to help a design team, for example testing an idea against the crime lifecycle model – see p. 233 – (Erol et al, 2002), or a sustainability model (Cooper & Boyko, 2010).

Designers are naturally trained to think in narratives or stories of use, and given the appropriate tools and reverences to agendas to be considered (for instance cyber-crime or sustainability) can narrate potential scenarios of use and emergent use. As Balsamo (2011) states 'Designers work the scene of technological emergence: they hack the present to create the conditions of the future'.

Balsamo illustrates the importance of deconstructing culture around the act of designing – that you must understand the cultural context that different people are working in, not because it is better or worse or essentially based on gender, race and so on, but because it is different depending on culture and experience, and individuality. Balsamo considers the need to cultivate imaginations as ingenious in creating new democratic cultural possibilities as they are in creating new kinds of technologies and digital media. For digital public space, the development of technical, user social and cultural imagination in a design team is critical to ensuring adoption and use of assets with digital public space that is not only beneficial to society and civilisation but that also does no significant damage.

It is also important for the designer to consider not just their own viewpoint, but that to truly understand the cultural context there must be inclusion at the design level. This is because, as discussed in Chapter 7, bias can be unconscious and unknown, and design decisions can be influenced not only by choices made but by omissions of knowledge or forethought; for example, that which led to Dr Louise Selby's changing room access issues. Diverse design teams need to include representation from different genders, races, cultures, levels of ability and so on. Where this is not possible in terms of the designers themselves, methods such as participatory design can include a diverse range of participants to work together to produce design outcomes that are as suitable as possible for the full variety of cultural contexts in which they might be used.

Designing for the environment

As described in Chapter 3, we are navigating multiple dimensions; the digital world offers simulations that replicate the physical world and completely new digital spaces, while augmented reality, pervasive computing and the internet of things bring 'intelligence' to us and our objects and systems in the physical world. Digital enhancement of these spaces such as GPS, mapping and sensors are already influencing our culture and society. The digital space is full of data flows across the physical digital boundary, connecting sensors, servers, devices and people, as Lindley et al (2017) suggest: 'in the unseen digital domain, where data swirls imperceptible to humans, the atmosphere is thick with the rapidly-moving data packets and content that constitute inter-machine chatter'.

Designing for this environment requires another level of social responsibility to ensure we are considering risk, safety and security. In order to understand and design in this hybrid world between the digital and physical spaces, designers are taking new approaches to the nature of objects. Coulton (2015) is concerned with how we classify objects in the internet of things in order to design for it. He draws upon the work of science fiction writer Bruce Sterling and in particular his description of spimes: networked objects with extensive and rich informational support that are designed on screens, fabricated on screens, and tracked in space and time throughout their lifespan (Sterling et al, 2005). We already have objects that are RFID tagged and managed in distribution and inventory systems, and the technologies are enabling further pervasiveness of spimes in our lives. Furthermore, Coulton refers to Sterling's category of biots; an entity that is both object and person, and which provides data to the network, with the example of the quantified self and personal health monitoring. If we see humans as simply another object in a constellation of objects applying an object-orientated ontology philosophy then designers may reimagine data, devices, and users, as equally significant actants in a flat ontology (Lindley et al, 2017). This will have an influence on the way in which designers undertake the design process, and how designers consider their role in terms of social, environmental and economic determinants in the system. What does it mean to be socially responsible when designing in a flat ontology? *Designers will need to consider the ethics of designing a constellation of objects and the effect on the overall environment in terms of human and global wellbeing.*

Design beyond now

The future is often designed but more often emerges, as humans respond and change their behaviours and values. Technologies will define new niches, as Clark (2004) explains: 'The idea would be to allow the technologies to provide for the kinds of interactions and interventions for which they are best suited, rather than to force them to (badly) replicate our original forms of action and experience'. Designers of digital public space will need to change their mindset in terms of how we respond to users and communities, providing not just solutions to current problems, needs and desires, but mechanisms to adapt to future opportunities, challenges and desires. As Holland and Stornetta (1992) put it:

> It is customary for a person with a broken leg to use crutches, but how odd it would be if they continued to use the crutches after their leg was restored to its natural condition. In contrast, one wears shoes because they provide certain advantages over our natural barefoot condition.
>
> … The crutch is designed specifically to make the best of a bad situation – to let someone hobble around until they are back in shape. On the other hand, shoes are to correct some of the problems of our natural condition, and, in the case of athletic shoes, to enhance our performance.

They argue that more telecommunications research should be geared to building shoes rather than crutches. Digital collaboration tools, in this analogy, can be considered a shoe rather than a crutch: as we described in Chapter 4, systems such as Google Docs allow joint working to happen when people are not physically co-located, but they also serve to enhance co-working even when people are in the same place. They function to replicate the work that can be done together with a pen and paper, but to do things that are not possible with that previous mode of working. Email was originally conceived as 'electronic mail', a replacement for written communication, memos and letters. In this sense the design process was creating a 'crutch', in that functions were considered only in terms of aspects that replicated previous functions. However, as email has developed and become embedded for most people as a fundamental part of working life, aspects have developed which function in ways that were not possible with paper based communication, such as the ease with which copies can be sent to a large number of correspondents. These capabilities enhance the way that we work in ways which were not necessarily considered in the original design, and turn email from a 'crutch' to a 'shoe'. This emergent process has taken a very long time and there are still aspects of email which, for no particular reason, remain because of the history of its development out of written communication. In designing for the future we should consider the shape of the problem and what the goal is, rather than updating existing solutions to what may be outdated questions.

As humans become the bridge between the physical and digital worlds, as we embody technologies, and as interfaces become invisible, we will not leave the body behind but will be 'extending embodied awareness in highly specific, local, and material ways that would be impossible without electronic prosthesis' (Hayles 1999). *Designing beyond now must consider the fact that humans will be in both the digital and the physical worlds at the same time; this will change how we view both spaces and what we design in them.*

Key principles for designing for digital lives

To design for digital public space means appropriating different paradigms or approaches, understanding the potential interaction between multiple systems of technology, use, and behaviour. These interactions stretch over many dimensions of geography, time, demographics, physical and digital domains. *Therefore, designing in digital public space demands a complex set of skills and expertise, emerging as hybrid versions out of traditional design fields to focus more holistically on products, people, places, technology moving between the digital and the physical.*

To this end, we should adopt the following five principles:

1 Interdisciplinarity

Design teams must be interdisciplinary; everyone who has insight is a designer to some degree. Balsamo (2011) talks about the importance of interdisciplinarity, and of bringing the humanities into discussions of technology development to address global and social problems. 'The kind of technological solutions that result from multidisciplinary research are not going to be (solely) complex technologies, but complex hybrids of technological objects, services, and applications, along with social and cultural implementation plans.' Such groups also need to include imagination of the arts and art practices. Furthermore, a design team must be diverse – in gender, race and discipline, and able to give a more effective design experience by bringing differences in life knowledge. This will help to neutralise to some degree the bias all designers bring, related to their cultural experience and presuppositions about their understanding of the design task, the world, and the end users. This is why co-design is important: building a shared meaning that includes many aspects so as to overcome these assumptions.

2 Co-design and co-creation

Co-creation and co-design allows buy in from users and consumers/customers – it enables them to buy into a venture, often to co-own it. In teams and communities this approach (Brand & Bevelo, 2003) tends to build commitment and trust in the brand and technology

(Linux for example was built by a community). The process is not just one of bringing in a few users and customers, but within digital public space, empowering audiences through co-design and making them co-owners to bring meaning into design and give it life beyond the original concept. This approach to co-design enables different user groups, varying in aspects including race, gender, sexuality and able-ness, to give different perspectives on use that might not be 'obvious', to counteract what have often been inherent biases of white male developers and designers. This involves using new and existing platforms in digital public space to re-appropriate for new design processes, for example taking into account functionality of social use of Facebook and new ways in which people inhabit digital public space. There are many examples of groups working together to co-design products, services and indeed policies. OpenIDEO is just one example of a global community working together to design solutions for the world's biggest challenges. Currently the emphasis is on health, education and economic issues in the developing world; nevertheless the internet and digital public space enables the global collaboration of designers to transform the physical and digital public space and how we navigate between them.

3 Tools for thinking

Obviously all the tools currently available to design teams such as user research, anthropology and ethnography as well as prototype development and testing are part of the portfolio of tools for when designing digital public space. There are methods that can be adapted from both science and social sciences to test the robustness of an idea in use. These tend to focus the design team on the issue to do with the life of the digital public space asset and how it will interface with the physical world. It is clear that understanding all the dimensions of digital public space becomes critical; it is important to consider how the design of tools, content and use of digital public space have implications for privacy, identity, trust, safety and security of the individual and society. For instance, there are some circumstances where visibility of some aspects of technology might be a desirable, or even a preferable quality in the design. Simple examples of this include ensuring the individual knows when and where their personal data is being accessed and used in digital public space, or how they understand who owns what in that space. The notion of cyber crime and how to think about it is one example where tools used in the design of the physical world can be adapted for digital public space:

Tool example 1: Design Against Crime

In the physical world, designers have developed mechanisms to think through potential crime-related attributes, for instance the use of the Design Against Crime cycle, based on criminology theory. This model, for instance, looks at the pre-crime issues, such as the offender's readiness to offend, the availability of resources to offend, the offender's presence in the situation, the design attributes of the product, the place, and the behaviour of others, all of which can be transferred into the digital domain; who is there, what skills will they have that they might need, and how is the behaviour of others offering opportunities for crime. This is something we encounter every day in digital and cyber crime, in cases of fraudulent websites, personas, and so on. Using an equivalent approach to thinking about crime in digital public space is one tool that designers who are not cyber crime experts might use in order to think through digital public space experience.

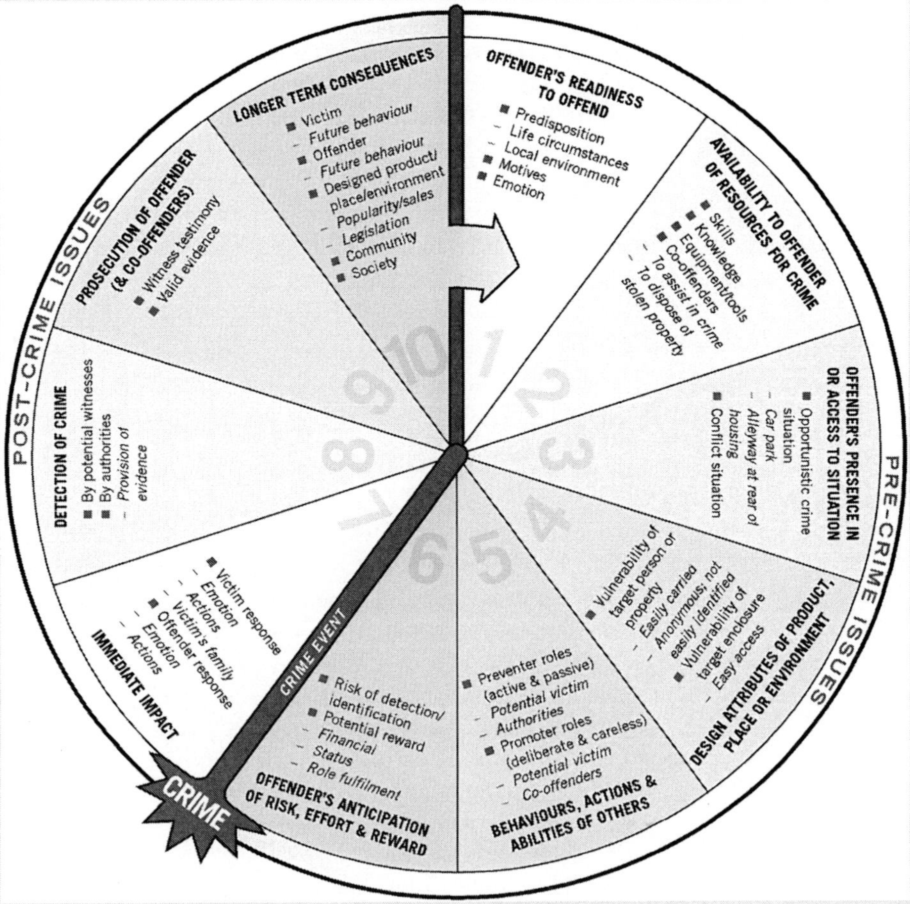

Figure 9.1 Design Against Crime model
Source: Cooper et al (2005), based on Eckblom (2001)

Another concern is privacy, for example we know that health related technologies in digital spaces such as health related internet of things can have far reaching implications in terms of health management, tele-monitoring patients, fitness and wellbeing monitoring and so on, but also raise concerns among users in terms of their privacy and of course inevitable trust of the system; similarly for finance systems. In the UK as in other countries we have bodies such as the Information Commissioner's Office whose remit is reactive in terms of upholding information rights in the public interest. In addition the EU 2016 General Data Protection Regulations (GDPR) will increasingly impact on how we deal with data and privacy. However, we have few proactive design guidelines aimed at privacy when designing for the digital public space. One rare example is Privacy by Design, developed by the Information and Privacy Commissioner of Ontario, Canada, Dr Ann Cavoukian in the 1990s:

Tool example 2: Privacy by Design

The objective of '*Privacy by Design*' is to advance the view that the future of privacy cannot be assured solely by compliance with legislation and regulatory frameworks; rather, privacy assurance must become an organization's default mode of operation.[1] This framework has subsequently been recognized globally by for instance the annual assembly of International Data Protection and Privacy Commissioners, the US Federal Trade Commission and the European Commission.

1. ***Proactive*** not Reactive; ***Preventative*** not Remedial
The *Privacy by Design* approach is characterized by proactive rather than reactive measures. It anticipates and prevents privacy invasive events before they happen. *Privacy by Design* does not wait for privacy risks to materialize, nor does it offer remedies for resolving privacy infractions once they have occurred – it aims to prevent them from occurring. In short, *Privacy by Design* comes before-the-fact, not after.

2. Privacy as the ***Default Setting***
We can all be certain of one thing – the default rules! *Privacy by Design* seeks to deliver the maximum degree of privacy by ensuring that personal data are automatically protected in any given IT system or business practice. If an individual does nothing, their privacy still remains intact. No action is required on the part of the individual to protect their privacy – it is built into the system, by default.

3. Privacy ***Embedded*** into Design
Privacy by Design is embedded into the design and architecture of IT systems and business practices. It is not bolted on as an add-on, after the fact. The result is that privacy becomes an essential component of the core functionality being delivered. Privacy is integral to the system, without diminishing functionality.

4. Full Functionality – ***Positive-Sum***, not Zero-Sum
Privacy by Design seeks to accommodate all legitimate interests and objectives in a positive-sum win-win manner, not through a dated, zero-sum approach, where unnecessary trade-offs are made. *Privacy by Design* avoids the pretense of false dichotomies, such as privacy vs. security – demonstrating that it is possible to have both.

5. End to End Security – ***Full Lifecycle Protection***
Privacy by Design, having been embedded into the system prior to the first element of information being collected, extends securely throughout the entire lifecycle of the data involved – strong security measures are essential to privacy, from start to finish. This ensures that all data are securely retained, and then securely destroyed at the end of the process, in a timely fashion. Thus, *Privacy by Design* ensures cradle to grave, secure lifecycle management of information, end-to-end.

6. ***Visibility*** and ***Transparency*** – Keep it ***Open***
Privacy by Design seeks to assure all stakeholders that whatever the business practice or technology involved, it is in fact, operating according to the stated promises and objectives, subject to independent verification. Its component parts and operations remain visible and transparent, to users and providers alike. Remember, trust but verify.

7. ***Respect for User Privacy*** – Keep it ***User-Centric***
Above all, *Privacy by Design* requires architects and operators to protect the interests of the individual by offering such measures as strong privacy defaults, appropriate notice, and empowering user-friendly options. Keep it user-centric.

Trilogy of Applications
Privacy by Design provides a method for proactively embedding privacy into information technology, business practices, and networked infrastructures.

Source: Information and Privacy Commissioner of Ontario, September 2013

Both the Design Against Crime and the Privacy by Design frameworks illustrate how when designing for digital public space, we have to keep in mind the behavioural aspects of the users from the outset, reconciling usability and access with use and misuse.

In addition to the thinking process that comes with co-design, to enable the design to be ingenious, robust, and innovate in a way that tests its use in the future, there are tools for taking designs into the realms of the imagination such as speculative design, or design fictions. These can enable the designers to consider not just the user but all the objects in the physical and digital spaces to generate new perspectives through an object-orientated ontology that we discussed above and view the design in this flat ontology. For instance, Lindley et al describe building a design fiction around an internet-connected kettle 'Polly':

> We designed two features… The first feature is a timeline inspired by the ubiquitous timelines we see on social networks and live news websites. Polly's timeline reports all data transactions that the kettle is involved in (i.e. every time the kettle communicates across the network, that communication appears as an entry in the timeline). The second feature is related to the first. Using a simple graphical display, Polly shows the user the relative volume of data that it sends to the Internet, receives from the Internet, and that is transferred within the kettle's local network (i.e. a user's home).
>
> (Lindley et al, 2017)

Their assumption is that in the future the pervasiveness and ubiquity of data collecting devices will grow. Therefore, in considering how to design for digital public space we can use such fictions to speculate on future places where objects and people move through the boundaries of physical digital dimensions and identify the potential opportunities and threats.

4 Allow for exploration, experimentation, emergence and appropriation

As we have made clear throughout, digital public space/s are not one-time objects, they have emergent qualities, and technology may amplify intended and unintended biases because it does not allow room for other interpretations; you cannot always use the technology in ways other than was intended by the designer and so good design must allow space for serendipity, for exploration, and 'breaking' things to find outcomes not anticipated; hence the trend to launch onto the market beta versions of software or one-off short runs of hardware for early adopters to use. This will affect the design process and how we conceive it as will be discussed in the next section.

5 Explore the moral issues

Socially responsible design has been a concern of the profession for a number of years; it has been of limited interest to business, especially large manufacturing companies, until regulatory measures have forced companies to address current moral issues. Digital public space is no different to the physical in this regard and indeed it is possibly more important because of the opaque nature of the space and the technologies that lead us into it. Design teams now need to move into domains of bias, ethics, trust, and privacy throughout the entire digital public space journey and experience, and in terms of how it has an effect on the physical public space.

These five essential principles of course need to be set within the evolving landscape of design and development, which due to the emergent and fluid nature of digital public space will not necessarily remain the same as previous design processes.

Design and development process for designing digital public space

Designers working in the traditional silos have predominantly worked alone or in small groups, following an iterative design process to develop concepts and detailed design. This is an internal process that is usually comprised of defining the problem, understanding the problem, thinking about the problem, developing an idea and creating a more detailed design. It is always somewhat iterative because in developing ideas the understanding of the problem is tested and revised.

However, over the past fifty years a number of factors have widened the designer's horizon and working methods. First is the change in the nature of the commercial product development process. For example, there has been a move from a conventional 'over the wall' process between professions, where marketing identifies the need and plans the concept, design produces the concept design and prototype, manufacturing produces it, and sales take it to market. It was recognised by large manufacturing companies in particular that this just did not work; the problems and solutions were just transferred from one department to another, increasing time and cost of new product development. Therefore the 'rugby approach' was introduced, essentially multidisciplinary teams working together from the beginning of the process. This not only speeded up the development process but enabled it to be accompanied by new philosophies of design, such as market led design, applying flexible manufacturing to respond to the flow of new information on customer demand and preferences, enabling products to be more tailored, adaptable and desirable to the customers (Evans 1985).

In 2005 the UK Design Council consolidated models of the design process through a study of global companies and arrived at the highly cited Double Diamond (Design

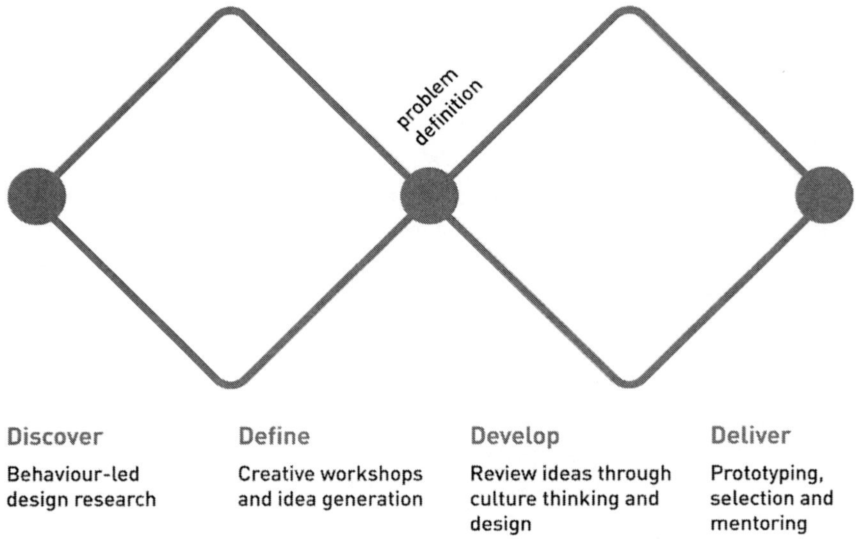

Figure 9.2 The Double Diamond

Source: www.thecreativeindustries.co.uk/uk-creative-overview/news-and-views/view-what-is-design-and-why-it-matters.

Council 2016) – see Figure 9.2. The Double Diamond divides the design process into four phases; Discover, Define, Develop and Deliver. It maps the divergent and convergent stages of the design process, showing the different modes of thinking that designers use. It is essentially focused on design but also illustrates the connection to the whole development team.

The Double Diamond does not illustrate the full scale of a new product development (NPD) process, a topic that has been researched significantly since the 1970s. There have been at least three generations of stage-gate new product development processes – the most common one has been that developed by Cooper (1994) which brings together the whole team at various stages throughout the NPD process to review and approve movement to the next stage. This has its critiques, a main one is that it slows down the process and inhibits creativity and innovation; therefore more fluid processes have been developed including Cooper's own third generation process. There has been a demand especially from the IT sector for more agile and flexible ways of developing new products. As Sommer et al discuss:

> Today, no less than nine different Agile methods have been described, including Scrum, Crystal, Extreme Programming, Adaptive Software Development, Agile Modeling, Dynamic Systems Development Method, Feature Driven Development, Internet Speed Development, and Pragmatic Programming. Each method offers different advantages and each comes with its own set of tools and approaches and its own development culture.
>
> (Sommer et al, 2015)

As the product becomes more complex so does the design and development process. The traditional 'designer' as we know them has become one of the multidisciplinary design team, comprised of all types of disciplines, but also all types of designers including product, software, services and interaction designers.

In product design in particular, complexity has come with the emergence of the notion of the product service system that brought together both the product and the system in which it is located, while in architecture the fusion of the service systems and the built form illustrate how vital designing holistically is to delivering effective products, places and services. Designing the experience has however expanded yet further, beyond the product, service, place, experience and towards designing the *total* experience 'the culture of use in which people negotiate, gain and express meaning from the designed world around them' (Press & Cooper, 2003). A model shown in Figure 9.3, adapted from Rhea (1992), aptly describes this process.

The complete design experience approach becomes critical when we combine the material and the digital world and start to think of the constellation of objects and people. *We have already seen how important it is to get the physical device, its interface and the digital journey right.*

If we combine the Double Diamond with aspects of the stage gate process and the design experience model, a framework for digital public space design process is revealed that can focus on the underlying principles and associated tools that must be taken into consideration when designing digital public space. This design process is not linear; it is a continuing and emergent process, whereby the *Discover phase* enables co-design and collaboration to discover the requirements and attributes necessary for the space, in other words the flexibility,

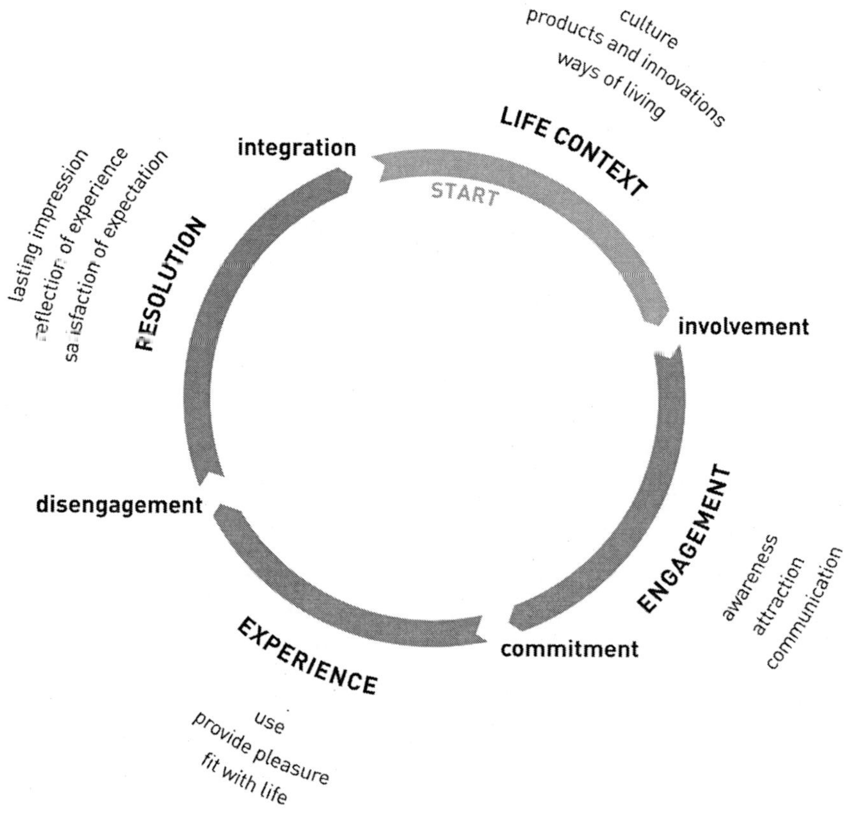

Figure 9.3 The Design Experience model

Source: Adapted from Rhea (1992) cited in Press & Cooper (2003). Redrawn by authors.

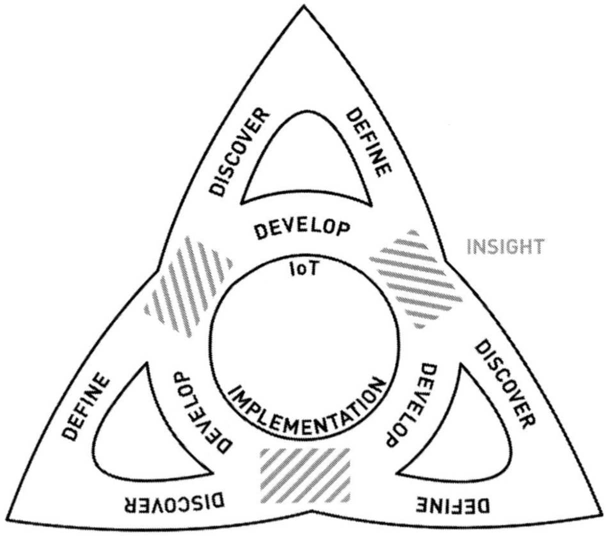

Figure 9.4 A new process for designing digital public space

robustness, and the safety and security needs. The *Define phase* uses narratives, scenarios and fictions to visualise and test the design idea before the *Develop phase*, through which the products and services of the space are developed with communities and lead adopters, and implemented, with in use insight revealing emergent and new qualities that feed another cycle of discover, define and develop, but which also inform policy and regulatory design related to digital public space.

Conclusion

The design of digital public space does mean we need to consider where we put people in the process, how we consider attributes of transparency that enables our plastic brains to adapt and adopt this world, against the concerns of privacy and security. Do we want the system to reveal itself, for us to understand what we are doing, how data is being used, and who we are really meeting; do we want to understand how we are objectified in digital public space and how useful that is to us, or how vulnerable it makes us. Designers have to consider what becomes visible and what is invisible, and what helps us navigate digital public space in a way that is valuable, safe and secure as they do in the physical world.

In summary, it is important to think about what is being designed; crucial to working in digital public space is recognising that digital public space is not something that is just created, but has developed as an emergent phenomenon from technology that has been designed for different, specific purposes. Specific examples of this include social media platforms, the open source movement, and data and archive facilities. Organisations and individuals appropriate and design with hardware and software to form digital public spaces that have varying dimensions of the geographical, temporal, physical and digital. Because the landscape of opportunity in digital public space is so vast and potentially complex, there is a need to consider how and what we design. As illustrated throughout this book, the relationship between humans, society and the digital phenomenon requires consideration of a number of factors that go beyond the user centred design, and therefore must be considered carefully by both public and private organisations. Designing in this space must cover skills and expertise between many traditional design fields: products, people, places, technology, and the digital and the physical; this can be assisted by adopting new design methods, processes and perspectives.

We are certainly designing in a new and emerging context, which offers both opportunities and challenges. It is wise therefore to understand where we came from, how our history informs the way we live in the digital world and how we will continue to create our futures. This book has attempted to set digital public space and indeed the digital world in context, to enable us to reflect on what type of world we are creating and how the human species will respond to it for good or ill.

Notes

1 For an illustration of this process, see the comic 'The long, slow death of Twitter' by Brad Colbow: www.theguardian.com/technology/picture/2016/mar/10/comic-long-slow-death-of-twitter
2 'Teens react to 90s Internet' www.youtube.com/watch?v=d0mg9DxvfZE

References

Abras, C., Maloney-Krichmar, D. and Preece, J., 2004. User-centered design. *In:* Bainbridge, W., ed. *Encyclopedia of Human-Computer Interaction*. Sage Publications.

Ageh T., 2012. *Speech by Tony Ageh, Controller, Archive Development, BBC*. Open lecture presented at Manchester Museum of Science and Industry. Available at: http://thecreativeexchange.org/CX_Launchpad_Tony_Ageh.pdf [Accessed 3 August 2016]

Balsamo, A., 2011. *Designing culture: the technological imagination at work*. Duke University Press, pp.129, 6 & 160.

Booth, P., 2010. *Digital fandom: New media studies*. Vol. 68. Peter Lang.

Brand, R. and Bevolo, M., 2003. Ambient intelligence and the climate for branding. *In:* Aarts, E. and Marzano, S., eds. *The new everyday: views on ambient intelligence*. 010 Publishers.

Clark, A., 2004. *Natural-born cyborgs: minds, technologies, and the future of human intelligence* Oxford University Press, p.109.

Cooper, R., 1994. Third generation new product process. *Journal of Product Innovation Management* 11(1), pp.3–14.

Cooper, R. and Boyko, C.T., 2010. How to design a city in five easy steps: exploring VivaCity 2020's process and tools for urban design decision making. *Journal of Urbanism,* 3(3), pp.253–273.

Cooper, R., Wootton, A.B., Davey, C.L. and Press, M., 2005. Breaking the cycle: fundamentals of crime-proofing design. *Crime Prevention Studies,* Vol 18, pp.179–201.

Coulton, P., 2015. Playful and gameful design for the internet of things. *In:* Nijholt, A., ed. *More playful user interfaces: interfaces that invite social and physical interaction*. Springer, pp.151–173.

De Michelis, Giorgio. 2003. Mixed objects. *Appliance Design Journal*.

Design Council, 2016. A study of the design process. [online] *Design Council.* www.designcouncil.org.uk/sites/default/files/asset/document/ElevenLessons_Design_Council%20(2).pdf [Accessed 1 August 2016]

Donath, J., 2014. *The social machine: designs for living online*. MIT Press.

Dourish P., 2001. *Where the action is: the foundations of embodied interaction*. MIT Press.

Ekblom, P., 2007. Going equipped: criminology, situational crime prevention and resourceful offender. *British Journal of Criminology, 40(3)*.

Ehn, P., 2006. Participation in interaction design: actors and artifacts in interaction. *In*: Bagnara, S. and Crampton Smith, eds. *Theories and practice in interaction design*. LEA, pp.1–356.
Erol, R., Press, M., Cooper, R. and Thomas, M., 2002. Designing out crime: raising awareness of crime reduction in the design industry. *Security Journal, 15(1)*, pp 49–61.
Evans, B., 1985. Japanese-style management, product design and corporate strategy, *Design Studies, 6(1)*, pp.25–33.
Floridi, L., 2007. A look into the future impact of ICT on our lives, The Information Society, 23(1).
Harrison, S., Tatar, D. and Sengers, P., 2007. The three paradigms of HCI, Alt. Chi. *Session of the SIGCHI Conference on Human Factors in Computing Systems*.
Hayles, N. K., 1999. *How we became posthuman: virtual bodies in cybernetics, literature, and informatics*. University of Chicago Press, p.290.
Holland, J. and Stornetta, S., 1992. Beyond being there. *Proceedings of the SIGCHI Conference on Human Factors in Computing Systems*. ACM, pp.119–125.
Lindley, J.G., Coulton, P. and Cooper, R., 2017. Why the internet of things needs object orientated ontology. *Proceedings of EAD 2017*. European Academy of Design.
Mashable.com, 2016. *38% of children under 2 use mobile media, study says.* [online] http://mashable.com/2013/10/28/children-under-2-mobile-media-study/#feBJP7f5ouqA [Accessed 1 August 2016]
Norman, D., 1998. *The invisible computer, why good products can fail, the personal computer is so complex, and information appliances are the solution.* MIT Press.
Norman, D., 2002. *The design of everyday things.* Basic Books.
Norman, D. A. and Draper, S. W., eds. 1986. *User-centered system design: new perspectives on Human-Computer Interaction*. Lawrence Earlbaum Associates.
Norman, D., 1988. *The psychology of everyday things*. HarperCollins, p.188.
Potter, N., 1969. *What is a designer: things, places, messages*. Studio Vista.
Press, M. and Cooper, R., 2003. *The design experience*: *the role of design and designers in the 21st century.* Ashgate, pp.12 & 74.
Rhea, D., 1992. A new perspective on design: focusing on customer experience. *Design Management Journal*, 9(4), pp.10–16.
Rheingold, H., 2002. *Smart mobs: the next social revolution*. Perseus Publishing.
Sommer, A. F., Hedegaard, C., Dukovska-Popovska, I. and Steger-Jensen, K., 2015. Improved product development performance through Agile/Stage-Gate Hybrids: the next-generation Stage-Gate process? *Research-Technology Management*, 58(1), pp.34–45.
Sterling, B., 2005. *Shaping things*, MIT Press.
Weise, S., Hardy, J., Agarwal, P., Coulton, P., Friday, A. and Chiasson, M., 2012. Democratizing ubiquitous computing: a right for locality. *Proceedings of the 2012 ACM Conference on Ubiquitous Computing*. ACM, pp. 521–530.

References

Aarts, E. and Marzano, S., 2013. *The new everyday: visions of ambient intelligence.* 010 Publishers.

Abbott, E.A., 1884. *Flatland; a romance of many dimensions, by A. Square.* Seeley & Co.

Abbruzzese, J., 2014. Amazon tensions with book publisher boil over. [online] *Mashable.* Available at: http://mashable.com/2014/05/23/amazon-blocks-hachette/#9B4f.0KREkq3 [Accessed 20 October 2016].

Abras, C., Maloney-Krichmar, D. and Preece, J., 2004. User-centered design. *In:* Bainbridge, W., ed. *Encyclopedia of Human-Computer Interaction.* Sage Publications.

Acampora, G., Cook, D.J., Rashidi, P. and Vasilakos, A.V., 2013. A survey on ambient intelligence in healthcare. *Proceedings of the IEEE, 101*(12), pp. 2470–2494.

Acquisti, A., Gritzalis, S., Lambrinoudakis, C. and di Vimercati, S., eds. 2007. *Digital privacy: theory, technologies, and practices.* CRC Press, p.348.

Adams, D., 1979. *The Hitchhiker's Guide to the Galaxy.* Pan Macmillan.

Adee, S., 2016. Will AI's bubble pop? *New Scientist, 231*(3082), pp. 16–17.

Ageh T., 2012. *Speech by Tony Ageh, Controller, Archive Development, BBC.* Open lecture presented at Manchester Museum of Science and Industry. Available at: http://thecreativeexchange.org/CX_Launchpad_Tony_Ageh.pdf [Accessed 3 August 2016].

Ageh, T., 2013. Why the digital public space matters. *In:* Hemment, D., Thompson, B., de Vicente, J.L. and Cooper, R., eds. *Digital Public Space.* FutureEverything Publications. Available from: http://futureeverything.org/publications/digital-public-spaces/ [Accessed 23 September 2016].

Ageh, T., 2015. *The BBC, the licence fee and the digital public space.* Open lecture presented at Royal Holloway University of London. Available from: www.royalholloway.ac.uk/harc/documents/pdf/tonyageh.pdf [Accessed 23 September 2016].

Aiello, L.C. and Dunbar, R.I., 1993. Neocortex size, group size, and the evolution of language. *Current Anthropology, 34*(2), pp. 184–193.

Alemseged, Z., Spoor, F., Kimbel, W.H., Bobe, R., Geraads, D., Reed, D., and Wynn, J.G., 2006. A juvenile early hominin skeleton from Dikika, Ethiopia. *Nature, 443*(7109), pp. 296–301.

Alexander, W., 2013. How to make thousands of pounds a month playing computer games. [online] *Vice.* Available at: www.vice.com/en_uk/read/i-make-thousands-of-dollars-a-month-from-playing-computer-games [Accessed 17 October 2016].

Aliaga, D.G., 1997. Virtual objects in the real world. *Communications of the ACM, 40*(3), pp. 49–54.

Allan, A. and Warden, P., 2011. Got an iPhone or 3G iPad? Apple is recording your moves. [online] *O'Reilly Radar.* Available at: http://radar.oreilly.com/2011/04/apple-location-tracking.html [Accessed 18 October 2016].

Allentoft, M.E., Sikora, M., Sjögren, K.G., Rasmussen, S., Rasmussen, M., Stenderup, J., Damgaard, P.B., Schroeder, H., Ahlström, T., Vinner, L. and Malaspinas, A.S., 2015. Population genomics of Bronze Age Eurasia. *Nature, 522*(7555), pp. 167–172.

Altman, I., 1975. *The environment and social behavior: privacy, personal space, territory, and crowding.* Brooks/Cole Publishing Co., p.2.

Andrews, A. and Hartevelt, M., 2003. Community, memory and ambient intelligence. *In*: Aarts, E. and Marzano, S., eds. *The new everyday: views on ambient intelligence*. 010 publishers, p.185.

Anglo-Filles, 2014. Alina's best day ever [podcast]. *Anglo-Filles. 27.* Available at: http://anglofilles.madeoffail.net/episodes/anglofilles-episode-27-alinas-best-day-ever/ [Accessed 11 October 2016].

Ayalon, O. and Toch, E., 2013, July. Retrospective privacy: managing longitudinal privacy in online social networks. *Proceedings of the Ninth Symposium on Usable Privacy and Security*. ACM, p.4.

Bakardjieva, M. and Gaden, G., 2011. Web 2.0 technologies of the self. *Philosophy and Technology*, *25*(*3*).

Baker, D.N., Balstad, R., Bodeau, J.M., Cameron, E., Fennel, J.F., Fisher, G.M., Forbes, K.F., Kintner, P.M., Leffler, L.G., Lewis, W.S. and Strachan, L., 2008. *Severe Space weather events–understanding societal and economic impacts: a workshop report*. National Academies Press.

Bakioglu, B.S., 2009. Spectacular interventions of Second Life: Goon culture, griefing, and disruption in virtual spaces. *Journal for Virtual Worlds Research*, *1*(3), p.5.

Bakshy, E., Messing, S. and Adamic, L.A., 2015. Exposure to ideologically diverse news and opinion on Facebook. *Science*, *348*(6239), pp. 1130–1132.

Balkan, A., 2015. Ethical Design Manifesto. [online] Available at: https://ind.ie/blog/ethical-design-manifesto/ [Accessed 19 October 2016].

Balsamo, A., 2011. *Designing culture: the technological imagination at work*. Duke University Press, pp. 4, 6, 134 &160.

Baltimore, D., Baylis, F., Berg, P., Daley, G.Q., Doudna, J.A., Lander, E.S., Lovell-Badge, R., Ossorio, P., Pei, D., Thrasher, A. and Winnacker, E.L., 2016. On human gene editing: International Summit Statement by the Organizing Committee. *Issues in Science and Technology*, *32*(3), p.55.

Banks, I. M., 1987–2012. *The Culture* (series). Macmillan.

Baraniuk, C., 2014. Digital doctors. *New Scientist*, *222*(2973), p.21.

Baraniuk, C., 2015. Caught before the act. *New Scientist*, *225*(3012), pp. 18–19.

Bargh, J.A. and McKenna, K.Y., 2004. The internet and social life. *Annual Review of Psychology*, *55*, pp. 573–590.

Bassett, C., Steinmueller, E. and Voss, G., 2013. Better made up: the mutual influence of science fiction and innovation. *Nesta Work. Pap*, *13*(07).

BBC, 2007. *75 years BBC World Service: A history. The 1930s.* [online] BBC World Service. Available at: www.bbc.co.uk/worldservice/history/story/2007/02/070123_html_1930s.shtml [Accessed 14 October 2016].

BBC, 2015. *Japan: Free wi-fi for Mount Fuji climbers.* [online] BBC News From Elsewhere. Available at: www.bbc.co.uk/news/blogs-news-from-elsewhere-33426338 [Accessed 14 October 2016].

BBC News, 2011. *150 officers warned over Facebook posts*. [online] Available at: www.bbc.co.uk/news/uk-16363158 [Accessed 18 October 2016].

BBC News, 2014. *Minecraft: all of Denmark virtually recreated.* [online] Available at: www.bbc.co.uk/news/technology-27155859 [Accessed 27 September 2016].

BBC News, 2016. *Did aerial photo of Michael McIntyre break privacy rules?* [online] Available at: www.bbc.co.uk/news/magazine-33535578 [Accessed 19 October 2016].

BBC News, 2016. *Plastic microbeads to be banned by 2017, UK government pledges*. [online] Available at: www.bbc.co.uk/news/uk-37263087 [Accessed 28 October 2016].

Beckett, C., 2011. *After Tunisia and Egypt: towards a new typology of media and networked political change.* [online] Polis. Available at: http://blogs.lse.ac.uk/polis/2011/02/11/after-tunisia-and-egypt-towards-a-new-typology-of-media-and-networked-political-change/ [Accessed 11 October 2016].

Beer, D. and Burrows, R., 2013. Popular culture, digital archives and the new social life of data. *Theory, Culture & Society*, *30*(4), pp. 47–71.

Bendell, J., 2015. What price a cashless life? *New Scientist*, 226(3024), pp. 24–25.

Benedikt, M.L., 2008a. *Human needs and how architecture addresses them.* University of Texas Press.

Benedikt, M.L., 2008. Cityspace, cyberspace, and the spatiology of information. *Journal for Virtual Worlds Research*, *1*(1), p.4.

Benkler, Y., 2006. *The wealth of networks: how social production transforms markets and freedom*. Yale University Press, p.9.

Ben-Ze'ev, A., 2003. Privacy, emotional closeness, and openness in cyberspace. *Computers in Human Behavior*, *19*(4), pp. 451–467.

Berger, L.R., Hawks, J., de Ruiter, D.J., Churchill, S.E., Schmid, P., Delezene, L.K., Kivell, T.L., Garvin, H.M., Williams, S.A., DeSilva, J.M. and Skinner, M.M., 2015. Homo naledi, a new species of the genus Homo from the Dinaledi Chamber, South Africa. *Elife*, *4*, p.e09560.

Berger, T.W., Hampson, R.E., Song, D., Goonawardena, A., Marmarelis, V.Z. and Deadwyler, S.A., 2011. A cortical neural prosthesis for restoring and enhancing memory. *Journal of Neural Engineering*, *8*(4), p.046017.

Beschizza, R., 2015. Google Glass chief "amazed" by privacy issues that helped kill his project. [online] Boing Boing. Available at: http://boingboing.net/2015/03/19/google-glass-chief-amazed.html [Accessed 27 September 2016].

Best, J., 2004. How eight pixels cost Microsoft millions. [online] *CNET.* Available at: www.cnet.com/uk/news/how-eight-pixels-cost-microsoft-millions/ [Accessed 20 October 2016].

Bezos, J., 2009. Announcement: an apology from Amazon. [online] *Amazon.com*. Available at: www.amazon.com/tag/kindle/forum/ref=cm_cd_ef_tft_tp?_encoding=UTF8&cdForum=Fx1D7SY3BVSESG&cdThread=Tx1FXQPSF67X1IU&displayType=tagsDetail [Accessed 14 October 2016].

Binder, J., Howes, A. and Sutcliffe, A., 2009, April. The problem of conflicting social spheres: effects of network structure on experienced tension in social network sites. *Proceedings of the SIGCHI Conference on Human Factors in Computing Systems*. ACM, pp. 965–974.

Booth, P., 2010. *Digital fandom: new media studies*. Vol. 68. Peter Lang Publishing.

Botvinick, M. and Cohen, J., 1998. Rubber hands 'feel' touch that eyes see. *Nature*, *391*(6669), pp. 756–756.

Bowyer, K.W., 2004. Face recognition technology: security versus privacy. *IEEE Technology and Society Magazine*, *23*(1), pp. 9–19.

boyd, d., 2007. Social network sites: public, private, or what. *Knowledge Tree*, *13*(1), pp. 1–7.

boyd, d., 2014. Is the Oculus Rift sexist? [online] *Quartz* Mar. 28. Available at: http://qz.com/192874/is-the-oculus-rift-designed-to-be-sexist/ [Accessed 26 September 2016].

Brand, R. and Bevolo, M., 2003. Ambient intelligence and the climate for branding. *In:* Aarts, E. and Marzano, S., eds. *The new everyday: views on ambient intelligence*. 010 Publishers.

Bridle, J., 2012. The library and the forum. *In:* Hemment, D., Thompson, B., de Vicente, J.L. and Cooper, R., eds. *Digital Public Spaces*. FutureEverything Publications. Available from: http://futureeverything.org/publications/digital-public-spaces/ [Accessed 23 September 2016].

Brietbart, P. and Vega, M., 2014. *Posthuman: introduction to transhumanism.* [video] Available at: www.youtube.com/watch?v=bTMS9y8OVuY Transcript available at: https://peterbrietbart.wordpress.com/2014/09/09/posthuman-introduction-to-transhumanism-script/ [Accessed 26 October 2016].

Brody, N., 2012. Modelling the digital public space: the new renaissance. *In:* Hemment, D., Thompson, B., de Vicente, J.L. and Cooper, R., eds. *Digital Public Space.* FutureEverything Publications. Available from: http://futureeverything.org/publications/digital-public-spaces/ [Accessed 23 September 2016].

Brody, N. and Fass, J., 2013. Digital Public Space and the Creative Exchange. Available from: http://researchonline.rca.ac.uk/1389/19/BrodyFass_DigitalPublicSpace_2013.pdf [Accessed 23 September 2016].

Brooks, 2007. *The outsourced brain*. [online] *The New York Times*. Available at: www.nytimes.com/2007/10/26/opinion/26brooks.html [Accessed 5 October 2016].

Burke, M. and Kraut, R.E., 2014, April. Growing closer on Facebook: changes in tie strength through social network site use. *Proceedings of the SIGCHI Conference on Human Factors in Computing Systems*. ACM, pp. 4187–4196.

Burns, C., 2014. Hear yourself happy. *New Scientist*, *224*(2996), pp. 40–43.

Bush, V., 1945. As we may think. *The Atlantic Monthly 176*(1), pp. 101–108.

Byrne, R. and Whiten, A., 1989. *Machiavellian intelligence: social expertise and the evolution of intellect in monkeys, apes, and humans*. Oxford Science Publications.

Campbell, H., 2014. *The Art of Neil Gaiman*. Harper Design, p.65.

Carr, N., 2011. *The shallows: what the internet is doing to our brains*. WW Norton & Company.

Cass, J., Goulden, L. and Koslov, S., 2003. Intimate media. Emotional needs and ambient intelligence. *In:* Aarts, E. and Marzano, S., eds. *The new everyday: views on ambient intelligence*. 010 Publishers.

Castells, M., 2011. *The rise of the network society: The information age: economy, society, and culture, volume 1*. John Wiley & Sons, p.1976.

Cat, Z., 2015. My name is only real enough to work at Facebook, not to use on the site. [online] Available at: https://medium.com/@zip/my-name-is-only-real-enough-to-work-at-facebook-not-to-use-on-the-site-c37daf3f4b03#.2b6wejs5c [Accessed 13 October 2016].

Ceurstemont, S., 2014. The virtual, in reality. *New Scientist*, *221*(2950), p.17.

Challis, C., 2015. Computer says no! Gym's sexist system locks female doctor out of women's changing room. [online] *BT.* Available at: http://home.bt.com/lifestyle/computer-says-no-gyms-sexist-system-locks-female-doctor-out-of-womens-changing-room-11363969434794 [Accessed 20 October 2016].

Chambers, P., 2010. My tweet was silly, but the police reaction was absurd. [online] *The Guardian: Comment is Free.* Available at: www.theguardian.com/commentisfree/libertycentral/2010/may/11/tweet-joke-criminal-record-airport [Accessed 18 October 2016].

Chan, A., 2009. If you think twitter is weird, you're not alone. [online] *Gravity 7.* Available at: www.gravity7.com/blog/media/2009/06/if-you-think-twitter-is-weird-youre-not.html [Accessed 11 October 2016].

Chinn, M.D. and Fairlie, R.W., 2010. ICT use in the developing world: an analysis of differences in computer and internet penetration. *Review of International Economics*, *18*(1), pp. 153–167.

Clark, A., 1996. *Being there: putting brain, body, and world together again*. MIT Press, p.180.

Clark, A., 2004. *Natural-born cyborgs: minds, technologies, and the future of human intelligence* Oxford University Press, p.109.

Clark, A. and Chalmers, D., 1998. The extended mind. *Analysis*, *58*(1). Oxford University Press, pp. 7–19.

Cohen, N., 2011. Define gender gap? Look up Wikipedia's contributor list. *The New York Times, January*, *30*(362), pp. 1050–56.

Collingwood, L. and Broadbent, G., 2015. Offending and being offended online: vile messages, jokes and the law. *Computer Law & Security Review*, *31*(6), pp. 763–772.

Collins, 2015. Fair use: copyright differences in the UK and US. [online] *The Design and Artists Copyright Society.* Available at: www.dacs.org.uk/latest-news/us-fair-use-uk-fair-dealing-differences-law?category=For+Artists&title=N [Accessed 14 October 2016].

Conti, M., Das, S.K., Bisdikian, C., Kumar, M., Ni, L.M., Passarella, A., Roussos, G., Tröster, G., Tsudik, G. and Zambonelli, F., 2012. Looking ahead in pervasive computing: challenges and opportunities in the era of cyber–physical convergence. *Pervasive and Mobile Computing*, *8*(1), pp. 2–21.

Cook, T., 2015. Always keep (nerd)fighting: Durkheim's collective effervescence and fandoms as reformative social movements. *PCA/ACA National Conference 2015.* Available at: http://ncp.pcaaca.org/presentation/always-keep-nerdfighting-durkheim%E2%80%99s-collective-effervescence-and-fandoms-reformative-so [Accessed 26 March 2016].

Cooper, R., 1994. Third generation new product process. *Journal of Product Innovation Management* 11(1), pp. 3–14.

Cooper, R. and Boyko, C.T., 2010. How to design a city in five easy steps: exploring VivaCity 2020's process and tools for urban design decision making. *Journal of Urbanism,* 3(3) pp. 253–273.

Cooper, R., Wootton, A.B., Davey, C.L. and Press, M., 2005. Breaking the cycle: fundamentals of crime-proofing design. *Crime Prevention Studies,* Vol 18, pp. 179–201.

Cooray-Smith, J., 2014. *I left my watch at home & it's really, really bothering me. I feel really, profoundly uncomfortable. Like I've lost a finger.* [Twitter]. 5 November. Available at: https://twitter.com/thejimsmith [Accessed: 1 October 2015].

Coulton, P., 2015. Playful and gameful design for the internet of things. *In:* Nijholt, A., ed. *More playful user interfaces: interfaces that invite social and physical interaction*. Springer, pp. 151–173.

Coulton, P., Murphy, E. and Smith, R., 2015. Live at Lica: collection access via augmented reality. *Digital R&D Fund for the Arts Research & Development Report.*

Cousins, J., 2012. Creating the backbone. *In:* Hemment, D., Thompson, B., de Vicente, J.L. and Cooper, R., eds. *Digital Public Space.* FutureEverything Publications. Available from: http://futureeverything.org/publications/digital-public-spaces/ [Accessed 23 September 2016].

Cowlishaw, T., 2015. The right to the network: radical urbanism of digital public space. [online] *Contributoria.* Available at: www.contributoria.com/issue/2015-05/551000321045c8eb71000132/ [Accessed 14 October 2016].

Crang, M., Crosbie, T. and Graham, S., 2006. Variable geometries of connection: urban digital divides and the uses of information technology. *Urban Studies*, *43*(13), pp. 2551–2570.

Crang, M., Crosbie, T. and Graham, S., 2007. Technology, time–space, and the remediation of neighbourhood life. *Environment and Planning A*, *39*(10), pp. 2405–2422.

Cuff, D., 2003. Immanent domain. *Journal of Architectural Education*, *57*(1), pp. 43–49.

Dalton, B., 2015. The Barrow Woodwose. *Royal College of Art Work in Progress 2015.* Available at: http://soc2015.rca.ac.uk/ben-dalton/ [Accessed 19 October 2016].

Dalton, B., and Fass, J., 2014. Work and wellbeing in digital public space. *In*: Myerson, J. and Gee, E., eds. *Time and motion: redefining working life*. Liverpool University Press, p.148.

Danezis, G. and Clayton, R., 2007. Introducing traffic analysis. *In*: Acquisti, A., Gritzalis, S., Lambrinoudakis, C. and di Vimercati, S. eds., 2007. *Digital privacy: theory, technologies, and practices*. CRC Press.

Davidson, R. and Poor, N., 2015. The barriers facing artists' use of crowdfunding platforms: personality, emotional labor, and going to the well one too many times. *New Media & Society*, *17*(2), pp. 289–307.

De Michelis, Giorgio. 2003. Mixed objects. *Appliance Design Journal.*

De Preester, H. and Tsakiris, M., 2009. Body-extension versus body-incorporation: is there a need for a body-model? *Phenomenology and the Cognitive Sciences*, *8*(3), pp. 307–319.

de Ruiter, J., Weston, G. and Lyon, S.M., 2011. Dunbar's number: group size and brain physiology in humans reexamined. *American Anthropologist*, *113*(4), pp. 557–568.

de Souza e Silva, A., 2006. From cyber to hybrid mobile technologies as interfaces of hybrid spaces. *Space and Culture*, *9*(3), pp. 261–278.

Dean, L.G., Vale, G.L., Laland, K.N., Flynn, E. and Kendal, R.L., 2014. Human cumulative culture: a comparative perspective. *Biological Reviews*, *89*(2), pp. 284–301.

Design Council, 2016. A study of the design process. [online] *Design Council.* www.designcouncil.org.uk/sites/default/files/asset/document/ElevenLessons_Design_Council%20(2).pdf [Accessed 1 August 2016]

DiMaggio, P., Hargittai, E., Neuman, W.R. and Robinson, J.P., 2003. Social implications of the internet. *In:* Nissenbaum, H., Price, M.E., and Lang, P., eds. *Academy and the Internet,* p.48.

Doctor Who, 1966. (First appearance of the Cybermen: *The Tenth Planet*). BBC Television.

Doctorow, C., 2008. *Little Brother*. Tor Books.

Doctorow, C., 2014. *You are not a digital native: privacy in the age of the internet.* [online] Available at: www.tor.com/2014/05/27/you-are-not-a-digital-native-privacy-in-the-age-of-the-internet/ [Accessed 26 September 2016].

Dodge, M. and Kitchin, R., 2001. *Mapping cyberspace*. London: Routledge.

Dodge, M. and Kitchin, R., 2009. Software, objects, and home space. *Environment and Planning A*, *41*(6), pp. 1344–1365.

Dodge, M., Kitchin, R. and Zook, M., 2009. How does software make space? Exploring some geographical dimensions of pervasive computing and software studies. *Environment and Planning A*, *41*(6), pp. 1283–1293.

Donath, J., 2014. *The social machine: designs for living online*. MIT Press.

Dourish P., 2001. *Where the action is: the foundations of embodied interaction.* MIT Press.

Dourish, P. and Bell, G., 2014. 'Resistance is futile': reading science fiction alongside ubiquitous computing. *Personal and Ubiquitous Computing*, *18*(4), pp. 769–778.

Downey, S., 2012. Visualizing a taxonomy for virtual worlds. *Journal of Educational Multimedia and Hypermedia*, *21*(1), p.53.

Dowthwaite, L., 2014. Getting paid for giving away art for free: the case of webcomics. [online] *CREATe.* Available at: www.create.ac.uk/blog/2014/02/25/webcomics-dowthwaite/ [Accessed 17 October 2016].

Dredge, S., 2014. Flickr takes flak for selling Creative Commons photos as wall-art prints. [online] *The Guardian.* Available at: https://www.theguardian.com/technology/2014/dec/02/flickr-creative-commons-photos-wall-art [Accessed 17 October 2016].

Drew, T., Võ, M.L.H. and Wolfe, J.M., 2013. The invisible gorilla strikes again: sustained inattentional blindness in expert observers. *Psychological Science*, *24*(9), pp. 1848–1853.

Duane, D., 2001. *High wizardry*. Houghton Mifflin Harcourt, p.179.

Dunbar, R., 2010. *How many friends does one person need?: Dunbar's number and other evolutionary quirks*. Faber & Faber.

Dunbar, R., 2014. *Human evolution: a Pelican introduction*. Penguin UK.

Dunbar, R.I., 1993. Coevolution of neocortical size, group size and language in humans. *Behavioral and Brain Sciences*, *16*(04), pp. 681–694.

Dunbar, R.I., 2012. Social cognition on the internet: testing constraints on social network size. *Philosophical Transactions of the Royal Society of London B: Biological Sciences*, *367*(1599), pp. 2192–2201.

Dunbar, R.I., Duncan, N.D.C. and Nettle, D., 1995. Size and structure of freely forming conversational groups. *Human Nature*, *6*(1), pp. 67–78.

Dunbar, R.I.M., 2004. Language, Music, and Laughter in Evolutionary Perspective. *In: Evolution of communication systems*, MIT Press, p.257.

Dunne, A. and Raby, F., 2013. *Speculative everything: design, fiction, and social dreaming*. MIT Press.

Ekblom, P., 2007. Going equipped: criminology, situational crime prevention and resourceful offender. *British Journal of Criminology*, *40(3)*.

Ehn, P., 2006. Participation in interaction design: actors and artifacts in interaction. *In*: Bagnara, S. and Crampton Smith, eds. *Theories and practice in interaction design*. LEA, pp. 1–356.

Ehrsson, H.H., Spence, C. and Passingham, R.E., 2004. That's my hand! Activity in premotor cortex reflects feeling of ownership of a limb. *Science*, *305*(5685), pp. 875–877.

Ellison, N.B. and Vitak, J., 2015. Social network site affordances and their relationship to social capital processes. *The Handbook of the Psychology of Communication Technology*, pp. 205–227.

Erol, R., Press, M., Cooper, R. and Thomas, M., 2002. Designing out crime: raising awareness of crime reduction in the design industry. *Security Journal*, *15(1)*, pp. 49–61.

European Commission, n.d. Factsheet on the 'Right to be Forgotten' ruling (c-131/12). Available at: http://ec.europa.eu/justice/data-protection/files/factsheets/factsheet_data_protection_en.pdf [Accessed 19 October 2016].

Evans, B., 1985. Japanese-style management, product design and corporate strategy, *Design Studies*, *6(1)*, pp. 25–33.

Evans, J.A., 2008. Electronic publication and the narrowing of science and scholarship. *Science*, *321*(5887), pp. 395–399.

Evers, S., 2015. Everyone says. *In*: Page, R., Amos, M. and Rasmussen, S., eds., 2015. *Beta-Life: stories from an A-life future*. Comma Press.

Farber, D., 2013. Philip Rosedale's second life with high fidelity. [online] Cnet. Available at: www.cnet.com/news/philip-rosedales-second-life-with-high-fidelity/ [Accessed 27 September 2016].

Featherstone, M., 2000. Archiving cultures. *The British Journal of Sociology*, *51*(1), pp. 161–184.

Featherstone, M., 2009. Ubiquitous media: an introduction. *Theory, Culture & Society*, *26*(2–3), pp. 1–22.

Fellett, M., 2011. Phone tech transforms African business and healthcare. *New Scientist*, 2833.

Fiesler, C. and Bruckman, A.S., 2014, February. Remixers' understandings of fair use online. *Proceedings of the 17th ACM Conference on Computer Supported Cooperative Work & Social Computing*. ACM, pp. 1023–1032.

Fiesler, C., Morrison, S. and Bruckman, A.S., 2016, May. An archive of their own: a case study of feminist HCI and values in design. *Proceedings of the 2016 CHI Conference on Human Factors in Computing Systems*. ACM, pp. 2574–2585.

Fischer, G., 2000. Symmetry of ignorance, social creativity, and meta-design. *Knowledge-Based Systems*, *13*(7), pp. 527–537.

Flood, A., 2017 'We're told to be grateful we even have readers': pirated ebooks threaten the future of book series. [online] *The Guardian*. Available at: https://www.theguardian.com/books/2017/nov/06/pirated-ebooks-threaten-future-of-serial-novels-warn-authors-maggie-stiefvater [Accessed 9 November 2017].

Floridi, L., 2007. A look into the future impact of ICT on our lives. *The Information Society*, *23*(1), pp. 59–64.

Forsythe, C., Kruse, A. and Schmorrow, D. 2005. Augmented cognition. In Forsythe, C., Bernard, M.L. and Goldsmith, T.E. eds. 2006. *Cognitive systems: human cognitive models in systems design*. Psychology Press.

Foster, J.K., 2011. Memory: how memories are made. *New Scientist*, *2841*.

Frenchman, D. and Rojas, F., 2006. Zaragoza's Digital Mile: place-making in a new public realm [Media and the city]. *Places*, *18*(2).

Fritsch, M., 2015. *The Story of Technoviking*. Available at: https://vimeo.com/140265561 [Accessed 18 October 2016].

Fuchs-Kittowski, K. and Krüger, P., 1997. The noosphere vision of Pierre Teilhard de Chardin and Vladimir I. Vernadsky in the perspective of information and of world-wide communication 1. *World Futures: Journal of General Evolution*, *50*(1–4), pp. 757–784.

Gaiman, N., 2008. The results of free. [online] Available at: http://journal.neilgaiman.com/2008/07/results-of-free.html [Accessed 17 October 2016].

Gallippi, A., 2013. The present and future impact of virtual currency. *Statement before the Subcommittee on National Security and International Trade and Finance and Subcommittee on Economic Policy of the United States Senate Committee on Banking, Housing and Urban Affairs*. Available at: www.banking.senate.gov/public/index.cfm/2013/11/the-present-and-future-impact-of-virtual-currency [Accessed 17 October 2016].

Garfield, S., 2012. *On the map: why the world looks the way it does*. Profile Books.

Garcia-Martinez, J., 2016. Here's what will happen when 30 billion devices are connected to the internet. [online] *World Economic Forum*. Available at: www.weforum.org/agenda/2016/06/nanosensors-and-the-internet-of-nano-things [Accessed 28 October 2016].

Garner, J., Wood, G., Pijnappel, S., Murer, M. and Mueller, F.F., 2013, September. Combining moving bodies with digital elements: design space between players and screens. *Proceedings of the 9th Australasian Conference on Interactive Entertainment: Matters of Life and Death*. ACM, p.17.

Ghena, B., Beyer, W., Hillaker, A., Pevarnek, J. and Halderman, J.A., 2014. Green lights forever: analyzing the security of traffic infrastructure. *8th USENIX Workshop on Offensive Technologies (WOOT 14)*.

Gibbs, S., 2014. Google Glass review: useful – but overpriced and socially awkward. [online] *The Guardian*. Available at: www.theguardian.com/technology/2014/dec/03/google-glass-review-curiously-useful-overpriced-socially-awkward [Accessed 27 September 2016].

Gibson, K.R., 1993. Tool use, language and social behaviour in relationship to information processing capacities. *In:* Gibson, K. R. and Ingold, T., eds. *Tools, language and cognition in human evolution*. Cambridge University Press, pp. 251–269.

Gibson, W., 2000. *Neuromancer*. Penguin.

Gillan, C., 2014. Why can't I stop? *New Scientist*, *223*(2980), pp. 28–29.

Gillen, J., 2016. How early picture postcards were the Edwardian equivalent of Instagram. [online] *The Conversation*. Available at: https://theconversation.com/how-early-picture-postcards-were-the-edwardian-equivalent-of-instagram-61870 [Accessed 11 October 2016].

Gillmor, D., 2011. WikiLeaks payments blockade sets dangerous precedent. [online] *The Guardian*. Available at: www.theguardian.com/commentisfree/cifamerica/2011/oct/27/wikileaks-payments-blockade-dangerous-precedent [Accessed 17 October 2016].

Glusman, A., 2010. *The Malkovich Bias*. [online] Available at: http://glusman.blogspot.co.uk/2010/02/malcovich-bias-over-years-ive-noticed.html [Accessed 26 September 2016].

Gonçalves, B., Perra, N. and Vespignani, A., 2011. Modeling users' activity on twitter networks: Validation of Dunbar's number. *PLOS One*, *6*(8), p.e22656.

Goodman, M., 2012. A vision of crimes in the future. [online] *TED Talks 2012*. Available at: www.ted.com/talks/marc_goodman_a_vision_of_crimes_in_the_future [Accessed 18 October 2016].

Goodwin, T., 2015. The battle is for the customer interface. [online] *TechCrunch*. Available at: https://techcrunch.com/2015/03/03/in-the-age-of-disintermediation-the-battle-is-all-for-the-customer-interface/ [Accessed 14 October 2016].

Google, 2015. Google Cultural Institute: Frequently asked questions. [online] Available at: www.google.com/culturalinstitute/about/ [Accessed 1 February 2015].

Gradinar, A., Burnett, D., Coulton, P., Forrester, I., Watkins, M., Scutt, T. and Murphy, E., 2015, September. Perceptive media – adaptive storytelling for digital broadcast. *In:* Abascal J., Barbosa S., Fetter M., Gross T., Palanque P. and Winckler M., eds. *Human–Computer Interaction*, Springer International Publishing, pp. 586–589.

Graeber, D., 2011. *Debt: the first 5000 years*. Melville House Publishing.

Graham, S., 2002. Bridging urban digital divides? Urban polarisation and information and communications technologies (ICTs). *Urban Studies*, *39*(1), pp. 33–56.

Graham, S. and Aurigi, A., 1997. Virtual cities, social polarization, and the crisis in urban public space. *The Journal of Urban Technology*, *4*(1), pp. 19–52.

Granovetter, M.S., 1973. The strength of weak ties. *American Journal of Sociology*, pp. 1360–1380.

Gray, R. 2013. The places where Google Glass is banned. [online] *The Telegraph*. Available at: www.telegraph.co.uk/technology/google/10494231/The-places-where-Google-Glass-is-banned.html [Accessed 27 September 2016].

Gray, W.D. and Fu, W.T., 2004. Soft constraints in interactive behavior: the case of ignoring perfect knowledge in-the-world for imperfect knowledge in-the-head. *Cognitive Science*, *28*(3), pp. 359–382.

Grayson, K. and Shulman, D., 2000. Indexicality and the verification function of irreplaceable possessions: a semiotic analysis. *Journal of Consumer Research*, *27*(1), pp. 17–30.

Green, J. 2003. Thinking the future. *In:* Aarts, E. and Marzano, S., eds. *The new everyday: views on ambient intelligence. 010 Publishers.*

Grinberg, R., 2012. Bitcoin: an innovative alternative digital currency. *Hastings Science & Technology Law Journal*, *4*, p.159.

Gulotta, R., Odom, W., Forlizzi, J. and Faste, H., 2013, April. Digital artifacts as legacy: exploring the lifespan and value of digital data. *Proceedings of the SIGCHI Conference on Human Factors in Computing Systems*. ACM, pp. 1813–1822.

Habermas, J., 1996. *Between facts and norms: contributions to a discourse theory of law and democracy.* Cambridge: MIT Press, p.360.

Hamilton, P.F., 2004. *Pandora's Star.* Macmillan.

Hamilton, P.F., 2004–05. *The Commonwealth Saga* (series). Macmillan

Hampton, K.N., Lee, C.J. and Her, E.J., 2011. How new media affords network diversity: direct and mediated access to social capital through participation in local social settings. *New Media & Society*, p.1461444810390342.

Hampton, K.N., Sessions, L.F. and Her, E.J., 2011. Core networks, social isolation, and new media: how internet and mobile phone use is related to network size and diversity. *Information, Communication & Society*, *14*(1), pp. 130–155.

Hampton, K.N., Sessions, L.F., Her, E.J. and Rainie, L., 2009. Social isolation and new technology. *Pew Internet & American Life Project, 4.*

Hardman, I., 2014. The menace of memes: how pictures can paint a thousand lies. [online] *The Spectator.* Available at: http://blogs.spectator.co.uk/2014/11/the-menace-of-memes-how-pictures-can-paint-a-thousand-lies/ [Accessed: 20 October 2016].

Harris, M., 2014. Are you looking at me? *New Scientist*, *224*(2998), p.21.

Harrison, S., Tatar, D. and Sengers, P., 2007. The three paradigms of HCI, Alt. Chi. *Session of the SIGCHI Conference on Human Factors in Computing Systems.*

Hawkins, J. and Blakeslee, S., 2007. *On intelligence*. Macmillan.

Hayles, N. K., 1999. *How we became posthuman: virtual bodies in cybernetics, literature, and informatics*. University of Chicago Press, p.290.
Hayles, N.K., 2002. Flesh and metal: reconfiguring the mindbody in virtual environments. *Configurations, 10*(2), pp. 297–320.
Hayles, N.K., 2009. RFID: human agency and meaning in information-intensive environments. *Theory, Culture & Society*, *26*(2–3), pp. 47–72.
Haythornthwaite, C., 2001. Introduction: the internet in everyday life. *American Behavioral Scientist, 45*(3), pp. 363–382.
Heinlein, R., 1942. Waldo. *Astounding Science Fiction*. USA: Street & Smith Publications Inc.
Helsper, E.J. and Eynon, R., 2010. Digital natives: where is the evidence? *British Educational Research Journal*, 36(3), pp. 503–520.
Henshilwood, C.S., 2007. Fully symbolic Sapiens behaviour: innovation in the middle stone age at Blombos cave, South Africa. *In:* Stringer, C. and Mellars, P., eds. *Rethinking the human revolution: new behavioural and biological perspectives on the origin and dispersal of modern humans*. Cambridge University Press, pp. 123–132.
Her, 2013. Film. Directed by Spike Jonze. USA: Warner Bros Pictures.
Hess, A., 2014. Why women aren't welcome on the internet. [online] *Pacific Standard.* Available at: https://psmag.com/why-women-aren-t-welcome-on-the-internet-aa21fdbc8d6#.e7k54g5wy [Accessed 13 October 2016].
Hill, R.A. and Dunbar, R.I., 2003. Social network size in humans. *Human Nature*, *14*(1), pp. 53–72.
Hillis, K., 1999. *Digital sensations: space, identity, and embodiment in virtual reality*. University of Minnesota Press.
Hinssen, P., 1995. Life in the digital city. [online] *Wired Magazine*, *3*(6) Available at: www.wired.com/1995/06/digcity/ [Accessed 26 September 2016].
Hodson, H., 2013. Online gamers harnessed to help disaster response. *New Scientist*, *219*(2928), p.21.
Hodson, H., 2013. Smartphone data could be used to influence what you do. *New Scientist*, *220*(2942), pp. 19–20.
Hodson, H., 2014. Welcome to the oracle. *New Scientist*, *223*(2983), pp. 18–19.
Hodson, H., 2015. Face recognition row over right to identify you in the street. [online] *New Scientist*. Available at: www.newscientist.com/article/dn27754-face-recognition-row-over-right-to-identify-you in the street [Accessed 19 October 2016].
Hodson, H., 2015. Nothing but the truth. *New Scientist*, *225*(3010), p.24.
Hodson, H., 2016a. Google knows your ills. *New Scientist*, *230*(3072), pp. 22–23.
Hodson, H., 2016b. Did Google's NHS patient data deal need ethical approval? [online] *New Scientist*. Available at: www.newscientist.com/article/2088056-did-googles-nhs-patient-data-deal-need-ethical-approval/ [Accessed 18 October 2016].
Hodson, H., 2016c. Printable electronics for all. *New Scientist*, *230*(3079), p.22.
Holland, J. and Stornetta, S., 1992. Beyond being there. *Proceedings of the SIGCHI Conference on Human Factors in Computing Systems*. ACM, pp. 119–125.
Holpuch, A., 2013. Damaged undersea internet cable causes widespread service disruption. [online] *The Guardian.* Available at: www.theguardian.com/technology/2013/mar/28/damaged-undersea-cable-internet-disruption [Accessed 29 September 2016].
Hongladarom, S., 2013. Ubiquitous computing, empathy and the self. *AI & Society*, *28*(2), pp. 227–236.
Hood, B., 2014. *The domesticated brain: a Pelican introduction*. Penguin UK.
Houghton, D.J. and Joinson, A.N., 2010. Privacy, social network sites, and social relations. *Journal of Technology in Human Services*, *28*(1–2), pp. 74–94.
Hulit, L.M., Fahey, K.R. and Howard, M.R., 1993. *Born to talk: an introduction to speech and language development*. Merrill.
Hunt, G.R., 1996. Manufacture and use of hook-tools by New Caledonian crows. *Nature*, *379*(6562), pp. 249–251.
Hutchins, E., 1995. *Cognition in the wild*. MIT Press.
Ikegami, E. and Hut, P., 2008. Avatars are for real: virtual communities and public spheres. *Journal of Virtual Worlds Research*, *1*(1), pp. 1–19.

Iron Man, 2008. Film. Directed by Jon Favreau. USA: Paramount Pictures.

Jackson, J., 2015. Less than half of UK adults are aware ads fund free content online. [online] *The Guardian.* Available at: www.theguardian.com/media/2015/jul/01/less-than-half-of-uk-adults-are-aware-ads-fund-free-content-online?CMP=share_btn_tw [Accessed 17 October 2016].

Jacobs, N. and Huck, J., 2017. Can we give ourselves extra senses? *In:* Heywood, I., ed. *Sensory arts and design*. Bloomsbury.

James, C., 2014. *Disconnected: youth, new media, and the ethics gap*. (Foreword by Jenkins, H). MIT Press, pp. 2, 27, 38 & 53.

Jeffries, S., 2014. Neil Harbisson: the world's first cyborg artist. [online] *The Guardian.* www.theguardian.com/artanddesign/2014/may/06/neil-harbisson-worlds-first-cyborg-artist [Accessed 26 October 2016].

Jenkins, H., 2006. *Convergence culture: where old and new media collide*. NYU Press, p.208.

Jenkins, H., 2012. *Textual poachers: television fans and participatory culture*. Routledge, p.xxxi.

Jenkins, H., Ford, S. and Green, J., 2013. *Spreadable media: creating value and meaning in a networked culture*. NYU Press.

Jinek, M., Chylinski, K., Fonfara, I., Hauer, M., Doudna, J.A. and Charpentier, E., 2012. A programmable dual-RNA-guided DNA endonuclease in adaptive bacterial immunity. *Science, 337*(6096), pp. 816–821.

Kahn, J.M., Katz, R.H. and Pister, K.S., 1999, August. Next century challenges: mobile networking for 'Smart Dust'. *Proceedings of the 5th Annual ACM/IEEE International Conference on Mobile Computing and Networking*. ACM, pp. 271–278.

Kanai, R., Bahrami, B., Roylance, R. and Rees, G., 2011. Online social network size is reflected in human brain structure. *Proceedings of the Royal Society of London B*. p. rspb20111959.

Kawamura, T., Fukuhara, T., Takeda, H., Kono, Y. and Kidode, M., 2003, July. Ubiquitous memories: wearable interface for computational augmentation of human memory based on real world objects. *Proceedings of 4th International Conference on Cognitive Science (ICCS2003)*, pp. 273–278.

Kelion, L., 2015. *Windows 10 to get 'holographic' headset and Cortana.* [online] BBC News. Available at: www.bbc.co.uk/news/technology-30924022 [Accessed 27 September 2016].

Kelly, A.M., 2013. *Google Maps as 21st C. Mercator projection: a map so engrained all other versions simply seem a bit strange.* [Twitter]. 12 December. Available at: https://twitter.com/a_m_kelly/status/411137681364512770 [Accessed: 20 October 2016].

Kennedy, J.B., 1926. When woman is boss: an interview with Nikola Tesla. *Colliers, Seattle.*

Kirsh, D. and Maglio, P., 1994. On distinguishing epistemic from pragmatic action. *Cognitive Science, 18*(4), pp. 513–549.

Kiss, J., 2013. BBC makes space for cultural history. *The Guardian,* 6 January 2013. Available from www.theguardian.com/media/2013/jan/06/bbc-digital-public-space-archive [Accessed 23 September 2016].

Kleinman, Z., 2015. Who's that girl? The curious case of Leah Palmer. [online] *BBC News.* Available at: www.bbc.co.uk/news/technology-31710738 [Accessed 17 October 2016].

Kosinski, M., Stillwell, D. and Graepel, T., 2013. Private traits and attributes are predictable from digital records of human behavior. *Proceedings of the National Academy of Sciences, 110*(15), pp. 5802–5805.

Koslowski, B., 2014. Screenscapes: of theatre, audiences and social media. *In*: Vicente, A. and Ferreira, H., eds. *Post-screen: device, medium and concept*. Faculdade de Belas-Artes da Universidade de Lisboa, 2014.

Krützen, M., Mann, J., Heithaus, M.R., Connor, R.C., Bejder, L. and Sherwin, W.B., 2005. Cultural transmission of tool use in bottlenose dolphins. *Proceedings of the National Academy of Sciences of the United States of America, 102*(25), pp. 8939–8943.

Kuhn, S. and Stiner, M.C., 2007. Cognitive perspectives on modern human origins. *In*: Mellars, P. ed. *Rethinking the human revolution: new behavioural and biological perspectives on the origin and dispersal of modern humans*. McDonald Institute for Archaeological Research.

Kurzweil, R., 2005. *The singularity is near: when humans transcend biology*. Penguin.

Lai, C.H. and Katz, J.E., 2012. Are we evolved to live with mobiles? An evolutionary view of mobile communication. *Periodica Polytechnica, Social and Management Sciences*, *20*(1), p.45.

Le Guin, U., 2004. A rant about "Technology". [online] Available at: www.ursulakleguin.com/Note-Technology.html [Accessed 26 September 2016].

Leckie, A., 2013–15. *Imperial Radch* (series). Orbit.

Lee, D., 2014. *What is the 'right to be forgotten'?* [online] Available at: www.bbc.co.uk/news/technology-27394751 [Accessed 26 September 2016].

Lerman, K., Yan, X. and Wu, X.Z., 2016. The 'majority illusion' in social networks. *PLOS One*, *11*(2), p.e0147617.

Lessig, L., 2008. *Remix: making art and commerce thrive in the hybrid economy*. Penguin.

Lindley, J.G., Coulton, P. and Cooper, R., 2017. Why the internet of things needs object orientated ontology. *Proceedings of EAD 2017*. European Academy of Design.

Lingel, J., 2013. 'Keep it secret, keep it safe': information poverty, information norms, and stigma. *Journal of the American Society for Information Science and Technology*, *64*(5), pp. 981–991.

Ljubič, K. and Fister Jr, I., 2014. How to store Wikipedia into a forest tree: initial idea. *Presented at First International Conference on Multimedia, Scientific Information and Visualization for Information Systems and Metrics.* Available at: www.iztok-jr-fister.eu/static/publications/35.pdf [Accessed 26 October 2016].

Ljubič Fister, K., 2016. Breaking the wall of data storage. *Presented at Falling Walls Lab Berlin 2015.* Available at: https://vimeo.com/148192811 [Accessed 26 October 2016].

lordoftheinternet, 2013. Some thoughts are so private that you only share them with a therapist or 17,000 people on the internet. [Tumblr post] Available at: http://lordoftheinternet.tumblr.com/post/44788412914/some-thoughts-are-so-private-that-you-only-share [Accessed 18 October 2016].

Lothian, A., 2013. Archival anarchies: online fandom, subcultural conservation, and the transformative work of digital ephemera. *International Journal of Cultural Studies*, *16*(6), pp. 541–556.

Lynch, J., 2015. EFF and eight other privacy organizations back out of NTIA face recognition multi-stakeholder process. [online] *Electronic Frontier Foundation.* Available at: www.eff.org/deeplinks/2015/06/eff-and-eight-other-privacy-organizations-back-out-ntia-face-recognition-multi. [Accessed 19 October 2016].

Ma, R., Kim, D.H., McCormick, M., Coleman, T. and Rogers, J., 2010, August. A stretchable electrode array for non invasive, skin-mounted measurement of electrocardiography (ECG), electromyography (EMG) and electroencephalography (EEG). In *2010 Annual International Conference of the IEEE Engineering in Medicine and Biology*. IEEE, pp. 6405–6408.

McCaffrey, A., 2012. *The ship who sang*. Random House, p.8.

McCarthy, J.F. and boyd, d. 2005. Digital backchannels in shared physical spaces: experiences at an academic conference. *CHI'05 Extended Abstracts on Human Factors in Computing Systems*. ACM, pp. 1641–1644.

McDonald, A.M. and Cranor, L.F., 2008. The cost of reading privacy policies. *Journal of Law and Policy for the Information Society*, *4*, p.543.

McPherson, M., Smith-Lovin, L. and Brashears, M.E., 2006. Social isolation in America: changes in core discussion networks over two decades. *American Sociological Review*, *71*(3), pp. 353–375.

Mäenpää, P., 2001. Mobile communication as a way of urban life. *In*: Mäenpäa, P., Warde, A. and Gronow, J., eds. *Ordinary consumption*. Routledge, p.111.

Manjoo, F., 2008. *True enough: Learning to live in a post-fact society*. John Wiley & Sons.

Malafouris, L., 2004. The cognitive basis of material engagement: where brain, body and culture conflate. *In:* DeMarrais, E., Gosden, C., and Renfrew, C., eds. *Rethinking materiality: the engagement of mind with the material world.* Cambridge: McDonald Institute for Archaeological Research, pp. 53–61.

Maravita, A. and Iriki, A., 2004. Tools for the body (schema). *Trends in Cognitive Sciences*, *8*(2), pp. 79–86.

Marescaux, J., Leroy, J., Gagner, M., Rubino, F., Mutter, D., Vix, M., Butner, S.E. and Smith, M.K., 2001. Transatlantic robot-assisted telesurgery. *Nature*, *413*(6854), pp. 379–380.

Marks, P., 2014a. Beams of sound immerse you in music others can't hear. *New Scientist*, 2954

Marks, P., 2014b. Messaging app lets you leave secrets on street corners. *New Scientist*, 2981

Marwick, A.E. and boyd, d., 2011. I tweet honestly, I tweet passionately: Twitter users, context collapse, and the imagined audience. *New Media & Society*, *13*(1), pp. 114–133.

Mashable.com, 2016. *38% of children under 2 use mobile media, study says.* [online] http://mashable.com/2013/10/28/children-under-2-mobile-media-study/#feBJP7f5ouqA [Accessed 1 August 2016].

Masters, S., 2010. *High tech kids cost parents £537 a year.* [online] Available at: www.parentallychallenged.co.uk/news/according-to-a-survey/high-tech-kids-cost-parents-537-a-year/ [Accessed 26 September 2016].

Matonis, J., 2012. The payments network as economic weapon. [online] *Forbes*. Available at: www.forbes.com/sites/jonmatonis/2012/03/27/the-payments-network-as-economic-weapon/#2b6cc33b2182 [Accessed 17 October 2016].

The Matrix, 1999. Film. Directed by The Wachowskis. USA: Warner Bros Pictures.

Mayer-Schoenberger, V., 2007. *Useful void: the art of forgetting in the age of ubiquitous computing*. KSG Working Paper No. RWP07-022. Available at: SSRN: https://ssrn.com/abstract=976541. [Accessed 7 November 2016].

Meier, P., 2015. *Digital humanitarians: how big data is changing the face of humanitarian response.* CRC Press, p.157.

Minkel, E., 2014. ICYMI: the internet has ruined our conception of time. [online] *New Republic*. Available at: https://newrepublic.com/article/117886/icymi-internet-acronym-destroying-our-conception-time [Accessed 13 October 2016].

Minority Report, 2002. Film. Directed by Steven Spielberg. USA: 20th Century Fox.

Minsky, M., 1980. Telepresence. *OMNI magazine*, June 1980. Available at: http://web.media.mit.edu/~minsky/papers/Telepresence.html

Mithen, S., 1996. *The prehistory of the mind: a search for the origin of art, religion and science.* Thames and Hudson.

Moggridge, B., 2006. *Designing interactions*. MIT Press.

Moggridge, B., 2006. Adopting technology. *In: Designing interactions*, MIT Press, pp. 237–317.

Müller, J., Alt, F., Michelis, D. and Schmidt, A., 2010, October. Requirements and design space for interactive public displays. *Proceedings of the 18th ACM International Conference on Multimedia*. ACM, pp. 1285–1294.

Müller, J., Wilmsmann, D., Exeler, J., Buzeck, M., Schmidt, A., Jay, T. and Krüger, A., 2009, May. Display blindness: the effect of expectations on attention towards digital signage. *International Conference on Pervasive Computing,* Springer Berlin Heidelberg, pp. 1–8.

Mullins, J., 2011. Squishybots: soft, bendy and smarter than ever. *New Scientist*, *212*(2838), pp. 48–51.

Mundie, C., 2014. Privacy pragmatism; focus on data use, not data collection. *Foreign Affairs*, *93*, p.28.

Murdock, G., 2004. Building the digital commons: public broadcasting in the age of the internet. *The 2004 Spry Memorial Lecture*, pp. 3–7.

Murphy, S., 2014. New virtual world put the real you on screen. *New Scientist*, *222*(2966), pp. 19–20.

Myerson, J. and Gee, E., eds. 2014. *Time and motion: redefining working life*. Liverpool: Liverpool University Press.

Myerson, J. and Ross, P., 2003. *21st century office*. Laurence King Publishing, p.10.

Nakamoto, S., 2008. Bitcoin: A peer-to-peer electronic cash system. [online] Available at: https://bitcoin.org/bitcoin.pdf

Nakamoto, M., Ukimura, O., Faber, K. and Gill, I.S., 2012. Current progress on augmented reality visualization in endoscopic surgery. *Current Opinion in Urology*, *22*(2), pp. 121–126.

NASA, 2014. Space Station 3-D printer builds ratchet wrench to complete first phase of operations. [online] Available at: www.nasa.gov/mission_pages/station/research/news/3Dratchet_wrench [Accessed 17 October 2016].

Nettle, D., Nott, K. and Bateson, M., 2012. 'Cycle thieves, we are watching you': impact of a simple signage intervention against bicycle theft. *PLOS One*, *7*(12), p.e51738.

New Scientist, 2014. Twitter health. *New Scientist*, 2994.

Newsweek, 2004. All eyes on Google. [online] *Newsweek*. Available at: http://europe.newsweek.com/all-eyes-google-124041 [Accessed 5 October 2016].
Nie, N.H., 2001. Sociability, interpersonal relations, and the internet reconciling conflicting findings. *American Behavioral Scientist*, *45*(3), pp. 420–435.
Nissen, B. and Bowers, J., 2015, April. Data-things: digital fabrication situated within participatory data translation activities. *Proceedings of the 33rd Annual ACM Conference on Human Factors in Computing Systems*. ACM, pp. 2467–2476.
Norman, D., 1998. *The invisible computer, why good products can fail, the personal computer is so complex, and information appliances are the solution.* MIT Press.
Norman, D., 1988. *The psychology of everyday things*. HarperCollins, p.188.
Norman, D., 2002. *The design of everyday things.* Basic Books.
Norman, D. A. and Draper, S. W., eds. 1986. *User-centered system design: new perspectives on Human-Computer Interaction*. Lawrence Earlbaum Associates.
Norman, K.L., 2008. *Cyberpsychology: an introduction to human-computer interaction*. Vol. 1. Cambridge University Press, p.327.
Ofcom, 2014. *The communications market 2014*. [online] Available at: www.ofcom.org.uk/__data/assets/pdf_file/0031/19498/2014_uk_cmr.pdf.
O'Neill, S., 2016. I plant memories in seeds. *New Scientist*, *229*(3056), p.27.
ONS, 2016. *Office for National Statistics. Statistical bulletin internet access – households and individuals, 2016*. Available at: www.ons.gov.uk/peoplepopulationandcommunity/householdcharacteristics/homeinternetandsocialmediausage/bulletins/internetaccesshouseholdsandindividuals/2016 [Accessed 26 September 2016].
Pacific Rim, 2013. Film. Directed by Guillermo del Toro. USA: Warner Bros Pictures.
Palmer, A., 2013. The art of asking. [online] *TED Talks 2013*. Available at: www.ted.com/talks/amanda_palmer_the_art_of_asking [Accessed 14 October 2016].
Pariser, E., 2011. *The filter bubble: what the internet is hiding from you*. Penguin UK.
Parks and Recreation (Season 7), 2015. NBC.
Peachey, K., 2015. *Q&A: universal credit and the benefits overhaul.* [online] Available at: www.bbc.co.uk/news/business-11735673 [Accessed 26 September 2016].
Pejovic, V. and Musolesi, M., 2015. Anticipatory mobile computing: a survey of the state of the art and research challenges. *ACM Computing Surveys (CSUR)*, *47*(3), p.47.
Person of interest, 2011–2016. TV series. CBS.
Pfeiffer, M., Dünte, T., Schneegass, S., Alt, F. and Rohs, M., 2015, April. Cruise control for pedestrians: controlling walking direction using electrical muscle stimulation. *Proceedings of the 33rd Annual ACM Conference on Human Factors in Computing Systems*. ACM, pp. 2505–2514.
Pine, K.J., 2014. *Mind what you wear: the psychology of fashion.* Kindle edn. [ebook].
Pinker, S., 1995. *The language instinct: the new science of language and mind* (Vol. 7529). Penguin UK.
Potter, N., 1969. *What is a designer: things, places, messages*. Studio Vista.
Prensky, M., 2001. Digital natives, digital immigrants part 1. *On the Horizon*, 9(5), pp. 1–6.
Press, M. and Cooper, R., 2003. *The design experience*. Ashgate, pp. 12 & 74.
Ptak, L., 2014. Wages for Facebook. *Presented at University Art Gallery, University of California* and available at: http://wagesforfacebook.com/
Ranner, V., 2013, December. UISilk: towards interfacing the body. *Proceedings of the Second International Workshop on Smart Material Interfaces: Another Step to a Material Future.* ACM, pp. 13–18.
Rapoport, M., 2013. Being a body or having one: automated domestic technologies and corporeality. *AI & Society*, *28*(2), pp. 209–218.
Reich, D., Green, R.E., Kircher, M., Krause, J., Patterson, N., Durand, E.Y., Viola, B., Briggs, A.W., Stenzel, U., Johnson, P.L. and Maricic, T., 2010. Genetic history of an archaic hominin group from Denisova Cave in Siberia. *Nature*, *468*(7327), pp. 1053–1060.
Reichelt, L., 2007. Ambient intimacy. [online] Available at: www.disambiguity.com/ambient-intimacy/ [Accessed 13 October 2016].

Rhea, D., 1992. A new perspective on design: focusing on customer experience. *Design Management Journal*, 9(4), pp. 10–16.

Rheingold, H., 2002. *Smart mobs: the next social revolution*. Basic Books.

Ritzer, G. and Jurgenson, N., 2010. Production, consumption, presumption: the nature of capitalism in the age of the digital 'prosumer'. *Journal of Consumer Culture*, *10*(1), pp. 13–36.

Roberts, A., 2014. A swarm of living robjects around us. *In*: Amos, M., Page, R., eds. *Beta Life*. Comma Press.

Robinson, H., 2013. How many teenagers are using Ask.fm to self-harm? [online] *New Statesman*. Available at: www.newstatesman.com/sci-tech/2013/08/how-many-teenagers-are-using-askfm-self-harm [Accessed 28 September 2016].

Ronson, J., 2015. How one stupid tweet blew up Justine Sacco's life. *New York Times*. Available at: www.nytimes.com/2015/02/15/magazine/how-one-stupid-tweet-ruined-justine-saccos-life.html [Accessed 18 October 2016].

Rothrock, K., 2016. Facial recognition service becomes a weapon against Russian porn actresses. [online] *Global Voices*. Available at: https://globalvoices.org/2016/04/22/facial-recognition-service-becomes-a-weapon-against-russian-porn-actresses/?platform=hootsuite [Accessed 19 October 2016].

Rutkin, A., 2014a. Bright lights, smart city. *New Scientist*, *223*(2981), p.17.

Rutkin, A., 2014b. Off the clock, on the record. *New Scientist*, *224*(2991), pp. 22–23.

Rutkin, A., 2014c. How data can save a city. *New Scientist*, *224*(2990), pp. 24–25.

Salinas, L., Coulton, P. and Dunn, N., 2016. Using game design as a frame for evaluating experiences in hybrid digital/physical spaces. *Architecture and Culture*, *4*(1), pp. 115–135.

Sato, Y., Nakamoto, M., Tamaki, Y., Sasama, T., Sakita, I., Nakajima, Y., Monden, M. and Tamura, S., 1998. Image guidance of breast cancer surgery using 3-D ultrasound images and augmented reality visualization. *IEEE Transactions on Medical Imaging*, *17*(5), pp. 681–693.

Scalzi, J., 2014. *Lock-In*. Tor Books.

Schmidt, E., 2005. Books of revelation. *Wall Street Journal*, *18*, p.A18.

Schwartz, L., 2006. Fantasy, realism, and the other in recent video games. *Space and Culture*, *9*(3), pp. 313–325.

Seefeld, B. and Chawathe, Y., 2013. Meet the new Google Maps: a map for every person and place [online] *Google Maps*. Available at: https://maps.googleblog.com/2013/05/meet-new-google-maps-map-for-every.html [Accessed: 21 October 2016].

Seife, C., 2014. *Virtual unreality: just because the internet told you, how do you know it's true?* Penguin.

Sense8, 2015. Netflix.

Seton, M., Williams, S., Zahirovic, S. and Micklethwaite, S., 2013. Obituary: Sandy Island (1876–2012). *Eos, Transactions American Geophysical Union*, *94*(15), pp. 141–142.

Seybert, H., 2012. Internet use in households and by individuals in 2012. *Eurostat*. Available at: http://ec.europa.eu/eurostat/documents/3433488/5585460/KS-SF-12-050-EN.PDF [Accessed 23 September 2016].

Shafy, S. 2008. An infinite loop in the brain. [online] *Spiegel International*. Available at: www.spiegel.de/international/world/the-science-of-memory-an-infinite-loop-in-the-brain-a-591972.html [Accessed 3 November 2016].

Shaw, I.G.R. and Warf, B., 2009. Worlds of affect: virtual geographies of video games. *Environment and Planning A*, *41*(6), pp. 1332–1343.

Shaw-Williams, K., 2014. The social trackways theory of the evolution of human cognition. *Biological Theory*, *9*(1), pp. 16–26.

Sheridan, B., 2016. Living in an old church means many things. Today it means my house is a Pokémon Go gym. This should be fascinating. [Twitter thread] Available at: https://twitter.com/boonerang/status/751849519407595520 [Accessed 27 October 2016].

Shirky, C. 2010. How cognitive surplus will change the world. [online] *TED Talks 2010*. Available at: www.ted.com/talks/clay_shirky_how_cognitive_surplus_will_change_the_world [Accessed 10 October 2016].

Simons, D.J. and Chabris, C.F., 1999. Gorillas in our midst: Sustained inattentional blindness for dynamic events. *Perception*, *28*(9), pp. 1059–1074.
The Six Million Dollar Man, 1973. TV series. ABC.
Smart, P. 2010. Cognition and the web. *1st ITA Workshop on Network-Enabled Cognition: The Contribution of Social and Technological Networks to Human Cognition*, USA.
Solove, D.J., 2007. *The future of reputation: gossip, rumor, and privacy on the internet*. Yale University Press, p.7.
Somebody's Going to Emergency, Somebody's Going to Jail, 2001. *The West Wing.* Season 2, episode 16. TV series. NBC.
Sommer, A. F., Hedegaard, C., Dukovska-Popovska, I. and Steger-Jensen, K., 2015. Improved product development performance through Agile/Stage-Gate Hybrids: the next-generation Stage-Gate process? *Research-Technology Management*, 58(1), pp. 34–45.
Southern, J., 2012. Comobility: how proximity and distance travel together in locative media. *Canadian Journal of Communication*, *37*(1), p.75.
Spooner, M.A., 2014. It's not a game anymore, or is it: virtual worlds, virtual lives, and the modern (mis)statement of the virtual law imperative. *University of St. Thomas Law Journal*, *10*, p.533.
Sposito, A., Bolognini, N., Vallar, G. and Maravita, A., 2012. Extension of perceived arm length following tool-use: clues to plasticity of body metrics. *Neuropsychologia*, *50*(9), pp. 2187–2194.
Statt, N., 2016. Apple says it doesn't know why iTunes users are losing their music files. [online] *The Verge.* Available at: www.theverge.com/2016/5/13/11674388/apple-music-itunes-file-deletion-bug-update [Accessed 14 October 2016].
Stephenson, N., 1992. *Snow crash*. Bantam Books.
Stephenson, N., 2011. *Reamde*. William Morrow & Co.
Sterling, B., 2005. *Shaping things.* Mediaworks Pamphlets, MIT Press, p.9.
Steuer, J., 1992. Defining virtual reality: dimensions determining telepresence. *Journal of Communication*, *42*(4), pp. 73–93.
Stevenson, S. 2014. Wish I were there: the beam telepresence robot lets you be in two places at once. [online] *Slate*. Available at: www.slate.com/articles/technology/technology/2014/05/beam_pro_telepresence_robot_how_it_works_and_why_it_is_strangely_alluring.html [Accessed 27 September 2016].
Stewart, I., and Cohen, J., 2009. *Figments of reality: the evolution of the curious mind*, revised ed. Cambridge University Press.
Stone, E.E. and Skubic, M., 2013. Unobtrusive, continuous, in-home gait measurement using the Microsoft Kinect. *IEEE Transactions on Biomedical Engineering*, *60*(10), pp. 2925–2932.
Stoop, E., 2003. Mobility: freedom of body and mind. *In*: Aarts, E. and Marzano, S., eds. *The new everyday: views on ambient intelligence*. 010 publishers, p.140.
Stout, D., Toth, N., Schick, K. and Chaminade, T., 2008. Neural correlates of Early Stone Age toolmaking: technology, language and cognition in human evolution. *Philosophical Transactions of the Royal Society of London B: Biological Sciences*, *363*(1499), pp. 1939–1949.
Strack, F., Martin, L.L. and Stepper, S., 1988. Inhibiting and facilitating conditions of the human smile: a nonobtrusive test of the facial feedback hypothesis. *Journal of Personality and Social Psychology*, *54*(5), p.768.
Streitfeld, D., 2014. Hachette says Amazon is delaying delivery of some books. *The New York Times,* May 9, 2014, B4. Available at: www.nytimes.com/2014/05/09/technology/hachette-says-amazon-is-delaying-delivery-of-some-books.html?_r=0 [Accessed: 20 October 2016].
Stross, C., 2011. *Rule 34*. Hachette UK.
Sturdee, M., Coulton, P., Lindley, J.G., Stead, M., Ali, H. and Hudson-Smith, A., 2016, May. Design fiction: how to build a Voight-Kampff machine. *Proceedings of the 2016 CHI Conference Extended Abstracts on Human Factors in Computing Systems*. ACM, pp. 375–386.
Suddendorf, T. and Corballis, M.C., 2007. The evolution of foresight: What is mental time travel, and is it unique to humans? *Behavioral and Brain Sciences*, *30*(03), pp. 299–313.
Susen, S., 2011. Critical notes on Habermas's theory of the public sphere. *Sociological Analysis*, *5*(1), pp. 37–62.

Sutherland, I.E., 1965. The ultimate display. *Proceedings of the International Federation of Information Processing Congress, 2*.

Swain, F., 2014. I can hear Wi-Fi. *New Scientist, 224*(2995), p.20.

Swain, F., 2016. Heard but not seen. *New Scientist, 231*(3080), p.20.

Swift, J., 1710. The art of political lying. *The Examiner, 14*.

Sykes, C.J., 1999. *The end of privacy: the attack on personal rights at home, at work, on-line, and in court*. Farrar, Straus, and Giroux, p.28.

Talbot, D., 2012. Given tablets but no teachers, Ethiopian children teach themselves. *Technology Review*. Available at: www.technologyreview.com/news/506466/given-tablets-but-no-teachers-ethiopian-children-teach-themselves/ [Accessed 23 September 2016].

Teng, S. and Whitney, D., 2011. The acuity of echolocation: spatial resolution in the sighted compared to expert performance. *Journal of Visual Impairment & Blindness, 105*(1), p.20.

Teperek, M., Simeone, A., Gaggioli, V., Miyamoto, K., Allen, G., Erkek, S., Peters, A., Kwon, T., Marcotte, E., Zegerman, P. and Bradshaw, C., 2016. Sperm is epigenetically programmed to regulate gene transcription in embryos. *Genome Research*, pp. gr-201541.

Thimm, C., 2012. Virtual worlds: game or virtual society? *In*: Fromme, J. and Unger, A. eds.

Computer Games and New Media Cultures. Springer Netherlands, pp. 173–190.

Thompson, C., 2008. Brave new world of digital intimacy. [online] *The New York Times Magazine*. Available at: www.nytimes.com/2008/09/07/magazine/07awareness-t.html?pagewanted=all&_r=1 [Accessed 13 October 2016].

Thompson, C., 2013. *Smarter than you think: how technology is changing our minds for the better*. Penguin.

Thompson, H. 2016. Being in a robot's shoes. *New Scientist, 3097*.

Tippmann, S. and Powles, J. 2015. Google accidentally reveals data on 'right to be forgotten' requests. [online] *The Guardian*. Available at: www.theguardian.com/technology/2015/jul/14/google-accidentally-reveals-right-to-be-forgotten-requests [Accessed 19 October 2016].

Titcomb, J., 2015. Asking Google 'What happened to the dinosaurs?' leads to a controversial answer. [online] *The Telegraph*. Available at: www.telegraph.co.uk/technology/google/11681955/Asking-Google-What-happened-to-the-dinosaurs-leads-to-an-controversial-answer.html [Accessed: 20 October 2016].

Tomasello, M., 1999. The human adaptation for culture. *Annual Review of Anthropology*, pp. 509–529.

Tomasello, M., Carpenter, M., Call, J., Behne, T. and Moll, H., 2005. Understanding and sharing intentions: the origins of cultural cognition. *Behavioral and Brain Sciences, 28*(05), pp. 675–691.

Toprak, C., Platt, J., Ho, H.Y. and Mueller, F., 2013, April. Cart-load-o-fun: designing digital games for trams. *CHI'13 Extended Abstracts on Human Factors in Computing Systems*. ACM, pp. 2877–2878.

Toups, M.A., Kitchen, A., Light, J.E. and Reed, D.L., 2011. Origin of clothing lice indicates early clothing use by anatomically modern humans in Africa. *Molecular Biology and Evolution, 28*(1), pp. 29–32.

Tsakiris, M. and Haggard, P., 2005. The rubber hand illusion revisited: visuotactile integration and self-attribution. *Journal of Experimental Psychology: Human Perception and Performance, 31*(1), p.80.

Tsakiris, M., Costantini, M. and Haggard, P., 2008. The role of the right temporo-parietal junction in maintaining a coherent sense of one's body. *Neuropsychologia, 46*(12), pp. 3014–3018.

Tsekleves, E., 2014. Wearable tech for Christmas? It probably won't help you get fit. [online] *The Guardian*. Available at: www.theguardian.com/media-network/2014/dec/01/wearable-technology-gadget-christmas-health-fitness [Accessed 28 September 2016].

Utrilla, P., Mazo, C., Sopena, M.C., Martínez-Bea, M. and Domingo, R., 2009. A palaeolithic map from 13,660 cal BP: engraved stone blocks from the Late Magdalenian in Abauntz Cave (Navarra, Spain). *Journal of Human Evolution, 57*(2), pp. 99–111.

Vaidya, J. and Atluri. V., 2007. Privacy enhancing technologies. *In*: Acquisti, A., Gritzalis, S., Lambrinoudakis, C. and di Vimercati, S., eds. *Digital privacy: theory, technologies, and practices*. CRC Press.

van Deursen, A.J. and Van Dijk, J.A., 2014. The digital divide shifts to differences in usage. *New Media & Society, 16*(3), pp.507–526.

Varga, S., 2011. Winnicott, symbolic play, and other minds. *Philosophical Psychology, 24*(5), pp. 625–637.

Verbücken, M., 2003. Towards a new sensoriality. *In:* Aarts, E. and Marzano, S., 2003. eds. *The new everyday: visions of ambient intelligence*, 010 publishers.

Vince, G., 2010. Who needs banks if you have a mobile phone? *New Scientist,* 2748.
Vinge, V., 1993. The coming technological singularity: how to survive in the post-human era. *Vision 21: Interdisciplinary Science and Engineering in the Era of Cyberspace*. Vol. 1. pp. 11–22.
Von Krogh, G., Ichijo, K. and Nonaka, I., 2000. *Enabling knowledge creation: how to unlock the mystery of tacit knowledge and release the power of innovation.* Oxford University Press on Demand.
Wainwright, M., 2010. Wrong kind of tweet leaves air traveller £1,000 out of pocket. [online] *The Guardian.* Available at: www.theguardian.com/uk/2010/may/10/tweeter-fined-spoof-message [Accessed 18 October 2016].
Waldron, S.M., Patrick, J., Morgan, P.L. and King, S., 2007. Influencing cognitive strategy by manipulating information access. *The Computer Journal*, *50*(6), pp. 694–702.
Wang, H. and Wellman, B., 2010. Social connectivity in America: changes in adult friendship network size from 2002 to 2007. *American Behavioral Scientist*, *53*(8), pp. 1148–1169.
Wang, L., Tan, T., Ning, H. and Hu, W., 2003. Silhouette analysis-based gait recognition for human identification. *IEEE Transactions on Pattern Analysis and Machine Intelligence*, *25*(12), pp. 1505–1518.
Watkins Jr, J.E., 1900. What may happen in the next hundred years. *Ladies Home Journal*, *8*.
Wayman, E., 2012. When did the human mind evolve to what it is today? [online] *Smithsonian Magazine.* Available at: www.smithsonianmag.com/science-nature/when-did-the-human-mind-evolve-to-what-it-is-today-140507905/ [Accessed 26 September 2016].
Weise, E., 2014. Amazon-Hachette: the war of the button. [online] *USA Today.* Available at: www.usatoday.com/story/tech/2014/11/13/amazon-hachette-preorder-publishing/18995643/ [Accessed: 20 October 2016].
Weise, S., Hardy, J., Agarwal, P., Coulton, P., Friday, A. and Chiasson, M., 2012, September. Democratizing ubiquitous computing: a right for locality. *Proceedings of the 2012 ACM Conference on Ubiquitous Computing*. ACM, pp. 521–530.
Weiser, M., 1991. The computer for the 21st century. *Scientific American*, *265*(3), pp. 94–104.
Weiser, M. and Brown, J.S., 1997. The coming age of calm technology. *In:* Denning, P.J. and Metcalfe R.M., eds. *Beyond calculation: the next fifty years.* Springer New York, pp. 75–85.
Weizenbaum, J., 1976. *Computer power and human reason: from judgment to calculation.* W.H. Freeman & Co, p.20.
Wellman, B., Haase, A.Q., Witte, J. and Hampton, K., 2001. Does the internet increase, decrease, or supplement social capital? Social networks, participation, and community commitment. *American Behavioral Scientist*, *45*(3), pp. 436–455.
Westin, A., 1967. *Privacy and freedom.* Atheneum Press.
Wikipedia, 2016. Wikipedia: systemic bias. [online] *Wikipedia.* Available at: https://en.wikipedia.org/wiki/Wikipedia:Systemic_bias [Accessed: 20 October 2016].
Wynn, T. and Coolidge, F.L., 2007. Did a small but significant enhancement in working memory capacity power the evolution of modern thinking. *In:* Mellars, P., Boyle, K., Bar-Yosef, O. and Stringer. C. eds. *Rethinking the human revolution.* McDonald Institute for Archaeological Research, pp. 79–90.
Yang, X.D., Hasan, K., Bruce, N. and Irani, P., 2013, October. Surround-see: enabling peripheral vision on smartphones during active use. *Proceedings of the 26th Annual ACM Symposium on User Interface Software and Technology*. ACM, pp. 291–300.
Yar, M., 2005. The global 'epidemic' of movie 'piracy': crime-wave or social construction? *Media, Culture & Society*, *27*(5), pp. 677–696.
Yee, N., Bailenson, J.N. and Ducheneaut, N., 2009. The Proteus effect: implications of transformed digital self-representation on online and offline behavior. *Communication Research. 36(2)*.
Yokota, T., Zalar, P., Kaltenbrunner, M., Jinno, H., Matsuhisa, N., Kitanosako, H., Tachibana, Y., Yukita, W., Koizumi, M. and Someya, T., 2016. Ultraflexible organic photonic skin. *Science Advances*, *2*(4), p.e1501856.
Zhong, C.B. and Leonardelli, G.J., 2008. Cold and lonely: does social exclusion literally feel cold? *Psychological Science*, *19*(9), pp. 838–842.
Zhou, W.X., Sornette, D., Hill, R.A. and Dunbar, R.I., 2005. Discrete hierarchical organization of social group sizes. *Proceedings of the Royal Society of London B: Biological Sciences*, *272*(1561), pp. 439–444.

Index